Embracing Transhumanism and Genomics in Human Resources Management

Jaspreet Kaur
Chandigarh University, India

A volume in the Advances in Human Resources Management and Organizational Development (AHRMOD) Book Series

IGI Global
Publishing Tomorrow's Research Today

Published in the United States of America by
IGI Global
Business Science Reference (an imprint of IGI Global)
701 E. Chocolate Avenue
Hershey PA, USA 17033
Tel: 717-533-8845
Fax: 717-533-8661
E-mail: cust@igi-global.com
Web site: http://www.igi-global.com

Library of Congress Cataloging-in-Publication Data

CIP Data Pending
ISBN: 979-8-3693-7668-3
eISBN: 979-8-3693-7670-6

British Cataloguing in Publication Data
A Cataloguing in Publication record for this book is available from the British Library.

All work contributed to this book is new, previously-unpublished material.
The views expressed in this book are those of the authors, but not necessarily of the publisher.

For electronic access to this publication, please contact: eresources@igi-global.com.

Table of Contents

Detailed Table of Contents

Amid the progress of the technoscientific revolution, which fundamentally alters nearly every sphere of people's existence, the idea of cyborg employees, people with artificial parts in their bodies, is no longer the stuff of fiction. This chapter is dedicated to the analysis of the relatively new philosophical movement called transhumanism and its possible manifestations in the contemporary work context. It may be important for cyborg employees to work in risky and exclusive areas of practice, and they include areas related to health, protection, and astronomy, among others. It concerns problems associated with privacy, security concerns, and the possibilities of losing the base of economically disadvantaged people. There is also evaluation of effects on the psychological aspects of the employees like their identity and self-pertaining view in addition to the social ramifications of implementing these types of technologies.

The study focuses on the impact of human enhancement technologies (HET) on job satisfaction from the human resources (HR) standpoint. It starts with the introduction and history of HET considering the types, cognitive, physical, and emotional booster, and recent developments. The chapter then provides details of the theoretical constructs used in the measurement of job satisfaction, the intrinsic and extrinsic incentive stemming from work, a favourable work environment, and the organizational culture. The consequence of the analyzed augmentation technologies include cognitive ones like brain-computer interface and the impact of memory, attention, and problem-solving on physical enhancements like exoskeletons and prosthetics, and emotional technologies for stress management and well-being.

The enhancement of human capability ushers in a new age of technology as people call for machinery to take over their work, increase human efficiency, and make the impossible become possible. But the major challenge organizations experience is the high level of resistance towards adoption of such innovations into the firm's operations. This study aims at discussing the dynamic factors behind technological resistance and how the necessary culture can be built. Some of these factors are perceived job loss, lack of knowledge on the new technology such as AI and biotechnology, culture and age differences in adoption of technology, ethics such as privacy and autonomy, bureaucratic resistance, psychological barrier to change, lack of resources, resistance to technology-based performance indicators, and fear of obsolescence of skills. Thus, there is a need to focus on such solutions as clear reporting by organizations, development of specific programs for effective training, ethical standards within the frameworks of the organization, considering employees, their views, ideas, and promoting widespread commitment and leadership from managers for stimulating innovation. Thus, proper management of change in organizations can prevent resistance and bring people and technologies to improve effectiveness, innovation, and strategic positioning in the age of high-tech environments. As for practical applications, this study calls for a partnership and future-oriented approach towards unveiling and addressing the challenges of technological implementation, and in turn, outlining the opportunities of human enhancement technologies to organizations as well as the society and ethics.

This chapter explores the integration of artificial intelligence (AI) in human resources (HR) processes, examining its evolution, current applications, and future implications. Beginning with a definition of AI in HR, including machine learning and predictive analytics, it highlights AI's role in enhancing efficiency, decision-making, and employee experiences. The chapter delves into specific applications across recruitment, onboarding, training, performance management, and employee engagement, showcasing AI's transformative impact. Ethical considerations such as bias mitigation, data privacy, and human-AI collaboration are also discussed. Ultimately, it emphasizes AI's potential to revolutionize HR practices while advocating for responsible adoption and ongoing professional development in HR professionals.

Genomics advancements have shaken the human resources (HR) field, giving insights into employee and organizational health and an understanding of the individualized approach to medicine. This study seeks to explore the evolving trends within human resource management (HRM) and genomics by reviewing past research. It also aims to identify key themes and outline potential directions for future investigations in these fields. The current research utilized electronic Scopus databases, which include all major prior studies. This bibliometric analysis utilizes content analysis to explore the methodology, annual contribution trends, and leading countries in the research area. VOS viewer software is also used for bibliometric analysis, incorporating bibliographic coupling, co-occurrence, and co-authorship analyses, which provide various graphical representations of the research network. Furthermore, thematic mapping is conducted to identify different clusters within the topic and highlight recent contributions, where new topics emerge, and new methodologies are employed to study consumer behavior. The current study aims to compile ideas for future research and explore how genomics can be pivotal and beneficial in comprehending HRM in the modern age.

The transhuman era has seen a significant shift in the public sector, with machine learning and artificial intelligence transforming HR analytics. This field has seen rapid growth, but its application in the public sector remains underexplored. This review explores the concepts and practices of HR analytics, proposing a five-step methodology for effective adoption. The study provides insights into workforce planning, recruitment, HR development, and performance management in the context of the transhuman era. It addresses critical issues such as privacy, data integrity, algorithmic bias, and public awareness. Key conditions for successful HR analytics implementation include robust data management, enhanced staff competencies, and broad acceptance of these technologies.

Genomic testing in workplace wellness programmes provides personalised health information, but it also gives rise to substantial ethical and privacy issues among employees. This study examines the viewpoints of employees, uncovering a combination of positive and negative sentiments regarding the potential advantages in comparison to the hazards associated with data security. HRM techniques must effectively manage these problems by prioritising transparency and obtaining permission. Efficient incorporation of genomic testing necessitates unambiguous communication, ethical principles, and strong data security measures. The findings of this research emphasise the importance of developing customised HRM methods that promote acceptance and engagement, while still protecting employee privacy.

When an employee is engaged in moonlighting, it means they are serving two different companies at the same time. There are different reasons behind the rise of this situation, such as the need to earn more to fulfill financial needs and dissatisfaction from existing pay. With the rise in work from home (WFH) culture, employees have access to multiple opportunities. With continued opportunities due to digitalization, work from home and moonlighting have made a significant impact on the overall well-being of employees. With that being said, this study is aimed to study the impact of moonlighting and work-from-home on employees' well-being in Delhi and Chandigarh and implications on moonlighting on their work-life balance. In order to fulfill the above objectives, this study is based on both primary and secondary data. Primary data is collected from an online survey conducted through self-structured questionnaire which is shared through Google Form. Survey data has been analyzed with Excel spreadsheet and statistical package for social sciences (SPSS) software. Secondary data has been collected from previous studies related to work from home and moonlighting in various sectors. This study seeks to explore the impact of work-from-home and moonlighting on employees' wellbeing and their understanding of work commitment. The study was conducted on Delhi and Chandigarh regions where participants were chosen with random sampling. The dependent variable for this study was "work from home" and independent variable was "employees' well-being". This study found significant impact of work from home and moonlighting on employees' well-being ($p<0.05$) by conducting one-sample T-test using SPSS software. The study provides a valuable insight to corporate leaders and managers to manage the issues of employee turnover and declining job security.

This research examines the connection between emotional intelligence, workplace PDP, employee learning, job happiness. The study employed review research papers and finds the role of EI in PDP initiative and job satisfaction. It was found that emotional intelligence is a key factor that influences how learning activities affect job satisfaction. These findings have important implications for organizational strategies including better job performance, increased employee retention, and a happier workplace environment, and have the potential to impact organizational practices significantly. To successfully fulfill the demands of a diverse workforce in changing business contexts, organizations should cultivate a culture that supports and fosters employee development and job satisfaction.

In the time of the genomics and transhumanism, protection of genetic and enhancement information represents an obvious necessity for human resources management (HRM). As identified in this chapter, there are ethical, legal, and practical issues in incorporating such information into the policies of an organisation. This increases the focus on the creation of sound protocols for the informed consent procedures, the standards of data access and control, and the policies against discrimination. This chapter shows how organisations are able to tackle the associated dilemmas of handling genetic and enhancement data responsibly by respecting employee privacy and encouraging truthful conversation. The goal of this chapter is to describe the key role of HR function in enhancing the level of trust in today's world of big data.

In a time portrayed by fast mechanical headways, the reconciliation of innovation into the working environment has fundamentally changed the idea of work and representative encounters. This exploration examines the effect of innovation upgraded workplaces on representative efficiency and occupation satisfaction, zeroing in on the jobs of different advancements like mechanization, man-made reasoning (artificial intelligence), coordinated effort apparatuses, and ergonomic developments. By looking at how these innovations impact representatives' work proficiency, commitment, and generally satisfaction, this study plans to give a far-reaching comprehension of the advantages and difficulties related with an innovation upgraded work environment. A blended strategies approach will be used, consolidating quantitative information from representative reviews and efficiency measurements with subjective experiences from meetings and contextual investigations. Reviews will evaluate workers' view of the innovative apparatuses they use, their effect on work execution, and levels of occupation satisfaction. Efficiency measurements will be dissected to gauge changes in work result and proficiency. Interviews with representatives, chiefs, and IT experts will give a more profound comprehension of the viable ramifications and individual encounters connected with innovation use in the work environment. Contextual investigations of associations known for their high-level utilization of working environment innovation will feature best practices and normal snags. Expected results incorporate recognizing the key advances that most altogether upgrade efficiency and occupation satisfaction, understanding the elements that add to fruitful innovation reception, and revealing potential downsides, for example, innovation related pressure or reliance. The exploration will likewise investigate the job of the executives in working with a smooth progress to an innovation improved workplace and keeping a harmony between mechanical coordination and human variables. The discoveries will offer significant bits of knowledge for associations expecting to streamline their innovation systems to further develop worker execution and prosperity. Vital suggestions will be created for directors to successfully carry out and oversee innovation in the work environment, guaranteeing that mechanical upgrades lead to positive results for both the association and its representatives. This study looks to add to the more extensive conversation on the fate of work, stressing the basic job of innovation in forming useful and fulfilling workplaces.

Preface

In an era where technological advancements are rapidly transforming every facet of our lives, the field of Human Resources Management is no exception. *Embracing Transhumanism and Genomics in Human Resources Management* delves into the intersection of cutting-edge technologies and HR practices, exploring how concepts from transhumanism and genomics are revolutionizing the way organizations recruit, develop, and manage talent.

This book provides a comprehensive overview of the theoretical foundations, practical applications, and ethical considerations of integrating these technologies into the workplace. By examining the latest innovations—from wearable technologies and biohacking to AI-driven talent acquisition and genomic-based employee development—this book aims to equip HR professionals with the knowledge and tools necessary to navigate and lead in this transformative era. Standing on the brink of a new age in human potential and organizational efficiency, understanding and embracing these advancements is crucial for building a future-ready workforce.

The book will be a comprehensive exploration aimed at HR professionals, researchers, educators, and students. It will delve into the integration of emerging technologies like transhumanism and genomics into HRM, aiming to deepen understanding and enhance various HR functions. Through practical insights, it will illustrate how these technologies can be applied in recruitment, talent management, and employee development, providing actionable strategies for implementation. Ethical considerations, privacy concerns, and societal impacts will be thoroughly examined, guiding readers towards responsible decision-making and policy implementation. By synthesizing existing literature and proposing innovative approaches, the book will contribute to advancing research in this field, fostering a deeper understanding of the intersection between technology, biology, ethics, and HRM. Ultimately, "Embracing Transhumanism and Genomics in Human Resources Management" will empower readers to navigate and embrace the transformative potential of these technologies in the evolving landscape of HRM.

The target audience of the book will be composed of HR professionals, researchers, technology enthusiasts, and practitioners. HR professionals will find valuable insights on integrating transhumanism and genomics into their practices to optimize recruitment, talent management, and employee well-being. Researchers and academics will benefit from the book's exploration of theoretical foundations, practical applications, and ethical considerations within the workplace. Technology enthusiasts will appreciate the in-depth examination of emerging technologies and their societal impacts. Practitioners will gain strategic insights to drive innovation, maximize workforce potential, and shape ethical guidelines and regulatory frameworks. This book aims to cater to a broad spectrum of professionals interested in understanding and leveraging the transformative potential of these technologies in HRM.

ORGANIZATION OF THE BOOK

Chapter 1: Cyborg Employees: Exploring the Benefits and Challenges of Transhumanism in Modern Work Environments

Authored by Roop Kamal from Chandigarh University, India, this chapter delves into the burgeoning realm of cyborg employees—individuals augmented with artificial parts. As the technoscientific revolution advances, the notion of cyborg employees is no longer confined to science fiction. This chapter critically examines transhumanism, a philosophical movement advocating for the enhancement of human capabilities through technology, and its practical manifestations in contemporary workplaces. Key areas of focus include the deployment of cyborg employees in high-risk and specialized fields such as healthcare, security, and space exploration. The discussion also navigates the intricate issues of privacy, security, and the socio-economic divide that may be exacerbated by these technologies. Additionally, the psychological implications for employees, including identity and self-perception, as well as the broader social consequences, are thoroughly explored.

Chapter 2: Can Human Enhancement Technologies Serve as Catalysts for Job Satisfaction in the Workplace? A Human Resources Perspective

This chapter, authored by Jaspreet Kaur, Noble Preet, and Navdeep Chahal, investigates the impact of Human Enhancement Technologies (HET) on job satisfaction from an HR perspective. Beginning with a historical overview and classification of HET—including cognitive, physical, and emotional enhancements—the chapter delves into the theoretical frameworks used to measure job satisfaction. It explores intrinsic and extrinsic motivations, favorable work environments, and organizational culture. The chapter evaluates the effects of cognitive technologies like brain-computer interfaces, physical enhancements such as exoskeletons, and emotional technologies for stress management and well-being, offering a comprehensive analysis of how these augmentations influence job satisfaction and overall workplace dynamics.

Chapter 3: From Resistance to Adoption: Promoting a Culture of Technological Acceptance in the Era of Human Enhancement

Authored by Noble Preet and Bhupinder Chahal, this chapter addresses the resistance to adopting human enhancement technologies in organizational settings. It examines the dynamic factors contributing to this resistance and proposes strategies for fostering a culture of technological acceptance. Emphasizing the importance of effective change management, the chapter outlines methods for integrating advanced technologies to improve organizational efficiency, innovation, and strategic positioning. By understanding and addressing the root causes of resistance, organizations can better prepare for the technological transformations of the future.

Chapter 4: AI Integration in HR Processes

Satya Subrahmanyam from Holy Spirit University of Kaslik, Lebanon, explores the transformative potential of AI in HR processes. This chapter provides an in-depth analysis of AI's evolution, current applications, and future implications in HR, covering areas such as recruitment, onboarding, training, performance management, and employee engagement. The chapter also addresses ethical considerations, including bias mitigation, data privacy, and the collaboration between humans and AI. By highlighting AI's capabilities in enhancing efficiency and decision-making, the chapter advocates for responsible adoption and continuous professional development in HR.

Chapter 5: Genomics in HR: A New Frontier for Employee Wellness and Organizational Growth

Authored by Dixit Kalra, Medha Gupta, Swati Gupta, and Harpreet Singh, this chapter investigates the integration of genomics in HRM, focusing on employee wellness and organizational growth. The chapter reviews existing research on genomics and its application in personalized medicine, analyzing trends and future directions. Utilizing bibliometric analysis, the authors identify key themes and emerging methodologies in the field. The chapter highlights how genomics can offer valuable insights into employee health and wellness, paving the way for innovative HR practices and policies.

Chapter 6: Transforming People Management: Harnessing HR Analytics in the Transhuman Era

Kapil Kumar Aggarwal and Ravneet Kaur examine the rapid growth of HR analytics in the Transhuman Era, particularly in the public sector. This chapter outlines a five-step methodology for effective HR analytics adoption, focusing on workforce planning, recruitment, HR development, and performance management. Critical issues such as privacy, data integrity, algorithmic bias, and public awareness are addressed. The chapter underscores the importance of robust data management, enhanced staff competencies, and widespread acceptance for the successful implementation of HR analytics.

Chapter 7: Genomic Testing in Workplace Wellness: Employee Perspectives, HRM Approaches, and Insights

Harwinder Singh and Aparna Saluja explore the ethical and privacy implications of genomic testing in workplace wellness programs. This chapter captures employee perspectives, highlighting the balance between the benefits of personalized health information and concerns over data security. Effective HRM approaches for integrating genomic testing are discussed, emphasizing the need for transparency, informed consent, and robust data protection. The chapter advocates for tailored HRM strategies to foster acceptance and engagement while safeguarding employee privacy.

Chapter 8: The Influence of Work From Home Policies on Employees' Well-Being and Moonlighting Behavior in the Transhuman Context

Authored by Diksha Taneja, this chapter examines the effects of work-from-home (WFH) policies on employee well-being and moonlighting behavior. It explores how WFH arrangements, driven by the digitalization and transhumanism trends, impact psychological well-being and work-life balance. The study includes recommendations for managing the challenges associated with moonlighting and leveraging WFH policies to enhance employee satisfaction and productivity in a technologically evolving workplace.

Chapter 9: Personal Development Plans (PDPs) Supporting Employee Learning and Job Satisfaction: Lightening Role of Emotional Intelligence

Trupti Patole, Swati Sharma, and Kshitiz Sharma investigate the relationship between emotional intelligence (EI), personal development plans (PDPs), and job satisfaction. This chapter highlights how EI influences the effectiveness of PDPs and contributes to employee learning and job satisfaction. The findings suggest that fostering a culture of EI can lead to better job performance, increased retention, and a more positive workplace environment. The chapter offers practical implications for organizational strategies aimed at supporting employee development.

Chapter 10: Ensuring Privacy in Genomics and Transhumanism: The HRM's Imperative

Gurdish Pal addresses the ethical and privacy challenges posed by genomics and transhumanism in HRM. This chapter provides a conceptual analysis of privacy concerns related to genomic data and human augmentation technologies. It proposes HRM solutions grounded in human rights principles, such as privacy, liberty, and informed choices. The chapter emphasizes the importance of ethical policies and practices to protect individuals' genetic information and ensure responsible use of transhumanist technologies in the workplace.

Chapter 11: The Impact of Technological Advancements on Corporate Culture and Employee Well-Being: A Comprehensive Analysis

Authored by Navpreet Sidhu, Swati Gupta, and Dixit Kalra, this chapter examines the dualistic impact of technological advancements on corporate culture and employee well-being. It explores how technologies like collaborative software, AI, and remote working platforms influence organizational development and employee health. The chapter reviews literature and case studies to discuss the benefits and challenges of integrating these technologies, highlighting their role in fostering a culture of lifelong learning and sustainability in complex business environments.

Chapter 12: Navigating the Future: A Comprehensive Study on Robo-Advisors in Wealth Management

Ravneet Kaur and Kapil Kumar Aggarwal explore the rise of robo-advisors in the wealth management industry. This chapter discusses the advantages of robo-advisors, such as cost-effectiveness, transparency, and competitiveness, compared to traditional human advisors. The study examines the current state and potential future of robo-advisors, emphasizing their ability to transform financial advising through effective decision-making processes and enhanced financial services.

Chapter 13: Enhancing Employee Engagement in Genetic Research Through Effective Communication Strategies: HRM Insights

Anupal Mongia, Aparna Saluja, and Harwinder Singh focus on the role of communication strategies in enhancing employee engagement in genetic research. This chapter leverages HRM insights to identify key factors influencing engagement, such as clear communication of research goals, transparent sharing of findings, and fostering a collaborative work environment. The chapter underscores the importance of aligning organizational objectives with individual aspirations through targeted communication efforts to promote innovation and productivity.

Chapter 14: Optimizing the Transhuman Workforce: How Technology-Enhanced Work Environments Affect Employee Productivity and Job Satisfaction

Dr. Ravinder Kaur investigates the impact of technology-enhanced work environments on employee productivity and job satisfaction. This chapter examines the roles of automation, AI, collaboration tools, and ergonomic innovations in shaping work efficiency and employee engagement. By analyzing the benefits and challenges of these technologies, the study provides insights into optimizing work environments to enhance employee experiences and overall satisfaction.

Chapter 15: Harnessing Predictive Analytics for Workforce Optimization in a Transhuman Age

Authored by Kapil Sharma and Bhupinder Chahal, this chapter explores the applications of predictive analytics in workforce management. It highlights how predictive analytics can enhance employee retention, reduce absenteeism, improve productivity, optimize recruitment processes, and boost employee engagement. By leveraging historical data and advanced statistical models, organizations can make proactive and informed decisions to optimize their workforce in the transhuman age.

Chapter 16: Building Workplace Resilience: Insights From Employee Engagement and Technological Enhancements

Medha Gupta, Rupali Arora, and Dixit Kalra delve into the concept of resilience in the modern workplace. This chapter provides a comprehensive understanding of resilience and its significance in organizational contexts. It explores the internal and external factors influencing resilience, drawing insights from existing literature and research. The study offers valuable recommendations for organizational leaders and policymakers to foster a resilient work environment, essential for navigating adversity and ensuring employee well-being.

IN CONCLUSION

As we conclude this comprehensive exploration of *Embracing Transhumanism and Genomics in Human Resources Management*, we stand at the cusp of a transformative era in HR practices. This book has endeavored to illuminate the profound impact that cutting-edge technologies, such as transhumanism and genomics, are having on the field. By bridging theoretical insights with practical applications, we aim to equip HR professionals, researchers, educators, and students with the knowledge and tools necessary to navigate this evolving landscape.

The integration of these emerging technologies promises to revolutionize how we approach recruitment, talent management, and employee development. Through this book, we have showcased the potential of AI-driven talent acquisition, wearable technologies, biohacking, and genomic-based employee development to enhance organizational efficiency and unlock human potential. At the same time, we have underscored the importance of addressing ethical considerations, privacy concerns, and societal impacts, guiding readers toward responsible and informed decision-making.

Our goal has been to provide a balanced perspective that not only celebrates the innovations reshaping HRM but also critically examines the challenges and implications they bring. By synthesizing existing literature and proposing forward-thinking approaches, we hope to contribute to the ongoing discourse and inspire future research in this dynamic field.

Ultimately, *Embracing Transhumanism and Genomics in Human Resources Management* serves as a call to action for HR professionals and stakeholders to embrace these technological advancements thoughtfully and strategically. By doing so, we can build a future-ready workforce that is not only more efficient and innovative but also ethically grounded and socially responsible.

We thank you for embarking on this journey with us and encourage you to continue exploring the transformative potential of these technologies in your own professional practice. Together, we can shape the future of Human Resources Management and unlock new possibilities for organizational success and human flourishing.

Jaspreet Kaur
Chandigarh University, India

Chapter 1

Cyborg Employees: Exploring the Benefits and Challenges of Transhumanism in Modern Work Environments

Roop Kamal
Chandigarh University, India

ABSTRACT

Amid the progress of the technoscientific revolution, which fundamentally alters nearly every sphere of people's existence, the idea of cyborg employees, people with artificial parts in their bodies, is no longer the stuff of fiction. This chapter is dedicated to the analysis of the relatively new philosophical movement called transhumanism and its possible manifestations in the contemporary work context. It may be important for cyborg employees to work in risky and exclusive areas of practice, and they include areas related to health, protection, and astronomy, among others. It concerns problems associated with privacy, security concerns, and the possibilities of losing the base of economically disadvantaged people. There is also evaluation of effects on the psychological aspects of the employees like their identity and self-pertaining view in addition to the social ramifications of implementing these types of technologies.

INTRODUCTION

A cyborg is defined as "a person whose physiological functioning is aided by or dependent upon a mechanical or electronic device." Just like the cyborgs from sci-fi novels and films, a successful organization is made up of people aided by machines (Khalifa, 2024). As businesses shift more responsibility and production

DOI: 10.4018/979-8-3693-7668-3.ch001

to the technology-led side of the equation, acquiring truly efficient software should take on the same importance as hiring good people (Adamczyk & Surdykowska, 2022). The term 'cyborg' arose as a short form of 'cybernetic organism,' which is an entity made up of both biological and technical elements. Initially it was used to describe any system of this mixed type; however, it has more recently been employed specifically for entities where the biology and technology are integrally attached (Palazzani, 2022). To be called a cyborg, it is normally the case that the entity has abilities above and beyond those exhibited by either its biological or its technological parts alone. This has possibly come about due to fictional stories in which the cyborg has mental and/or physical capabilities that are far beyond those of humans (Sadat Ahadzadeh et al., 2024). A cyborg is essentially a man-machine system in which the control mechanisms of the human portion are modified externally by drugs or regulatory devices so that the being can live in an environment different from the normal one (Kennedy, 2024).

Evolving technologies are rapidly transforming the nature of services, service customer experiences and the very essence of relationships between customers and service providers. In terms of applications, there has been much focus on the 'infusion' of smart technologies, Artificial Intelligence (AI) and/or the assimilation of robots into service delivery processes. However, the nature of emergent technologies and their application within service delivery systems is increasingly blurring the boundaries between 'physical, digital and biological spheres' (Knell & Rüther, 2024). For example, car rental companies equip frontline representatives with headsets so they can communicate with supervisors and artificial intelligence (AI) agents and obtain detailed, accurate information about available vehicles or local recommendations, which they then can provide immediately to customers (Almahameed et. al., 2021). Overall then, technological innovations may help firms enhance customer– employee interactions and create a richer customer experience, with meaningful implications for customer loyalty to the firm (Pelegrín et al., 2020). Although arguably, they might also prioritize benefits to the firm, rather than customers, raising potential dark-side ethical concerns about modern technologies (as we address in the General Discussion), our primary focus is on how technology can improve frontline interactions. The use of innovative technologies and applications already has revolutionized the frontline interface, but a more recent development involves human enhancement technology (HET), which comprises tools that proactively improve human functions beyond what is "necessary for life or well-being" (Sharma & Makhija, 2024).

Transhumanism

Transhumanism is a philosophical and scientific movement that advocates the use of current and emerging technologies—such as genetic engineering, cryonics, artificial intelligence (AI), and nanotechnology—to augment human capabilities and improve the human condition. Transhumanists envision a future in which the responsible application of such technologies enables humans to slow, reverse, or eliminate the aging process, to achieve corresponding increases in human life spans, and to enhance human cognitive and sensory capacities. The movement proposes that humans with augmented capabilities will evolve into an enhanced species that transcends humanity—the "posthuman." Companies today are strategizing about future investments and technologies such as artificial intelligence, the internet of things, or growth around new business models (Murata et. al., 2024). While many of these trends will make for solid investments for the next 5-10 years, fewer companies are considering the revolutionary convergence of disparate trends pulled from technology, behavioral and societal changes, and medical advances to understand how they will converge to transform society (Das & Chanda, 2023). This transformation will be messy, complex, and sometimes scary, but signals already point to a future of humanity that will blur our identities into "transhumanism." In other words, a transhuman is a being that resembles a human in most respects but who has powers and abilities beyond those of standard humans. These abilities might include improved intelligence, awareness, strength, or durability. Transhumans sometimes appear in science-fiction as cyborgs or genetically-enhanced humans (Khalifa, 2024).

Unlike some other philosophical positions and social movements (e.g., feminism, liberalism, communism, and libertarianism), transhumanism is not best characterized by a particular dogma, principle, or prediction. Transhumanists vary in their political doctrines, their positions on ethical issues, and their religious outlooks (Ireni-Saban & Sherman, 2022). What distinguishes transhumanism as a movement and ideology is instead a set of shared core values and the conviction that those values can be pursued using technology. Even in terms of values and methods, however, there are differences among transhumanists (Knell & Rüther, 2024). To better understand the position then, it is useful to distinguish three levels of transhumanist culture and thought. The first level is a set of values and beliefs necessary for being considered a transhumanist. The second level is a set of values and beliefs that are consistent with sympathetic to, and historically co-occurrent with transhumanist movements and thought – sometimes so much so that they are seen as essential components of transhumanism (Modliński & Gladden, 2022). However, these values and beliefs are not logically implied by core transhumanist attitudes, and one may be a transhumanist without them. The third level is a set of technologies and futurist ideas that are strongly affiliated with transhumanism. Some of these technologies embody transhumanist beliefs about the metaphysical nature of human beings, consciousness, and life; some are philosophically aligned with common transhumanist values; and some are seen as technical means necessary for achieving transhumanist goals. Transhumanism is the movement and belief system promoting the use of technology to enhance human bodily function and cognition (Sharma & Makhija, 2024). A transhuman, however, is not a person who subscribes to transhumanism. That person would simply be called a transhumanist (Martins & Martins, 2024). Instead, a transhuman would be a human being who has undergone enough technological modification that he or she is no longer easily categorized as an ordinary human (possessing, for example, extended life span, enhanced cognitive abilities, and prosthetic implants) but is still human enough in essential outlook, cognition, and biology to be identifiable as member of the human species (Pelegrín et al., 2020). Although obviously a very imprecise term, and part of a fuzzy continuum, there are probably no human beings sufficiently modified at this time to merit being called transhuman. It is worth remembering, then, that transhumanism is a human ideology (Toker et al., 2023).

TRANSHUMNISM AT MODERN WORKPLACE

Techno-progressive ideas such as transhumanism where human beings are supposed to be improved through technologies has very significant consequences with regards to present day workplaces (Modliński & Gladden, 2022). In light of this, it has been observed that with the future developments of biotechnological solutions, implementation of these solutions within the working environment is possible. Transhumanism issues are therefore defined by productivity, efficiency, ethical concerns, and social implications (Das & Chanda, 2023).

Another possible benefit is the boost in productivity that is likely to be experienced in organizations that enforce and or adopt the policy. Segmented cyborgs with neural implants, developed prostheses or enhanced sensory systems are more effective at handling certain specified tasks than the non-cybernetic employer might be (Knell & Rüther, 2024). For example the neural implant may help them to capture and analyze information much faster enhancing on decisiveness and problem solving areas of an employee. Technological limbs may enable people to work hard physically without any exhaustion or difficulty, therefore let alone getting injured. Sensory enhancements like the ability to see or hear beyond the normal human range could prove useful in aspects like more efficient surveillance or communication, especially when Lives are at risk for example during medical procedures, or in the military (Ireni-Saban & Sherman, 2022).

Additionally, one can state that transhumanism might influence job satisfaction and employee well-being considerably (Adamczyk & Surdykowska, 2022). If lifting the physical and mental barriers restricts the employees, then it can be assumed that their work might no longer be as difficult or as tiring as it used to be. Thus, enhanced productivity and performance may result in a feeling of accomplishment and increase the level of employees' power. Transhumanism can further contribute to providing more opportunities on the job market for disabled people: with the help of given technology even such people with disabilities can perform work that has always been impossible for them in practice (Murata et. al., 2024). This inclusiveness could make work places more diverse and sustainable thus improving the working conditions for everyone. (Almahameed et. al., 2021)

Yet, there are some stumbling blocks on the way of integration of transhumanism in the world of work. That is why ethical issues are so important, especially focusing on the aspects of privacy and security. Neural implants and other forms of biotechnological enhancements creates issues over ownership of data and the possibility for an external entity to gain access to a person's mind or to their data. The requisite of the employers would be to balance the use of these technologies in order to reap efficiencies while ensuring that their employees' right to privacy and self-determination is not violated (Grewal et al., 2020).

Another major difficulty is the issue of social justice that is closely connected with socioeconomic inequality. Indeed, biotechnological enhancements can be expensive for many people, particularly if they have to pay for enhancements of themselves and of their future children, which means that such enhancements can create a population of enhanced and non-enhanced individuals (Murata et. al., 2024). This could lead to emergence of new form of 'improved' or 'better' employees thus increasing employment inequalities. Society has learned that these technologies have to be made available to the less privileged in an endeavor to close the gap between the haves and the have nots thus a challenge that policymakers and business persons have to address (Modliński & Gladden, 2022).

HUMAN PSYCHOLOGY AND TRANSHUMANISM

Transhumanism as another factor also has its consequences in the field of psychological effects on the employees. This paper explores the factors, both physical and psychological, that disturb an individual's self-identity through the insertion of artificial products into the human system (Sadat Ahadzadeh et al., 2024). Another possible necessary condition for the emergence of psychological stress is the existence of a feeling of dehumanization or rather a feeling of being an object among the personnel. These are indeed some of the grey areas that the employers have to ensure that they offer the necessary support and tools that would assist the employees to manage on the complex emotions and keep their mental health well patronized (Forlano & Glabau, 2024).

Moreover, there are the social consequences of the spread of transhumanism in organizations and their everyday functioning. Culture could change with time with the blurriness of the human and mechanical zone that has been created (Sharma & Makhija, 2024). This shift could in the long run lead to a shift in the orientation to work and employment and the relationship that exists between employers and employees. The society is left with dilemmas that arise out of transhumanism that hinders the answer to what it means to be human in the society (Galloway, 2020).

Regarding transhumanism the consequences for contemporary working environments are manifold: While very positive outcomes are to be expected, innovative opportunities as well as remarkable risks are attached to them as well. Improvement in efficiency, organizational commitment and employee relations and increased diversity should consider ethical issues, inequitable distribution of work, consequences on people's psychological states and social transformations. Now, at the eve of this technological dawn, it is necessary to raise the question of how to incorporate transhumanism into the working environment in a way that is both efficient and fair (Das & Chanda, 2023).

ISSUES RELATED TO PRIVACY AND SECURITY

Figure 1. Security Issues in transhumanism

Source: https://www.linkedin.com/pulse/business-intelligence-analytics-amongst-data-privacy-security-kamath/

Considering transhumanism as a part of enterprises and working environments it is important to consider the basic, yet crucial concerns of privacy, security, and distribution of opportunities. When it comes to Huxley's vision of the evil use of biotechnology, privacy invasion does not start with neural implants for employees but rather is the logical extension of the availability of advanced prosthetics, sensory augmentations, and other tools of biotechnological enhancements available to the developing society to employees (Kennedy, 2024). These enhancements which we can used to store and process large amount of personal data requires great measures to guard against unauthorized access. The employers and third parties can possibly exploit these technologies to track down the thoughts, behaviors, and personal information which would result into compromising of personal autonomy and privacy. The social effects of such surveillance are quite complex as this raises a lot of ethical questions on the rights to privacy of individuals in the course of their duty (Adamczyk & Surdykowska, 2022).

Security issues are also a burning question. Amalgamation of cybernetic enhancements leaves the employees open to cyber bullying whereby the hackers can control and even switch off those technologies. For instance, neural implants that improve functions such as memory, or control prosthetic devices may be attacked and cause the impairment of body parts or lead to adverse effects on the mind. What Jaeger (2014) observes is that the security of such devices is an issue that cannot be addressed without having to incorporate effective cybersecurity measures and constantly watch out for danger signs. To shield the employees from the tangible and intangible threats posed by cyber risks, the society needs to invest in the creation of secure and tamper-proof technologies (Knell & Rüther, 2024).

The last issue is the concerns arising from inequalities due to the level of income and social status in the case of application of transhumanism in the workplace. The fact is that biotechnological enhancements are expensive which will make it possible for only the privileged to have access to such enhancements in the initial stages hence increasing inequality in the society (Khalifa, 2024). Such a situation could generate a polarized labor force: enhanced individuals would reap benefits that put them far ahead of their non-enhanced counterparts regarding output, job prospects, and wages. Such a divide could possibly aggravate stratification with an emergence of a new set of people who have superior capacities and preferential access to the resources (Martins & Martins, 2023).

To tackle these disparities, one needs to devise certain prevention strategies that would help to make installation of biotechnological enhancements more egalitarian. There is this guiding rationale that employers need to work hand-in-hand with government officials as well as other keyshakers to come up with strategies that can be used to underwrite prices of such technologies and thus increase the access of their employees to such systems (Reichel et al., 2024). Also, at the same time, it implies that there should be certain measures against discrimination of non-enhanced people at the workplace. This could be done in ways such as developing positions that need not involve augmentations or providing other ways of progression in one's job (Ireni-Saban & Sherman, 2022).

The deficiencies have major ethical and societal consequences. Hence, if not prevented, the imbalance in the enhancements will escalate social tension and conflict (Martins & Martins, 2023). However, perceiving the enhanced individuals as superior to the others can create new social problems and jeopardise the idea of fair and equal treatment in the work environment (Kennedy, 2024).

PSYCHOLOGICAL IMPACT OF CYBORG ON EMPLOYEES

Figure 2. Impact of cyborg on employee psychology

Source: https://www.aigantic.com/ai-jobs/ai-employment-trends/ai-psychological-impact/

While the integration of transhumanism has the potential to supply major advantages into today's operating environments, it also has associated risks that include privacy, security, and the risk for people's socioeconomic status. The implementation of further improvements in enhanced employees entails strict policies and IT solutions against cyber risks and comparable risks. That is why fighting social injustice requires approaches aimed at making bio-technological improvements available for all and prevent the stratification of the workforce (Galloway, 2020). Addressing these challenges, society is able to benefit from the advancements of transhumanism when eliminating the risk of violation of employees' rights and abuses.

Cyborg or transhumanism in the conceptualization of employees' identity pertains to many ramifications that can be categorized into several major clusters of psychological concerns within the work setting (Kiess, & Schmidt, 2024). In view of this, where employees are subjected to enhancements via biotechnological interventions like implants in the brain or neural prosthesis, bionic limbs, or surgical procedures to alter the human senses, these changes have certain side effects in regards to the conceptualization and perception of self by the employee (Pelegrín et al., 2020).

The major psychological effects that arise are those of self-identity being at risk. It has been postulated that the integration of artificial components may cause a reappraisal of one's humanity (Martins & Martins, 2023). The employees may face issues with perceived aloneness or loss of one's spontaneous personality as they seek to understand that some of the parts of the body or the way they think do not fully belong to them. This identity crisis results in a lot of anxiety, depression or what some individuals refer to as state of being depersonalized (Kennedy, 2024). The mentioned combination of human and mechanical aspects can create a feeling that humans are mere tools or parts of technological systems in the workplace, where their mental well-being could be negatively impacted (Sharma & Makhija, 2024).

Moreover, the transhumanism leads to the improvement of capabilities, which inevitably can become a source of pressure and expectations. Superiority may become a burden to the enhanced employees in the sense that they are expected to deliver better outcomes because of enhanced capability (Thomas, 2024). This means that one may be forced to work longer hours to meet the set production rates thus increasing stress and even burnout as it becomes difficult to differentiate between where man ends and machine begins. The pressure forced on employees in aspiring organizations to work at their best may in the long run reduce their morale andEmpty diligences as well as lead to depression (Grewal et al., 2020).

Working relationships with peers can also be distorted beyond recognition due to social relations at the workplace (Murata et. al., 2024). As for the non-enhanced employees, the counterparts who are experienced an increase in employees' performance may feel jealous or resentful, or suffer from inadequacy, knowing that similar technologies are available to their colleagues. Without adequate strategies put in place to oversee the relationships between these divisions, it is possible that it creates a culture of dominance and childhood rivalry instead of sister sisterhood and comradeship. Superiority is also associated with social isolation if people with enhancements have reasons to think that they are essentially different from their co-workers. (Almahameed et. al., 2021)

Moreover, general views of enhancements as those that are essential to have in order to progress in one's career breed coercion. It is always a vast pressure for employee to follow the same trend just because they feel that they need to go through the enhancement whether they have a change of heart towards the procedures or what they mean to them (Karayaman, 2024). This pressure can bring about feeling of lost decision-making, and Personal choice, which in turn affects the mental health and job satisfaction the employees have.

Transhumanism is not just a negative movement from a psychological standpoint, nevertheless. Increased capacities implies that an individual will enjoy the feeling of being more effective at doing tasks. Original coded enhancements: integrated Enhancements include computer literacy, team work, communication skills, managerial skills, etc employees who successfully integrate enhancements may have boost in self-esteem and confidence because their skills have been boosted. The freedom from some form of disability whether physical or intellectual brings happiness at the work place as well as a deeper meaning to the work done (Martins & Martins, 2023).

To avert such impacts on the employees, the employer needs to ensure that there are mechanisms that help in the development of improved employees (Palazzani, 2022). These preventive measures are inclusive of providing accessibility to mental health services, reviewing counseling procedures and creating platforms that will allow the employees to share and possibly seek ways on how to deal with such instances and feelings or enhancements that they might be encountering. Additional, involving people with an enhancement or disability into the workplace as all other employees and promoting their talents to be utilized to the maximum is something that can go hand in hand (Khalifa, 2024).

Consequently, psychological affects of cyborg or transhumanism employees is a bit larger and involves identity, mental position, social relation and job satisfaction (Ireni-Saban & Sherman, 2022). Mitigating these impacts calls for an effort that entails psychological evaluation and treatment, employment of policies that encourage equal treatment of all employees, and acknowledging the worth of the diversity of the human resource (Grewal et al., 2020). In doing so, the organization will thus work to derive the best of the transhumanism while avoiding its psychological consequences (Karayaman, 2024).

CHALLENGES OF CYBORG AT WORKPLACE

Cyborg in the workplace encompasses various difficulties that must be resolved so that the implementation of transhumanist ideas remains both ethical and efficient. The solutions to these problems lie in areas such as ethical issues, privacy, security, organizational relationships, legal matters, and population in general.

Figure 3. Challenges of cyborg at workplace

1. Ethical Considerations

Perhaps the greatest of them is the issue of ethics and the enhancement of human abilities as a result of bionics. This notion of building on the body of human capital with biotechnological solutions like neural chips, super limbs, and other bionics is problematic when one is philosophically answering what it means to be a human (Thomas, 2024). Evaluative conflicts center such concerns as consent, control over one's body, and coercion. loans, there is likely to be a form of pressure exerted on employees to go for these enhancements in order to meet requirements of employment hence creating some sort of ethical issue with regards to issues of voluntarism and coercion. On the same note, issues of equity and how it would affect if some employees are enhanced while others are not enhanced raises the concern of rights (Forlano & Glabau, 2024).

2. Privacy Concerns

Privacy is a major concern because pro-levens usually require the input, processing and storage of one's personal information. Neural implants for instance can record thoughts and memories of an individual through implantation in the subject's head. Because this data is personal, it should be guarded from informational or identity pirates and it should be consumed in a proper manner (Almahameed et. al., 2021). There is a need to design effective data security procedures to protect enhanced employees' privacy, and there is a need for regulatory organizations to start implementing strict privacy measures.

3. Security Risks

The invulnerability of biotechnological enhancements is another grave consideration (Ireni-Saban & Sherman, 2022). These are exploitable through cyber attacks with disastrous effects as has been witnessed in the past. For example, one can interfere with or even destroy neural implants or prosthetic limbs and control an individual's vital body functions or alter a person's thinking processes. There is need to employ high end security on these devices because they are prone to threats and therefore, should be monitored regularly for any gave threats. Besides, the prospects of improved employees require secure and antitamper technologies to counter possible threats to their invulnerability (Martins & Martins, 2023).

4. Workplace Dynamics

The concept of introducing enhanced employees is a plausible one which can drastically change dynamics of relations in workplace. Enhanced individuals may be seen as 'better' or 'fitter' for a task than other non-enhanced individuals, thus, creating resentment, jealousy, and feelings of inadequacy among the subordinates. Such division may bring about unhealthy competition leading to rivalry and conflict and hence the colleagues do not work together (Das & Chanda, 2023). Organizations should ensure that people with enhancements are accepted and encouraged to make their best input to the company's productivity without any prejudice.

5. Regulatory Issues

Technological progress in transhumanism is fast, and very often the rate of development surpasses the speed of creating legislation (Ireni-Saban & Sherman, 2022). Man and his governments and regulating authorities are on the horns of the dilemma of coping up with the existing technologies and at the same time guarantee that they are being applied safely and legally. This concerns standards for the use of enhancements, employee's freedom, and guarantee that enhancements do not cause inequity or abuse of workers. Establishing the coherent policy for transhumanism that will contain major integral regulation of its usage is needed for the further introduction of the phenomenon into work environment (Kiess, & Schmidt, 2024).

6. Societal Impact

Effects of transhumanism on the broader society especially within the workplace must also be taken into consideration. Use of enhancements in general populace may worsen the societies social injustices by creating a clear line between haves and have nots in terms of enhancements. This could give a genesis to a range of new types of class clash, wherein enhanced persons have stark advantages over non-enhanced persons indeed (Martins & Martins, 2024). Thus, enhancing and developing society, writers underline the importance of collaboration between policymakers and businesses to avoid the creation of a society regarde, where certain individuals and companies receive enhancements while others do not (Pelegrín et al., 2020).

7. Psychological Impact

The socialization process of improved employees can also be another main issue which is accompanied by the psychological effects experienced. Surgically incorporating technology into an individual's body can have psychological effects such as identity issues, stress and the feeling of being an outsider. Thus, there might be the issue of diminished naturalness and the problem of dependent employees losing their individuality. Organizations require offering mental health assistance and tools to the workforce to manage such sentiments and improve their mental health status (Reichel et al., 2024).

To this end, the points outlined below demonstrate how the endorsement of transhumanism in workplace comes with several issues that need to addressed (Mehta et. al., 2020). The potential ethical issues include, privacy, security, interpersonal relationships, the formulation of policies, legal boundaries, the general impact on society and the well-being of employees to ensure any posthuman benefits are gained whilst minimizing drawbacks. Taking all these into consideration, people in businesses and policy-making authorities can manage these issues and build a future in which the transhumanism is a positive addition to the workplace that does not favor any employee or group over the other (Galloway, 2020).

FUTURE OF CYBORG AT WORKPLACE

The prospects of cyborg employees into the workplace is set to revolutionize the mode of working, the manner in which work is accomplished and productivity and efficiency as influenced by technological developments (Khalifa, 2024). In the light of advanced and widely available biotechnological improvements, the depolymerization of cyborg employees as mentioned earlier is bound to present more requirements relatives to the future of various professions. Here are several key aspects to consider:Here are several key aspects to consider:

Figure 4. Future of Cyborg at workplace

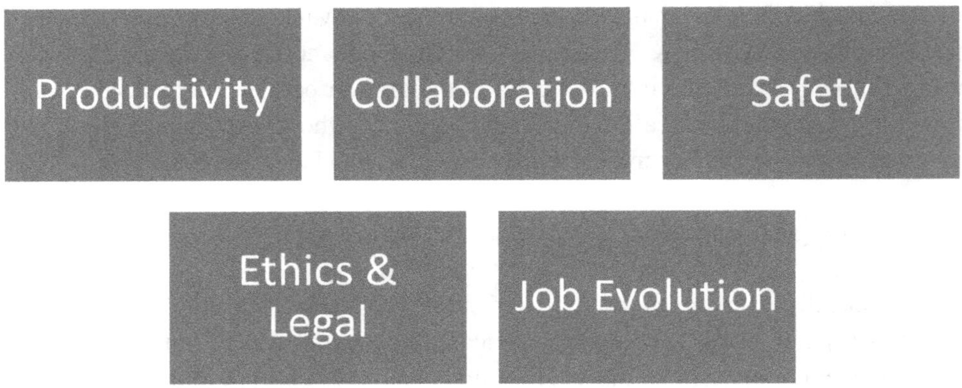

1. Enhanced Productivity and Efficiency

Neural enhancements, proper prostheses, and bioengineering of the sensory input of the employee-where the employer is ahead of the cyber-cyborg employee. Neural implants could provide fast data processing and decision aiding which, in turn, can enhance the speed and effectiveness of employees in complex procedures (Das & Chanda, 2023). Prosthetic enhancements may enhance people with the ability to perform strenuous tasks with relative ease, and so, improve efficiency. All abilities that are aimed at extending sensory organs could be beneficial: better vision or hearing aids could enhance perception of environment and communication in conditions like healthcare and defense (Karayaman, 2024).

2. Specialization and High-Risk Jobs

Cyborg employees can be seen fit for vocation types where specialized and hazardous work is required. In such areas as in medicine, this means that through the use of enhanced surgeons with better precision in surgeries would mean that the chances of surgeries yielding positive results would be much improved (Karayaman, 2024). In military and defense, cyborg soldiers with improved physical and mental prowess could take up assignments that are strength, stamina and perceptiveness intensive.

In the same way, in space exploration, those people who have had biotechnological augmentation of their body could experience less deterioration in the conditions of space environment (Kiess, & Schmidt, 2024).

3. Inclusivity and Accessibility

It was also seen that transhumanism could lead to more acceptance and accommodate persons with disabilities in the workforce (Thomas, 2024). Disabled people may have lot to gain from, improvements that make lose bodily and mental abilities and help them get back to their usual activities which may have been humiliating to do before. This inclusion could help in addressing difficulties that may be faced by women and minorities in the work environment by allowing them to work in equality with other people in the organization (Martins & Martins, 2024).

4. Ethical and Privacy Considerations

As cyborg employees become more prevalent, ethical and privacy considerations will be paramount. Ensuring that enhancements are used ethically and that employees' autonomy and consent are respected will be crucial (Martins & Martins, 2024). Privacy concerns related to the data generated by biotechnological enhancements will need to be addressed through robust regulatory frameworks and advanced cybersecurity measures. Employers must navigate the fine line between leveraging these technologies for productivity gains and protecting employees' rights to privacy and autonomy (Grewal et al., 2020).

5. Psychological and Social Impact

The psychological and social impact of integrating cyborg employees into the workplace cannot be overlooked (Karayaman, 2024). Enhanced employees may face identity challenges and feelings of alienation, as the distinction between human and machine becomes blurred. Employers will need to provide mental health support and foster an inclusive culture that values the contributions of all employees, regardless of their enhancement status. Social dynamics within the workplace will also need to be managed to prevent divisions and ensure collaboration and mutual respect among all employees (Toker et al., 2023) .

6. Socioeconomic Disparities

The potential for socioeconomic disparities is a significant challenge that must be addressed. The high cost of biotechnological enhancements could create a divide between those who can afford these advancements and those who cannot, leading to inequality in job opportunities and earning potential (Kiess, & Schmidt, 2024). Policymakers and businesses must work together to ensure equitable access to enhancements and prevent the emergence of a bifurcated workforce. Strategies such as subsidies, insurance coverage, and employer-sponsored programs could help bridge this gap.

7. Regulatory and Legal Frameworks

The integration of cyborg employees will require comprehensive regulatory and legal frameworks to ensure safety, ethics, and fairness. Governments and regulatory bodies will need to develop guidelines for the use of enhancements, protect employee rights, and address issues such as consent, data privacy, and cybersecurity. Employers will also need to establish policies and practices that comply with these regulations and promote ethical use of biotechnological enhancements (Mehta et. al., 2020).

8. Continuous Innovation and Adaptation

The future of cyborg employees in the workplace will be characterized by continuous innovation and adaptation. As technology evolves, so too will the capabilities and applications of biotechnological enhancements. Employers will need to stay abreast of these advancements and adapt their strategies and practices accordingly. Lifelong learning and reskilling will become essential for employees to keep pace with the rapid changes brought about by transhumanism (Forlano & Glabau, 2024).

The future of cyborg employees in the workplace holds immense promise for enhancing productivity, efficiency, and inclusivity. However, it also presents significant challenges that must be carefully navigated to ensure ethical and equitable integration (Mehta et. al., 2020). By addressing ethical considerations, privacy concerns, psychological and social impacts, socioeconomic disparities, and regulatory requirements, society can harness the potential of transhumanism to create a more innovative and inclusive future of work (Grewal et al., 2020).

CONCLUSION

Cyborg employees are becoming a part of working environments of contemporary societies providing distinguishing opportunities for changes as well as presenting profound issues. On one hand, augmentation with the help of cybernetic technology may increase the work rate or the efficiency with which things can be done or the ability of the performer; In all these cases, augmentations offer businesses an edge over others. Members of staff who have been genetically modified through technology like having prosthetics that are technologically enhanced and neural interfaces can work past the limits of ordinary humans by coming up with more creative ideas and doing their work more efficiently. But this technological incorporation poses highly sensitive ethical, social, and pragmatic questions. The matters of privacy, self-ownership, and consent are immediately relevant since the human-MP relationship becomes indistinct. Possible challenges of the future state include more surveillance and control over employees, which would result in exploitation of an individual and reduction of freedom.

However, drawbacks from these technologies are that they are relatively expensive; therefore, it may lead to emergence of inequality in the workforce to enhance where only a few people or companies can afford such technologies. All in all, this type of relations between people and cyborg employees is useful for increasing productivity and stimulating innovations, but it requires a significant reflection upon its ethical and juridical aspects. It is important that technological progress does not overshadow the preservation and enforcement of human rights and human beings' dignity. These are the challenges that societies and businesses must collectively face to make sure that the opportunities of transhumanism in the workplace can be achieved sustainably and fairly.

REFERENCES

Adamczyk, S., & Surdykowska, B. (2022). Cyborgs in the Workplace? Some Preliminary Considerations. *E-Journal of International and Comparative Labour Studies*.

Almahameed, A., Oliva, M. A., & Borondo, J. P. (2021). Cyborg acceptance in healthcare services: The use of cyborg as a surgeon. In *Moving technology ethics at the forefront of society, organisations and governments* (pp. 127–144). Universidad de La Rioja.

Das, A., & Chanda, D. (2023). To trust or not to trust cybots: ethical dilemmas in the posthuman organization. In *New Horizons for Industry 4.0 in Modern Business* (pp. 189–208). Springer International Publishing. 10.1007/978-3-031-20443-2_9

Forlano, L., & Glabau, D. (2024). *Cyborg*. MIT Press. 10.7551/mit-press/13864.001.0001

Galloway, K. (2020). The COVID Cyborg and Protecting the Unaugmented Human. *Alternative Law Journal*, 45(3), 162–167. 10.1177/1037969X20930431

Grewal, D., Kroschke, M., Mende, M., Roggeveen, A. L., & Scott, M. L. (2020). Frontline cyborgs at your service: How human enhancement technologies affect customer experiences in retail, sales, and service settings. *Journal of Interactive Marketing*, 51(1), 9–25. 10.1016/j.intmar.2020.03.001

Ireni-Saban, L., & Sherman, M. (2022). Cyborg ethics and regulation: Ethical issues of human enhancement. *Science & Public Policy*, 49(1), 42–53. 10.1093/scipol/scab058

Karayaman, S. (2024). The Alienating Effect of Technology: Does Technological Innovation Cause Work Alienation? *Journal of Management and Economic Studies*, 6(2), 204–218. 10.26677/TR1010.2024.1418

Kennedy, T. (2024). Transhuman Existentialism: Purpose in the Changing Human (Doctoral dissertation, Macquarie University).

Khalifa, E. The Cyborg Civilization: Bridging The Gap Between Humans and Robots.

Kiess, A., & Schmidt, A. L. (2024, January). Hospitality in the Cyborg Age: The Power of Brain-Computer Interfaces in a Field-Experiment. In *ENTER e-Tourism Conference* (pp. 140–145). Springer Nature Switzerland. 10.1007/978-3-031-58839-6_14

Knell, S., & Rüther, M. (2024). Artificial intelligence, superefficiency and the end of work: A humanistic perspective on meaning in life. *AI and Ethics*, 4(2), 363–373. 10.1007/s43681-023-00273-w

Martins, A., & Martins, I. (2023). Seeking Truer Reality: Quark Influence on the Transhuman Future. Managerial Challenges of Industry 4.0, 15.

Martins, I., & Martins, A. (2024). Back to the future: multidimensional transhuman influence on creativity and innovation. In *Artificial Intelligence in Production Engineering and Management* (pp. 193–206). Woodhead Publishing. 10.1016/B978-0-12-819471-3.00007-0

Mehta, A., Sharma, J., & Chahal, D. (2020). Overview of Cyborg Technology. [ijtsrd]. *Published in International Journal of Trend in Scientific Research and Development*, 4(3), 932–935.

Modliński, A., & Gladden, M. E. (2022). An organizational metaphor for the 4th industrial revolution: The organization as Cyborg. *World Futures*, 78(6), 372–391. 10.1080/02604027.2021.1996187

Murata, K., Orito, Y., Adams, A. A., Arias-Oliva, M., & Fukuta, Y. (2024). The ethics of body modification: Transhumanism in Japan. In Ethics and Sustainability in Digital Cultures (pp. 112-140). Routledge.

Palazzani, L. (2022). The future of work in the era of emerging technologies. Pensamiento. Revista De Investigación E Información Filosófica, 78(298 S. Esp), 777-784.

Pelegrín-Borondo, J., Arias-Oliva, M., Murata, K., & Souto-Romero, M. (2020). Does ethical judgment determine the decision to become a cyborg? Influence of Ethical Judgment on the Cyborg Market. *Journal of Business Ethics*, 161(1), 5–17. 10.1007/s10551-018-3970-7

Reichel, P., Bassler, C. T., & Spörrle, M. (2024). Embracing the enhanced self now and in the future: The impact of temporal focus, age, and sex on cyborg products use intention. *Personality and Individual Differences*, 225, 112665. 10.1016/j.paid.2024.112665

Resimić, M. (n.d.). Changing of the Roles and Behaviour of Employees from the Aspects of Christian Spirituality and Ethical Leadership-From People Through Human Resources to Cyborg. *Academic Journal*, 8.

Sadat Ahadzadeh, A., Wu, S. L., Lee, K. F., Ong, F. S., & Deng, R. (2024). My perfectionism drives me to be a cyborg: Moderating role of internal locus of control on propensity towards memory implant. *Behaviour & Information Technology*, 43(5), 862–875. 10.1080/0144929X.2023.2190821

. Sharma, K., & Makhija, T. K. (2024). Transhumanism, AI, and Global Business: Navigating the Intersection of Augmentation, Intelligence, and Enterprise. Eduzone: international peer reviewed/refereed academic multidisciplinary journal, 13(1), 90-98.

Thomas, A. (2024). *The Politics and Ethics of Transhumanism: Techno-Human Evolution and Advanced Capitalism*. Policy Press.

Toker, K., Afacan Fındıklı, M., Gözübol, Z. İ., & Görener, A. (2023). To be a cyborg or not: Exploring the mechanisms between digital literacy and neural implant acceptance. *Kybernetes*. Advance online publication. 10.1108/K-07-2023-1297

Chapter 2
Can Human Enhancement Technologies Serve as Catalysts for Job Satisfaction in the Workplace?
A Human Resources Perspective

Jaspreet Kaur
https://orcid.org/0000-0002-3587-6841
Chandigarh University, India

Noble Preet
Conestoga College, Canada

Navdeep Chahal
https://orcid.org/0000-0002-0960-7883
Granville College, Canada

ABSTRACT

The study focuses on the impact of human enhancement technologies (HET) on job satisfaction from the human resources (HR) standpoint. It starts with the introduction and history of HET considering the types, cognitive, physical, and emotional booster, and recent developments. The chapter then provides details of the theoretical constructs used in the measurement of job satisfaction, the intrinsic and extrinsic incentive stemming from work, a favourable work environment, and the organizational culture. The consequence of the analyzed augmentation technologies include cognitive ones like brain-computer interface and the impact of memory, attention,

DOI: 10.4018/979-8-3693-7668-3.ch002

and problem-solving on physical enhancements like exoskeletons and prosthetics, and emotional technologies for stress management and well-being.

INTRODUCTION

Human Enhancement Technologies (HET) are defined as the approaches and tools used in redefining the capability of the human body and the brain to achieve an improved level. "HET" the wider category of the technologies that represent enhancement of human traits and abilities for the superior consist of the pharmacy-therapies, genetic enhancements and surgeries, external gadgets, nanotech enhancements and implants (Döbler et al., 2024).The primary goal of HET is to enhance people's capabilities over and above their abilities for the sake of enhancing the standard of living, raised production and overall human happiness of the common citizen. Modern IT and mathematics related technologies developed very lately because of the development of other related technologies like biotechnology, nanotechnology and so on, and the formation of HET (Bloomfield and Dale, 2015). All these developments, however, have not only enhanced the efficiency and accessibility of enhancements but along with it brought out ethical, social and legal issues that needs consideration. Now, the integration of HET in to the work, health, learning and social aspects of people, is defining a drastic shift the society's interaction with technology and each other.

Based on the analysis of this chapter, the objectives are as follows:

- ✓ To explain on the new technology Human Enhancement Technologies
- ✓ To deliberate on effects of Human Enhancement Technologies on the issue of job satisfaction at the place of work.

The introduction of HET in professional discourses can produce definite transformations of the attitudes toward and of the ways of working. Alternatively, in terms of the dynamics of the integration of HET with job satisfaction, the information provided in this chapter should serve as one that presents the beneficial as well as the harmful consequences that comes with these technologies (Bloomfield and Dale, 2015). From the above, one may be in a position to define organizational satisfaction as the state and general attitude that an individual holds with regard to the activities to be carried out within an organization. It, therefore, involves the level of satisfaction the employees have towards themselves, their duties, the workplace, etc The benefits of self-organizing work environments may include the following;

- ✓ The productivity level is high, the turnover rate of employees and health is good because contented workers rarely show symptoms of ill health.
- ✓ On the other hand, low job satisfaction attracts organizational ill health where employees experience burn out, truancy, and high turnover, which is inappropriate in the achievement of the goals of the organization.
- ✓ Therefore, understanding need of positive organizational climate stabilization and increasing the degree of employees' satisfaction is crucial. As HET asserts, it becomes explicitly crucial for the onsite observer to notice how these technologies affect: the state of perceived accomplishment, the state of perceived autonomy, and job satisfaction level of employees.

RESEARCH QUESTIONS AND OBJECTIVES

This chapter seeks to answer several key research questions regarding the intersection of Human Enhancement Technologies and job satisfaction:Thus, this specific chapter seeks to answer several main research questions touching on the HT&E and job satisfaction, which include:

1. What are the impressions of different categories of HET and their relationship with level of job satisfaction of the employees?
2. Consequently, the objective of this paper is to explain the positive and negative aspects of HET in the perception of the employee as well as the employer.
3. On the basis of individual provision of Information Technological Possibilities which are offered by contemporary world, it is possible to determine, which norms of professional ethics and regulation should be fulfilled, to minimize losses and to raise part which is filled in by usage of HET by representatives of different professions.
4. To what extent the organizations can opt for implementation of measures in order to integrate HET for promotion of job satisfaction without compromising with the ethical practices?

DEFINING HUMAN ENHANCEMENT TECHNOLOGIES

Human Enhancement Technologies (HET) can be well-described as technologies aimed at bringing human's capabilities beyond his/her limits or biology power. These technologies include improvements in different aspects or can also be described as cognitive, physical and even emotional boosts (Grewal et al., 2020). In contrast to conventional approaches to clinical therapy intended to bring the performance back to standard baseline levels, HET strive to provide more than the characteristics possessed by an ordinary person, increase effective work capacity, and improve the quality of life. HET can involve pharmacological treatments that enhance cognitive performance, robotic limbs as well as other devices that provide or enhance the abilities of the human body and gadgets that modify or enhance psychological and emotional conditions. However, the classification of HETs is constantly expanding based on new technologies which appear due to developments in biotechnologies, nanotechnologies, artificial intelligence, and cybernetics. These innovations foretell dramatic alterations in how people live in their own skin and heads, provocations concerning who we are and what we can do, or perhaps more fundamentally, what it means to be human (Döbler et al., 2024).

Categories of Human Enhancement Technologies

HET can be categorized into three primary types based on the specific aspect of human function they target: The first is cognitive enhancements, the second one is physical enhancements and the last one is the enhancement of emotional and psychological state (Grewal et al., 2020). Each of the categories contain a number of technologies and applications that are aimed at enhancing human performance in specific manners (figure 1).

Figure 1. Working of cognitive enhancement technology and their potential applications in improving brain functions and mental health

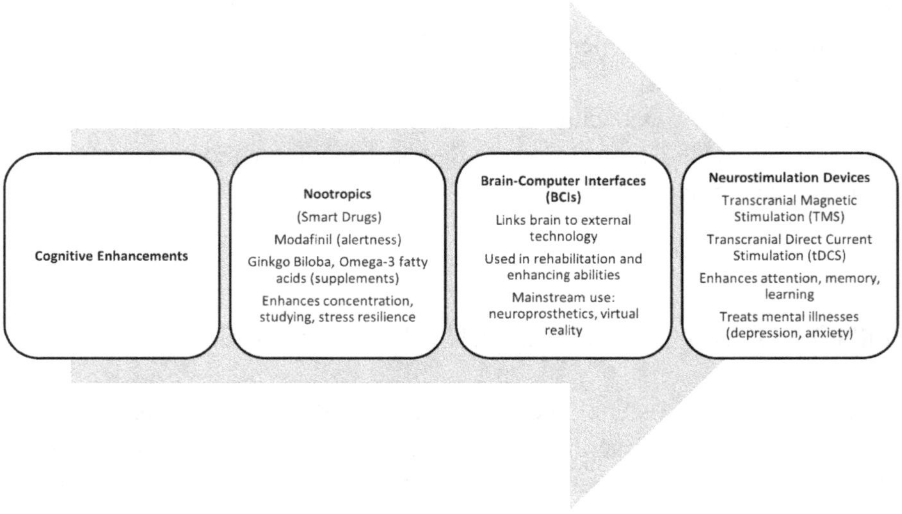

1. **Cognitive Enhancements**

The strategy of cognitive improvement implies the better condition of memory, focus, decision-making, and imagination. Some of these technologies are nootropics; BCIs, and Neuro stimulating devices (Pariseau-Legault et al.,2019).

- **Nootropics:** Otherwise known as smart drugs, nootropics are compounds that improve the efficiency of operations of the human brain. It could be drugs such as modafinil that is used to treat excessive sleepiness and to improve alertness, or supplements such as ginkgo biloba and Omega 3 fatty acids that are said to be beneficial to the brain. Everyone wants a better brain, that is why people consume nootropics: for better concentration, effective studying, and a stronger brain that can cope with stress in a high-pressure job (Döbler et al., 2024).
- **Brain-Computer Interfaces (BCIs):** BCIs are devices and tools that allow individual's brain and the outside world to interface. It can be employed in the process of the rehabilitation of patients with certain disabilities when an organism lost some functions or used in increase of apparent abilities in healthy people. BCIs function based on the identification of the signals in the nervous system and then issuing command to computers or equipment. The mainstream use of this technology might apply to neuroprosthetics, virtual reality and it might even alter how humans interface with computers and other technology (Saritas, 2019).
- **Neurostimulation Devices:** Some of these include the transcranial magnetic stimulation (TMS), and the transcranial direct current stimulation (tDCS) which are used to make electric or magnetic shocks in the brain to boost the desired cognitive properties. There are some evidences on the applicability of neurostimulation in increasing the attention, memory, and learning system and it is effectively being used to treat certain mental illnesses such as depression and anxiety (Pustovrh et al., 2018).

2. **Physical Enhancements**

Physical enhancements are more related to improving or providing body capabilities beyond those normally inherent to human beings through such things as advanced limb replacements, external frameworks for the body such as exoskeletons, or even genetic engineering (Hogle, 2005).

- **Exoskeletons:** These are devices that have to worn by a person and are designed to help him or augment his strength. By the use of exoskeletons, strength and endurance can be improved in healthy persons, particularly where a lot of physical work is involved. They are also useful specifically in medical and rehabilitation environments where people with mobility disorders can find the means to move again (Pustovrh et al., 2018).

- **Prosthetics:** Modern prosthetics do not only include the ordinary looking fake limbs but they are built to connect with the user's nervous system in order to inform the limb. These prosthetics can regain missing activities and, at times, even superior ordinary tasks. For instance, the advance in technology where artificial limbs for users which was formerly developed to enable a prosthetic limb user to feel touch and pressure is a major improvement as it enhances a prosthetic limb user's ability to a great extend to carry out his/her normal activities in society (Döbler et al., 2024).
- **Genetic Modifications:** Essentially, technologies type CRISPR-Cas9 could be used to perform selective changes within genes that are responsible for diseases, thus fixing the disorders or improving an individual's desirable traits. Organisms can be changed genetically in a way that allows them to have enhanced muscle pull, disease immunity among other values (Pustovrh et al., 2018).

Emotional and Psychological Enhancements

Affective and psychological upgrades involve boosting one's positive feelings or interacting with devices in order to elevate emotional healing and mental health (Coeckelbergh, 2017).

Mood-Regulating Devices: These are equipment's like Deep Brain Stimulation (DBS) and neurofeedback equipment. DBS comprises of implantation of electrodes in the brain to control what is regarded as abnormal neural activity, with the ailments that it is used to treat including Parkinson's disease as well as depression. Neuro feedback devices provide the enhancement of training the individuals concerning the demands of their brain in terms of relaxation, or for concentration and also in controlling emotions (Brey, 2009).

Virtual Reality (VR) Therapy: Applying VR to therapy, one can turned it into stress, anxiety, and post- traumatic stress disorder therapy. It refers to engaging patients in realistic virtual environments representing situations that are either soothing or evoking symptoms of the patient's condition to guide exposure therapy and gradually master the primary emotional problem (Döbler et al., 2024).

Psychopharmacology: It entails the use of drugs to control and also improve the various feelings and mood of a person. Current drugs target the specific diseases of depression, anxiety, and other related disorders and whereas new exert molecules for optimum mental performance focusing on the overall population of an aforementioned nation (Saritas, 2019).

CURRENT STATE AND ADVANCEMENTS IN HET (FIGURE 2)

Figure 2. Current developments and applications of Human Enhancement Technologies (HET)

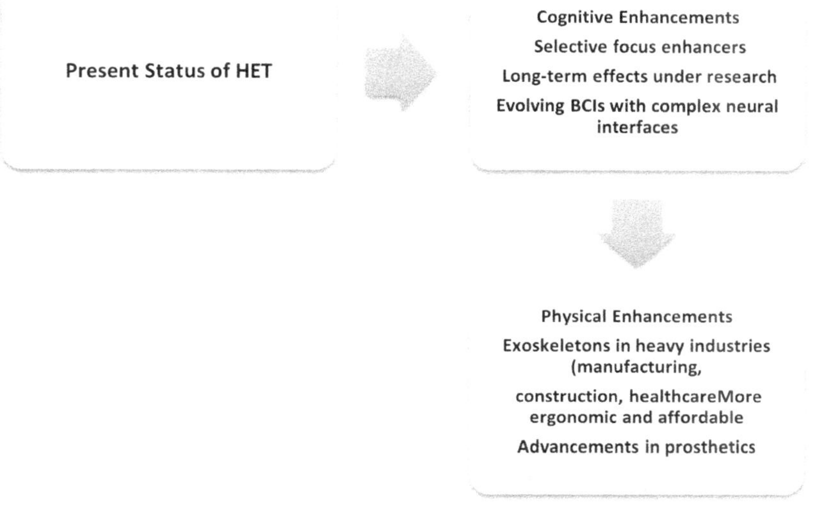

The present status of HET is characterized by continuous development and expanding applications across different facets of society. In the cognitive area, much has discussed, and more and more people use the selective focus enhancers, but the effect of their long-term use is still widely being researched. Currently, BCIs are evolving step by step with more complex interfaces of neural interaction and enhanced processing of the signals which are encouraging us towards fulfilling the idea of man-machine merger. Techniques in neurostimulation are also improving and more promising methods of usage are beginning to appear (Grunwald, 2007).

3. Physical Enhancements

In the case and classification of physical enhancements, exoskeletons are no longer now considered as novelty robotic suits, but are becoming tangible robust technologies to augment heavy industries like manufacturing, construction, and healthcare. These devices are also more ergonomic and relatively cheaper; anybody can afford to have them these days. The field of prosthetics is improving due to the changes in supply materials in the materia science and advanced robotics that makes them more functional and user friendly. Genetic treatments are relatively untested in humans but present promising outcomes in medical and cosmetic applications while being considered risky and permitting only limited liberal use and/or requiring extensive governmental approval (Grewal et al., 2020).

Emotional and psychological changes are further being underlined as augmenting at a fast pace. Smart lightning and virtual reality therapy are considered efficient for handling mental disorders at the moment. The area of pharmacological examine related to the mind remains on the rise, and more and new drugs and chemical compounds are being advanced as ways to improve mood and cognition (Döbler et al., 2024).

ETHICAL AND LEGAL CONSIDERATIONS

The development and use of HET provoke a series of social, ethical and legal issues that need to be examined so that HET can be used properly, and in a fair way. These are consent, fairness and possibly coercion. For example, the use of HET should be voluntary meaning that no one should force a person to use it or refrain from using it. The access to HET could even enhance the social inequalities if only some particular groups will be able to afford or benefit from these technologies. There is also an implication of negative signification or the creation of new discrimination where HET tends to put a premium on enhanced individuals and those who decide or are unable not to enhance place themselves in a position of jeopardy in competitive areas such as at the place of work. Security of data is highly valued because the data collected and analyzed in therapeutic processes by using BCIs and neurostimulation devices are highly confidential and personal. People's data should be protected and processed in a correct way, this is important to trust and avoid abuse (Lin & Allhoff, 2008).

Thus, speaking of the current regulations, it is possible to conclude that they need strengthening in order to effectively combat the elements that can be linked to HET. This also consists of guidelines that assist in defining the components of development, implementation and evaluation of the mentioned technologies; that is innocent in development, secure in implementation, and respects the dignity and rights of humans. this has a dimension of integrating the international players in the regulation and standardization processes as the international environment demands harmonization of such laws across different jurisdictions. In conclusion, it is possible to notice that Human Enhancement Technologies seems to be one of the most significant and the most rapidly evolving branches, which can predict significant shifts in the perspectives of different aspects of human life including work, health, and the quality of the individual's life (Bostrom & Roache, 2008).

In any case, it is clear that the benefits of HETs are apparent; nonetheless, it is equally indicative that there exist questions of ethic and law associated with their use; therefore, the discussions of advantages of such matters should be consulted carefully to ensure that technologies will enhance the quality of human life as is ethical, legal, and more importantly, moral (Ziesche & Yampolskiy, 2020).

JOB SATISFACTION: THEORETICAL FRAMEWORK

The satisfaction with the job is the overall level of contentment an employee has concerning his or her job.It is described as openness that may have components related to that employees' perception of their job description, responsibilities, physical context and organizational interactions with other people. Job satisfaction can be understood through various dimensions: emotional, the member's intellectual personality and the motor (Kaur & Arora, 2023).This affective aspect of work is workers' attitude toward their job; those emotions are the feelings of pleasure, zeal or irritation. The cognitive component of the employees' attitude includes their views on rights and responsibilities, the role of the job, the perceived justice of organizational decisions and perceived job relevance. Lastly, the behavioral implication covers the outcomes of job satisfaction or dissatisfaction, such as; productivity, performance, attendance, and turnover. Knowledge of these dimensions aids organizations to come up with methods of addressing certain areas in order to promote the general satisfaction of employees in their assigned working stations (Judge et al., 2020).

Factors Influencing Job Satisfaction

There are a number of variables, which have significant roles to play in the determination of the employee job satisfaction. These factors may be grouped under contents factors and process factors.

Intrinsic vs. Extrinsic Factors

Intrinsic Factors: Intrinsic factors, on the other hand, refer to the Holland's conception of the job and the self-fulfillment that the employee gets from the job he or she is assigned to do. These include aspects such as the content of the work, employee's developmental opportunities, choice, and the achievement of organizational goals. Those employees whose jobs are naturally satisfying, they are going to experience high level of job satisfaction because the job itself provides satisfaction. For instance, a software developer who solves problems and develops unique software is likely to be content with his or her position as it involves the strength of interest (Kaur & Arora, 2023).

Extrinsic Factors: Hypotheses related to extrinsic aspects of the job refer to factors that are outside the job and these are- Pay, employment security, physical environment and circumstances of work, organizational policies/standards, and relationship with supervisors or fellow workers. Such factors give pain to the job satisfaction as these factors respond to the material and social requirements of the employees. For instance; attractive wages, secure employment, considerable support from supervisors and organization, and pleasant workplace all make job satisfaction possible by offering those job resources and communities by which can enable the employees to carry out their tasks proficiently. On the other hand, poor extrinsic factors like low wages, said environment, etc. will lead to dissatisfied and hence the employees do not perform well (Judge et al., 2020).

EMPLOYING WORK ENVIRONMENT AS WELL AS ORGANIZATION CULTURE

The study seeks to appreciate the fact that culture of work or organizational environment influences job satisfaction. Organizational climate or work environment can thus be defined as the 'space or context within which organizational members carry out their tasks and deals with the characteristic facets of working context including physical structure of the working place, physical tools employed and the noise level, safety and security. Comfort, safety and duration taken by the employees to perform their tasks are known Worthy work environment since if these values are provided, chances of job satisfaction are high (Acquah et al., 2021).

Closely related to the development of organizational culture is the concept stating that organizational culture refers to specific standards for action and decision-making as generated by the aspiration of the organization toward uniformity among employees. If, within the framework of an organizational culture, managers and other employees at all organizational levels understand and behave in a way that supports working together with others, providing help accepting recognition for achievements, then satisfaction at work will increase.

If the employees receive those things, it is certain that they accept the organizations as their own, and therefore they will have high job satisfaction, organizational commitment and personal / organizational goals' congruence. On the other hand an organizational culture which is vice as adopted with parameter like conflict, no recognition and poor communication leads to more dissatisfaction and high turnover rates (Thant & Chang, 2021).

THEORIES OF JOB SATISFACTION

The construct of job satisfaction in addition to the variables that makeup the concept is rather some what complicated, and therefore there are several theories out there. Other theories are; the need hierarchy theory by Maslow, the two factor theory by Herzberg and the job characteristics or model (figure 3).

Figure 3. Key components of Theories of Job satisfaction

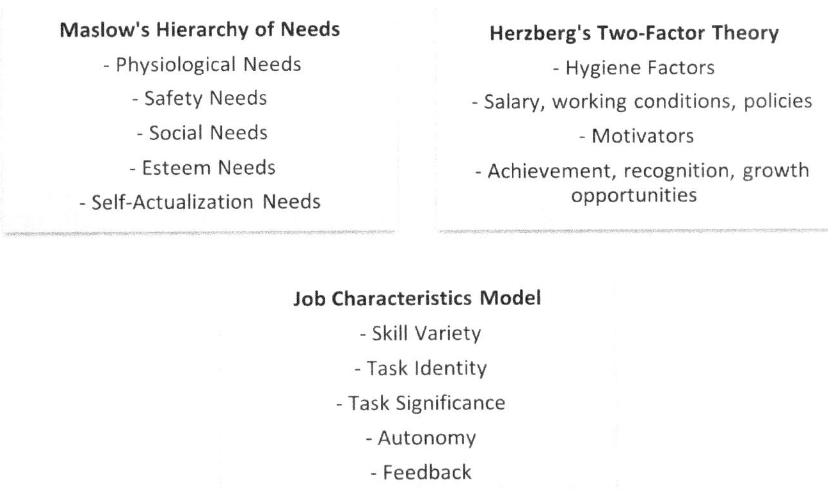

Maslow's Hierarchy of Needs: According to Maslow need hierarchy theory, people's requirements are divided into 5 categories from the most elemental ones to the most sophisticated. In a scenario where job satisfaction is the focal point of this theory, then the organization will endeavor to satisfy these needs in a prescribed order (Adamopoulos, 2022). The hierarchy consists of five levels: such as physiological requirements, including the need for pay and proper certification, in the second category we have safety requirements such as safety from injury or disease, followed by affiliation needs such as interaction requirements with other employees in the organization, recognition needs that are requirements for recognition by other employees or employers, and finally the actualization needs which are the requirements that allow the individuals to grow and accomplish his or her potential in the work place. Maslow stated that In order to achieve the satisfaction or to motivate people, there are certain conditions or premises that must be fulfilled or met; the lower order needs must be satisfied to permit the sixth level needs to surface (Thant & Chang, 2021).

Herzberg's Two-Factor Theory: Herzberg has distinguished between what he referred to as hygiene factors and motivators (Patricia & Asoba, 2021). Hygiene factors into focus include salary, working conditions, company policies, and supervision among others are extrinsic factors and lack of them is likely to cause dissatisfaction among employees. Nevertheless, enhancing these variables does not lead to job satisfaction; instead, it only acts to negate it. Whereas, the pull factors are primarily internal factors associated with the actual character of the job, such as; challenge, satisfying, autonomy, and promotion. If present these factors have the potential of improving job satisfaction. Herzberg's two-factor theory reveals that motivators, which only increase job satisfaction if properly attended to hygiene factors, should be the priorities of management (Rokeman et al., 2021).

Job Characteristics Model: Among the many elements of work in context theory, Hackman and Oldham's job characteristics model asserts five core job dimensions-generating job satisfaction and motivation including: skill variation, task totality, task importance, independence and feedback. Skill variety defines the extent to which an occupation entails various types of activities and skills. Task interdependence focuses on the requirements of carrying out a whole and identifiable job. Task significance is the degree and the extent to which the job affects other people. In this case, autonomy is the amount of control or amount of choice that is available for choosing the method for completing the job. A performance feedback is the communication of results that show the degree of task accomplishment. In studies that are based on this model, jobs which have high scores on these dimensions are presumed to result in higher job satisfaction, because they represent challenging, interesting, meaningful, and fulfilling work that gives people a feeling of accomplishment (Ahmad et al., (2021).

Altogether, the job satisfaction is a complex concept which depends on numerous motives intrinsic and extrinsic aspects of the job, pay, work conditions, and organizational climate. Knowledge of the conceptual models of job satisfaction including the Maslow's hierarchy of needs, Herzberg's two-factor theory of motivation, and the job characteristics model will assist organizations in formulating ways to improve this aspect to ultimately increase the standards of the welfare of their workers and organizational outcomes. It therefore becomes clear that there many ways employers can satisfy employees' needs in a work environment, reduce turnover and increase organizational productivity by improving intrinsic and extrinsic motivations as well as the organizational culture (Korir & Ndegwa, 2020).

IMPACT OF COGNITIVE ENHANCEMENT TECHNOLOGIES ON JOB SATISFACTION

These are technologies that seek to advance some or several areas of the human brain or components that could easily be understood like memory or attention or even aspects that propel problem solving or decision making abilities which are essential to meet organizational undertakings successfully and enthusiastically. Therefore, it is possible to improve these cognitive functions for the sake of increasing the employees' efficiency, attaining challenging goals, as well as receiving desired positions. For instance, modafinil, and piracetam are stationed under the broad title of smart drugs through which people ingest some substances with an aim of enhanced thinking abilities, memory and focus (Acarturk & Mucen, 2022).

Another way of enhancing the human brain is through; fast and efficient decision making and also the ability to intentionally switch from one activity to the other. Other methods of brain stimulation which are considered to be non invasive include TMS and tDCS and they have also been noted to assist in enhancing attention and flexibility of thought. The above technologies can operate to change a workspace by improving the people's cognitive features that are essential to achieve a certain job and, therefore, the satisfaction level that the people have at the workspace. As found out from the above-mentioned identified literature, some empirical research studies along with case studies can be sourced to expand the investigation on over-hypothesized effect of CET on specific facets of employees' job satisfaction. For instance, one of the major experiments that relate to the nootropic utilization in the employees doing stressful jobs like trading, Analysts, architects, etc. is that the nootropic compounds may help enhance the co-operative concentration, mental exhaustion, and other various mental toppling abilities (Pustovrh et al., 2018).

The integration of cognitive enhancement technologies into the workplace presents numerous potential benefits and drawbacks. On the positive side, these technologies can lead to improved job performance, higher productivity, and greater job satisfaction. Enhanced cognitive functions such as memory, attention, and problem-solving can help employees handle their tasks more efficiently, reducing stress and increasing their sense of accomplishment. This, in turn, can lead to greater job satisfaction as employees feel more competent and valued in their roles. Additionally, cognitive enhancements can foster innovation and creativity, as employees are better equipped to think critically and generate new ideas (Bloomfield & Dale, 2015).

However, there are also significant drawbacks and challenges to consider. The long-term effects of cognitive enhancers like nootropics are not yet fully understood, raising concerns about potential health risks and dependency. The use of BCIs and neurostimulation devices also raises ethical and privacy issues, as these technologies involve the collection and manipulation of neural data. There is a risk that employees may feel pressured to use cognitive enhancements to keep up with their peers, leading to ethical dilemmas around consent and autonomy. Moreover, reliance on cognitive enhancers could diminish natural cognitive abilities over time and create a workforce that is overly dependent on technology for basic cognitive functions (Acarturk & Mucen, 2022).

IMPLICATIONS FOR WORKPLACE PRODUCTIVITY AND CREATIVITY

Cognitive enhancement technologies are highly topical with regard to aspects of efficiency and innovation at the workplace. In that aspect, the technologies may actually be beneficial to the organization in boosting the performance of the employees and in also offering a better approach to do things and better manner of operating. It will be easier for employees to remember newly acquired information and long term data while they also exhibit better focal attention on work-related tasks, they/ are also capable of coming up with better decisions and solving problems. This can result in increased output as the employees can complete their work faster and with fewer mistakes. Cognitive enhancement technologies can also augment creativity in the work place to a great degree (Marois & Lafond, 2022).

Mental rigidity and fatigue is minimized in these workers, and hence they are able to come up with new ideas on how to solve various problems in the organization hence an innovation workplace is fostered. This can lead to new ideas and concepts in the organization which in turn broadens it and hence improves on the competitiveness of the organization. On one hand it is a very good thing that organizations take a critical responsibility on the advocates to use those technologies responsibly and deal with health and privacy effects (Dubljević et al., 2020).

Thus, it could be suggested that technology for the cognitive enhancement can highly impact the factor of job satisfaction, based on the presumed improvements in areas of memory, attention, and various problem-solving skills. The study shows that the mentioned technologies can enhance the job performance and satisfaction simultaneously it is evident that these technologies are accompanied with risks and ethical concerns. What has been established is the fact that companies have to give much attention to the questions that relate to advantages and disadvantages of using cognitive enhancements and that there is a need for the formulation of rules that do apply to the ethical use of the enhancements in corporations. Hence, they may benefit from the opportunities of the cognitive technologies and getting at the same time organizational effectiveness, creative work place, pleasant or healthy employee and his freedom (Weekes & Eskridge, 2022).

PHYSICAL ENHANCEMENT TECHNOLOGIES AND THEIR ROLE IN JOB SATISFACTION

Physical enhancement technologies aim to augment human physical abilities, thereby increasing strength, endurance, and dexterity. These technologies encompass a range of innovations, from exoskeletons and advanced prosthetics to wearable devices and genetic modifications. Exoskeletons, for instance, are wearable robotic suits that enhance the wearer's strength and endurance by providing mechanical assistance to muscles and joints. These devices are particularly useful in industries that require heavy lifting, prolonged physical activity, or repetitive tasks, such as manufacturing, construction, and logistics. Advanced prosthetics, which can restore or even enhance physical capabilities, integrate with the nervous system to provide more natural control and sensory feedback, improving the user's dexterity and fine motor skills. Wearable devices, such as smart gloves and augmented reality (AR) systems, can enhance hand-eye coordination and precision, making them valuable tools in fields like surgery, craftsmanship, and high-tech assembly. Genetic modifications, though still in the experimental stage, hold potential for enhancing physical attributes such as muscle strength and stamina. By improving physical capabilities, these technologies can significantly impact job satisfaction by enabling employees to perform their tasks more effectively and with less physical strain (Langton & Ramseur (2001).

EMPLOYEE PERCEPTION AND ACCEPTANCE OF PHYSICAL ENHANCEMENTS

It is important to understand and obtain favourable attitudes toward the technologies for the enhancement of the physical body among employees because this determines the success of the implementation of the technologies for the improvement of job satisfaction. HRM can incorporate the following tenets: each of the identified technologies carries certain benefits and risks, employees' attitudes, and perceived usefulness, as well as alignment with both personal and cultural values. According to research, subgroups of the population are typically more accepting of modifications that provide visible and tangible returns for the business, its workers, and customers in terms of efficiency, stress alleviation, and safety (Duan et al., 2024).

For example, employees in lifting-heavy occupations will appreciate technologies such as exoskeletons and enhanced prostheses since they would ease their work and reduce the risk of injury. However, there could also be some issues associated with the possible long-time health or/and ethical consequences and risks of applying such technologies. People can be concerned with the configuration of these devices to their bodies, that their occupations will be automated, and their levels of independence and individuality. Thus, to solve the indicated concerns, the leaders of employing organizations have to discuss with their employees the distinctive characteristics of enhancement physical technologies and their potential advantages and negative facets. Actually, training and education could also let the employees know how to properly implement these technologies thereby enhancing the technologies acceptation and satisfaction from the employees (Bação et al., 2023).

LONG-TERM EFFECTS AND SUSTAINABILITY

The jobs, career treatment and welfare of the employees are the major issues at play so the impact and future-ability of enhancement tools are major considerations. Despite providing short-term solutions that are full of immense benefits, it is important to look at top level effects these technologies would have on health, job satisfaction and the organizations in the long-run. For instance, the device like exoskeletons that are frequently used in industries as well as other advanced prosthetics have the possibility to have negative impacts on health among the users, which may include muscle weakness or joint pains if the periods of using the devices are not well regulated. Long-term and periodic surveillance and appraisal of people's physical and mental conditions are critical to guarantee the prudent and efficient use of the specified technologies (Kaur, 2024).

However, the costs and overall maintainability of the enhancing technologies, which are primarily fleshly, might be huge, provoking the issue of the economic feasibility. It is necessary for an organization to look into the cost of the equipment, the costs of maintaining it, and even the costs of acquiring a new piece of equipment if the current one is defective, as well as the opportunities that come with a decreased number of injuries and boosted productivity. The other consideration here is thereby the environmental sustainability whereby the manufacturing, application as well as degradation of these technologies affects the environment. Aim at the conservation of the environment while designing and implementing technological appliances for physical improvement. Last but not the least, the sustainability of these technologies in the long-run can only be possible if there is a constant evolution and enhancement of same. Additionally, organizations require investing on researches and development as a way of adapting to the new technologies while at the same time providing employees with the most adequate and efficient means of mitigating the identified issues and concerns (Duan et al., 2024).

EMOTIONAL AND PSYCHOLOGICAL ENHANCEMENTS IN THE WORKPLACE

The matter of stress and equality of the employees in the context of demands on the contemporary workplace has now become a significant factor because most working positions entail fast decisions and exerts considerable pressure on the employee. Problematic behaviors which these technologies can effectively manage, can notably contribute to the solution of the listed problems and the overall well-being of personnel, as well as orderliness and effectiveness within working environments. Other LTTs are portable stress monitoring equipment, biosensor feedback, stress controlling applications, such as meditation, and Virtual Reality (VR) based therapy, and other digital mental health solutions. Smart eyeglasses and fitness and health wearable devices will thus be able to record the levels of stress through some parameters such as heart rate variability and cortisol and give recommendations to the client depending on the condition (Bloomfield & Dale, 2015).

Biofeedback devices are one of the nuts-and-bolts technologies, which teach individuals various functions of their bodies using signals in the form of sound and light and, therefore, can assist the employees who suffer from such ailments as anxiety and emotional instability. Online applications for mobiles like head space and calm offer sessions that are positive and productive which can be integrated in user's working hours. Clients' perception is introduced to the concept of virtual reality therapy to help in developing an environment to reduce stress and improve client's mental treatment. In addition, the digital mental health platforms deliver the professionals' counseling and support, which the workers can access, and seek in secret, and at any given place. As a result, the employers, while introducing such technologies into the workplace, provided the conditions of support (Grewal et al., 2020).

SOCIAL RELATIONSHIP AND TEAMWORK CONSEQUENCES

Emotional and psychological improvements are of similar significance and they work on the relationships between people and overall teamwork in organizations. Further, when the employees learn ways to cope with pressure and stress and get a healthy psychological state, then they treat their fellow workers in a good manner hence, creating a healthy environment for teamwork. Technologies that nurture well-being can boost employees' abilities, especially in different aspects of close cooperation, such as empathy, interpersonal skills, and conflict management within groups (Kaur, 2024).

For example, through biofeedback and mindfulness, personnel will be in a position to manage how they feel, think and react hence, controlling outbursts during interpersonal communications. The authors also note that the increased self-awareness may decrease the number of times people misunderstand or misinterpret one another, thus preventing a lot of conflict in the relations. Moreover, using VR therapy and available digital mental health applications, team building and classes on emotional intelligence can also be conducted improving interpersonal dynamics and team collaboration. Thus, it can be stated that with the help of these technologies, a positive organization culture is created where each worker is happy, supported, and willing to perform tasks cooperatively (Saritas, 2019).

Part Played in Alleviating Burn Out Syndrome and Improving Work-Life Balance

Where workplace stress is left unaddressed, over time the combination of physical, mental, and emotional exhaustion that defines burnout becomes a major concern that impacts workers positively. In this context, it should be noted that technologies that help to enhance the employee's emotional and psychological state can be objectively effective against burnout and for maintaining the healthy life-work balance. One of the commonly applied technological solutions is the implementation of stress monitoring gadgets and applications that are aimed at raising awareness of such early signs of burnout and prompt actions to prevent them (Cvenkel, 2021).

For instance, exercises such as those provided by mindfulness and meditation smartphone applications lead to lower stress levels and build up emotional strengthening thus decreasing the employee's likelihood of broker. Electronic suits are bracelets that constantly report on the current levels of stress and suggest that the employee relax and take a break from work. Mental health applications and other related online platforms can help stressed employees deal with anxiety or other mental disorders and can act, therefore, as an opportunity to seek help. Of course, such technologies will benefit in addressing one's stress and having a positive outlook, thus aiding an employee in balancing his/her life as a full- time employee and a family man/woman. This balance is very helpful in avoiding burnout, because even the health of an employee is conserved thus resulting to more efficiency and joy in their work (Kaur & Arora, 2023).

ETHICAL IMPLICATIONS AND PRIVACY CONCERNS

The disclosures of the human body's emotional and psychological improvement technologies at the workplace come with other importance considerations such as ethical issues of privacy to consider. On the same note, there is always the risk that more information about the employees will be exploited or the data treated poorly. Stress and mental health technologies collect personal information on people and for this reason, people's privacy and equality can be infringed. Employers need to make sure that through the mentioned technologies, data is protected and only utilized for a specific purpose. Another important component is the freedom to choose, implying that the employees that are targeted by these technologies should be allowed to decide whether or not they want to partake in them with no penalties for refusing. Also, more worrying is the downside potential of these technologies, that is if they will be employed in a forceful manner (Lin & Allhoff, 2008).

Employers should not give their workers the feeling that they have to engage in the usage of the emotional enhancement tools or mental health programs. However, these resources should be provided as optional employee assistance where one can opt to make use of a particular assistance scheme or not depending on their situation. In addition, it is critical to address the possible long-term psychological effects of outsourcing people's ability to care for their emotional needs to technology. Even though these applications can deliver certain benefits in the short time, it is important not to rely on them instead of consulting with a person and using the services of a mental health worker (Bostrom & Roache, 2008). Managers should ensure the availability of human support programs for the employees' emotional health, for example, counseling and peer support. Thus, understanding and solving these ethical and privacy problems, organizations can use employee's emotional and psychological improvement technologies, hence boosting employee performance, morale and satisfaction without trampling on their rights and freedom.

Lastly, it could be suggested that most of the SEL and psychological boosting technologies present a vast opportunity to enhance the efficacy of stress and emotion regulation as well as to increase workers' psychological well-being and job satisfaction. Thus, the technologies aimed at assisting people to build an effective stress coping and providing proper emotional support could also improve the relations between people, encourage the teamwork, and decrease the number of employees who could become susceptible to burnout. However, the effectiveness of such technologies in organizations must be balanced with the ethical and privacy implications so that employees' rights won't be violated and it remains more of an option and a help. It is therefore necessary for organizations to be open, ensure that participants understand the steps being taken and secure data in the process of updating emotional and psychological needs of the employees. In that way, organizations can reap the advantages of these technologies to fashion a healthier, more productive, and more fulfilling work environment for their personnel (Ziesche & Yampolskiy, 2020).

FUTURE TRENDS AND PROSPECTS

New technologies are expected to deliver integrated advancements in the human intellectual, bio-mechanical, and psychological functioning. Without a doubt, one of the most appealing fields relates to the enhancement of contemporary neuro-technology and the introduction of further refined systems of brain-computer interfaces (BCIs). Such BCIs are anticipated to progress from today's models, which are designed to aid people with disabilities, to universal mechanisms for boosting cognitive processes like memory, focus, and learning in the workforce population (Bloomfield & Dale, 2015).

On the same note, other technologies based on gene editing such as CRISPR-Cas9 are also rapidly evolving with possible uses in improving the prowess of muscles and their endurance, brain function, and ability to resist diseases. Thus, the further development of exoskeleton technology and prosthetics are prospective in the sphere of physical enhancements. Subsequent generation exoskeletons also tend to be more lightweight, easier to control and provide even better assistance and freedom of movement. Such enhancements can increase the usage of exoskeletons in sectors that involve manual labor, thus cutting down the rates of workplace accidents, and increasing effectiveness. Prosthetics should also become improved with the use of AI and machine learning algorithms and enhanced movement and sensation of the limbs. Here it could be pointed out that these improvements will be effective for people with such disability as limb loss and they will be able to accomplish tasks with less difficulties, and with more accuracy.On the emotional and psychological side, the stakeholders expect that virtual reality (VR) and augmented reality (AR) will be significant. The use of VR and AR makes it possible to develop mental health treatment and opportunities for recreational and calming scenarios (Grewal et al., 2020).

As for the future of HET at the workplace, this field is virtually inexhaustible and it is only a matter of time before it changes the way we work as well as communicate. Cognitive enhancements could mean that the workers can be smarter in processing information at work and dealing more efficiently with their tasks, as well as yielding better solutions to problems. For instance, better memory and improved learning ability would mean that training took lesser time and could help employees quickly develop new skills; consequently, a more flexible labour force would be created. Perhaps these enhancements could the generator of new ideas in knowledge-based industries for example in the financial industry or technology, legal sector among others. Physical augmentations will surely revolutionize industries which involve optimization of muscle strength and physical strength generally. Powered suits and better prosthetics could mean that physically strenuous professions are less likely to lead to physical harm and diseases. It would also increase production and the quality of job done because workers would not be over-fatigued or pulled out of position by constantly having to stand or sit. In construction, manufacturing, and logistics industries, utilization of such technologies could result in vast enhancements of productivity and employees' health (Kaur, 2024).

The advancements based on emotions and psychology will equally impact the organizational culture and entitle employees as well. They are the components of the framework that in link, with technologies to support mental health and emotional regulation, contribute to improving the work environment. For example, VR therapy sessions or mindfulness applications that can be incorporated in the workplace help employees build the stress management and focus as well as the EI skills. Such interventions can foster improvement of the organizational working culture and climate, where employee is recognized and appreciated.

CHANGES IN THE FUNCTIONALITY OF HR IN REGARDS TO MANAGEMENT OF HET

✓ Consequently, as more applications of HET enter the workplace, Human Resource (HR) for your company will be revolutionized. An appreciation of these technologies' capacities and ramifications will be expected from HR professionals for those wanting to handle their adoption and utilization. This will encompass coming up with new policies and practices in relation to the ethical, legal, and realistic use of HET. A notable task of HR will be to guarantee that HET is being used willingly and the workers understand the advantages and drawbacks of this measure. This entails developing elaborate informed consent procedures and coming up with awareness creation and sensitivity to enable employees make constructive decision on the technologies of enhancement. For this reason, legal and compliance teams will be useful for HR to consult when implementing and incorporating HET into the organization because the utilization of HET probably falls under some kind of regulation.

✓ Data privacy and security will be major issues as more people and sensitive information will be involved due to HET. Also, for effective implementation of the change, HR will need to introduce strict procedures to protect the data of the employees and ensure that they do not lose trust in the organization. This include following up on things such as, how data collected from, neuro technology and other enhancements are put to proper use and for the right reasons such as from wearable devices.

- ✓ Another responsibility is also felt in the area of diversity, inclusion and equity, as HET gains ground; this, under the coordination of the HR department. One of the most crucial tasks that will need to be accomplished is to have a non-discriminatory access to the enhancement technologies to avoid exacerbation of prejudicial difference or reinforcement of existing ones. This may entail offering HET in connection with the company's health insurance programs and making sure improvements are not only accessible to employees, but also open to all personnel without restrictions to rank or job title.
- ✓ However, what one should do is to consider if the policies and strategies that are being implemented will impact on the present job content and process particularly to the HET incorporation at HR as well as the career opportunities that are explored at this firm. This poses problems to HR as the varied tasks will gradually be made easier or outsourced to development technologies, and the employees will need management's help in searching for new opportunities for training the development of the specific skills. This will require name spearhead to develop courses that meet the new skills that workers and employees need to acquire in a better workplace.
- ✓ Lastly, it will be the role of the initial hosting Human Resource department to evaluate the degree of HET concerning the health and productivity of the employees. As such, it involve the collection of data regarding the effects of these technologies on satisfaction, output, and wellness. This is another reason on which, concurrently with the application of HET, HR should assess the outcome in order to identify the most suitable approaches for increasing its effectiveness with targets' gain on the side of the employee and the organization.

CONCLUSION

Thus, the general perspectives concerning the enhancement of human capabilities that would define the future of work environment is optimistic. Advanced Neurotechnology, Gene Editing, Exoskeletons and Emotional Well-being are the emerging technologies which has tons of potential to enhance neural and physical capacities along with well-being. Such possibilities mean that the improvements could enhance the productivity, quality and satisfaction of working experiences for people in different fields of work. Yet, setting up for a successful implementation of HET in workplace implies addressing ethical, legal as well as practical issues. Yet, enhancement technologies should be used sparingly, voluntarily, fairly, and in a way that respects the employees; HR will be central in this matter. Through culture change, the provision of the right training, and an ongoing appraisal of HET, HR can play the most significant role of managing these tools to ensure that the advantages are availed while managing the potential negative implications.

Therefore, managers should pay attention to literature advancements in HET and adapt to challenges while utilizing opportunities in the future. Incorporating these new technologies sensibly and ethically, organizations can build the future where evolving human capabilities result in improved satisfaction and well-being together with an organization's success.

REFERENCES

Acarturk, C., & Mucen, B. (2022). Performance in the workplace: A critical evaluation of cognitive enhancement. *NanoEthics*, 16(1), 107–114. 10.1007/s11569-021-00407-6

Acquah, A., Nsiah, T. K., Antie, E. N. A., & Otoo, B. (2021). Literature review on theories of motivation. *EPRA International Journal of Economic and Business Review*, 9(5), 25–29. 10.36713/epra6848

Adamopoulos, I. P. (2022). Job satisfaction in public health care sector, measures scales and theoretical background. *European Journal of Environment and Public Health*, 6(2), em0116. 10.21601/ejeph/12187

Ahmad, M., Khan, A., & Arshad, M. (2021). Major theories of Job Satisfaction and their use in the field of Librarianship. Library Philosophy and Practice (e-journal), 6385.

Bação, P., Gaudêncio Lopes, V., & Simões, M. (2023). AI, demand and the impact of productivity-enhancing technology on jobs: Evidence from Portugal. *Eastern European Economics*, 61(4), 353–377. 10.1080/00128775.2022.2064307

Bloomfield, B., & Dale, K. (2015). Fit for work? Redefining 'normal'and 'extreme'through human enhancement technologies. *Organization*, 22(4), 552–569. 10.1177/1350508415572507

Bostrom, N., & Roache, R. (2008). Ethical issues in human enhancement. New waves in applied ethics, 120-152.

Brey, P. (2009). Human enhancement and personal identity. In *New waves in philosophy of technology* (pp. 169–185). Palgrave Macmillan UK. 10.1057/9780230227279_9

Coeckelbergh, M. (2017). Cyborg humanity and the technologies of human enhancement. Philosophy: Technology, Macmillan Interdisciplinary Handbooks, 141-160.

Cvenkel, N. (2021). Work-life balance and well-being at work: Employees' perspective to promote a psychologically healthy workplace. The Palgrave handbook of corporate social responsibility, 429-451.

Döbler, N. A., Carbon, C. C., & Schaub, H. (2024). Human Enhancement Without Organizational Knowledge and by Organizational Order. *Journal of Cognitive Enhancement: Towards the Integration of Theory and Practice*, 8(1), 170–183. 10.1007/s41465-023-00278-7

Duan, S. X., Deng, H., & Wibowo, S. (2024). Technology affordances for enhancing job performance in digital work. *Journal of Computer Information Systems*, 64(2), 232–244. 10.1080/08874417.2023.2188497

Dubljević, V., McCall, I. C., & Illes, J. (2020). Neuroenhancement at work: Addressing the ethical, legal, and social implications. Organizational neuroethics: Reflections on the contributions of neuroscience to management theories and business practices, 87-103.

Grewal, D., Kroschke, M., Mende, M., Roggeveen, A. L., & Scott, M. L. (2020). Frontline cyborgs at your service: How human enhancement technologies affect customer experiences in retail, sales, and service settings. *Journal of Interactive Marketing*, 51(1), 9–25. 10.1016/j.intmar.2020.03.001

Grunwald, A. (2007). Converging technologies for human enhancement. Assessing societal implications of converging technological development, 11, 271-88.

Hogle, L. F. (2005). Enhancement technologies and the body. *Annual Review of Anthropology*, 34(1), 695–716. 10.1146/annurev.anthro.33.070203.144020

Judge, T. A., Zhang, S. C., & Glerum, D. R. (2020). Job satisfaction. Essentials of job attitudes and other workplace psychological constructs, 207-241.

Kaur, J. (2024). Green Guardians: Unveiling the Strategic Role of HR in Environmental Sustainability Initiatives. In Building Sustainable Human Resources Management Practices for Businesses (pp. 125-143). IGI Global.

Kaur, J., & Arora, R. (2023, June). Exploring the dimensionality of employee silence in healthcare sector. In AIP conference proceedings (Vol. 2782, No. 1). AIP Publishing. 10.1063/5.0154178

Korir, C. K., & Ndegwa, P. (2020). Job Satisfaction and Employee Performance at Finlays Kenya Limited in Kericho County, Kenya. International Journal of Business Management. *Entrepreneurship and Innovation*, 2(2), 82–97.

Langton, A. J., & Ramseur, H. (2001). Enhancing employment outcomes through job accommodation and assistive technology resources and services. *Journal of Vocational Rehabilitation*, 16(1), 27–37.

Lin, P., & Allhoff, F. (2008). Untangling the debate: The ethics of human enhancement. *NanoEthics*, 2(3), 251–264. 10.1007/s11569-008-0046-7

Marois, A., & Lafond, D. (2022). Augmenting cognitive work: A review of cognitive enhancement methods and applications for operational domains. *Cognition Technology and Work*, 24(4), 589–608. 10.1007/s10111-022-00715-1

Pariseau-Legault, P., Holmes, D., & Murray, S. J. (2019). Understanding human enhancement technologies through critical phenomenology. *Nursing Philosophy*, 20(1), e12229. 10.1111/nup.1222930378247

Patricia, N. M., & Asoba, S. N. (2021). Theories of job satisfaction in the higher education context. *Academy of Entrepreneurship Journal*, 27(2), 1–16.

Pustovrh, T., Mali, F., & Arnaldi, S. (2018). Are better workers also better humans? On pharmacological cognitive enhancement in the workplace and conflicting societal domains. *NanoEthics*, 12(3), 301–313. 10.1007/s11569-018-0332-y

Rokeman, N. R. M., Kob, C. G. C., & Yaacob, M. N. M. (n.d.). A Review of The Job Satisfaction Theory for Vocational College Education Perspective. Journal of Modern Education, 6(20), 522-532.

Saritas, O. (2019). Emerging technologies, trends and wild cards in human enhancement. Emerging Technologies for Economic Development, 243-259.

Thant, Z. M., & Chang, Y. (2021). Determinants of public employee job satisfaction in Myanmar: Focus on Herzberg's two factor theory. *Public Organization Review*, 21(1), 157–175. 10.1007/s11115-020-00481-6

Weekes, T. R., & Eskridge, T. C. (2022, June). Responsible human-centered artificial intelligence for the cognitive enhancement of knowledge workers. In International Conference on Human-Computer Interaction (pp. 568-582). Cham: Springer Nature Switzerland. 10.1007/978-3-031-21707-4_41

Ziesche, S., & Yampolskiy, R. (2020, December). Introducing the concept of ikigai to the ethics of AI and of human enhancements. In *2020 IEEE International Conference on Artificial Intelligence and Virtual Reality (AIVR)* (pp. 138-145). IEEE. 10.1109/AIVR50618.2020.00032

Chapter 3
From Resistance to Adoption:
Promoting a Culture of Technological Acceptance in the Era of Human Enhancement

Noble Preet
Conestoga College, Canada

Bhupinder Pal Singh Chahal
https://orcid.org/0000-0002-8106-7121
Yorkville University, Canada

ABSTRACT

The enhancement of human capability ushers in a new age of technology as people call for machinery to take over their work, increase human efficiency, and make the impossible become possible. But the major challenge organizations experience is the high level of resistance towards adoption of such innovations into the firm's operations. This study aims at discussing the dynamic factors behind technological resistance and how the necessary culture can be built. Some of these factors are perceived job loss, lack of knowledge on the new technology such as AI and biotechnology, culture and age differences in adoption of technology, ethics such as privacy and autonomy, bureaucratic resistance, psychological barrier to change, lack of resources, resistance to technology-based performance indicators, and fear of obsolescence of skills. Thus, there is a need to focus on such solutions as clear reporting by organizations, development of specific programs for effective training, ethical standards within the frameworks of the organization, considering employees, their views, ideas, and promoting widespread commitment and leadership from managers

DOI: 10.4018/979-8-3693-7668-3.ch003

for stimulating innovation. Thus, proper management of change in organizations can prevent resistance and bring people and technologies to improve effectiveness, innovation, and strategic positioning in the age of high-tech environments. As for practical applications, this study calls for a partnership and future-oriented approach towards unveiling and addressing the challenges of technological implementation, and in turn, outlining the opportunities of human enhancement technologies to organizations as well as the society and ethics.

INTRODUCTION

The traditional era way of doing things is not sustainable in the rapidly changing environment as the world's geopolitics and economies are experiencing a revolution, brought about by technology, which becomes a necessity in everyone, institutions, and societies. Global challenges can be solved, new industries can emerge and communication can be improved by making technologies increase the efficiency of many processes. But these potentials do not occur automatically, hence from the idea to permeate the society, there are always challenges and one of the biggest is resistance. This is a kind of resistance to change and adoption of technologies that is evident in various ways and across various domains of application; this resistance, poses a stiff challenge to development and actualization of change sparked by technologies (Abbasi et al., 2015).

The reasons for the RA can therefore be viewed as a complex entity, which consists of psychological, organizational, and sociocultural factors that influence the technology adoption process. Fundamentally, it is a constituent of restrictiveness or conservatism regarding new technology, even though the new technology in question certainly has positive implications and capability to bring change. This is not a simple phenomenon that can be ascribed to the individuals' reluctance to change or their orientation toward the status-quo but it's an intricate interrelation between feelings, behaviors, beliefs, and structures within individuals, organizations, and societies (Anandarajan et al., 2000)

As applied to the psychological facet, the subject exhibits technology hesitation stemming from factors including but not limited to the following: This psychological inertia is accompanied by organizational inertia such as end-customers' attachment to legacy systems, the fear of disruption of normal business operations, and the costs of technological change. Furthermore, organizational cultures that do not support risk-taking and innovation reinforce these structures thus resulting in celebration of adoption of new technologies as being defensive and slowing rather than as agents of change (Carmen & María José, 2008).

Thus, it is evident that culture and social factors are the other influential factors as far as adoption of technology is concerned. Therefore, disillusionment, socioeconomic disparity's construct, perceived understand-ability of technology, and, more generally, cultural perceptions and beliefs about the role of technology in addressing diverse challenges all determine definitively how attainable the above mentioned and offered technological systems are to the openness of the various communities. Other factors include data privacy, ethic issues on use of technology and the overall trust on the digital platforms affects the perception and the adoption behaviour (Chau & Hu, 2002). The big picture of not being willing to accept new technologies in this particular case is grim and disastrous in general. At the individual level, such thinking means it becomes possible not to learn new abilities, switch the job and enhance the chosen sphere of action. Organizations fail again to introduce technological changes and might be left out in the cold in terms of competition flour and be blind to efficiency 'kickers' and consumers' wants. At societal level, recurrent refusal to embrace technology keeps digital divide unrecognized, socioeconomic inequalities to increase, and slows down society to address the set sustainable development goals, not to mention delaying solutions to arising international challenges such as climate change and COVID-19 pandemic (Kaur & Arora, 2023).

The kinds of organizations that need to identify psychological, organizational, and socio-political factors likely to inhibit change are those that seek to introduce new technologies or improve the usage of the already existing ones. For an acceptable technological culture to prevail, it is imperative that intentional work should be undertaken by all the stakeholders starting from the government, through the industry, universities and other civil bodies. Reforms of this kind should seek to establish a climate favourable to learning and risk-taking, actively and a climate in which the ability to transfer learning from the idea stage to the implementation phase is encouraged positively. In addition, there are still policies planned to eliminate hindrances to adapt to technology these include funding incentives for technologies, those policies governing data protection, polices on access to technologies for everyone are very important in eliminating inequalities in the distribution of technology. This way, the resistance to technological evolution can be influenced freeing new opportunities for societies and their inclusion and sustainable development in the digital economy (Chen & McDonough, 2012).

Finally, it can be stated that the issues presented in this study address the challenges of resistance to adopt new technologies by showing that it is not only a matter of implementing new tools and solutions but also changing the fixed mindset, practices, and organizational culture that prevent innovation and diffusion process. In essence, only when investors, producers, and consumers embrace technology and welcome new solutions, innovative technology will be able to shape the future of the world that is efficient and effective for everybody's benefit (Ciganke, Mao, & Srite, 2008).

ERA OF HUMAN ENHANCEMENT

Organizations are experiencing new and diverse changes due to the phenomenon of human enhancement that involves artificial intelligence and robotics, biotech cum cybernetic augmentation. That is why many experts believe that these innovations can greatly increase the capabilities inherent in humans: increase intelligence, physical strength, lifetime, and quality of life. As for the organizations, it implies reconsideration of the specifics of roles, organizational processes and overall strategies in order to incorporate these innovative tools to the best advantage (Coenen, 2024). Such change entails premising a company's efforts on strong change management tactics, the usage of training initiatives, ethical implementations, and promotion of a compelling culture that embraces the human and artificial aspects of a given process (Blok,2024). Through timely changes and progressive stances, employing enhancement technologies, organizational profit, innovation, and business edge increase while the moral and social impact of efficiency's new concepts are taken care of. Adopting this post human condition does not only change how organizations are run but also recalculates the essence of work and possibility of humans in business (figure 1 & 2) (Coenen, 2024).

Figure 1. Human enhancement

Human Enhancement

- AI
- Robotics
- Biotech
- Cybernetic Augmentation

Benefits:

- Increased Intelligence
- Enhanced Physical Strength
- Extended Lifetime
- Improved Quality of Life

Figure 2. Organizational Impact of Human Enhancement

Human Enhancement

Organizational Impact:
- Reconsider Roles and Processes
- Adapt Strategies for Integration

Strategies:
- Change Management
- Training Initiatives
- Ethical Implementations
- Promote Culture of Integration

Outcomes:
- Increased Profit
- Enhanced Innovation
- Competitive Edge

Implications:
- Ethical and Social Considerations
- Redefinition of Work

FACTORS THAT WILL LEAD TO TECHNOLOGICAL RESISTANCE DURING ENHANCEMENT OF HUMAN RESOURCE IN ORGANIZATIONS

This post humanistic elements of artificial intelligence, bio genetics and cybernetics already define mainly the post human epoch which means significant shifts in the carrying out of work and organizations. However, this potential is normally confronted with forces that lead to conflict. It is therefore necessary that organizations which are in various development stages, wishing to take advantage of the benefits that come with the innovations of these technologies understand the forces that entails resistance to change. The following are the reasons why organizations do not accept and adapt to technology advances during this transition period:

¬**Fear of Job Displacement:** The notion that employment is easy to lose to the advances of technology is one of the major barriers to why acceptance is sure to persist. Since AI and automation technologies are being applied more frequently in organizations, the employees' concern has changed towards the chances of being replaced by these technologies. This fear cannot be said to be baseless as percentages obtained by various polls done by different surveys indicate that sundry occupations are likely to be subcontracted to robotic methods (Demetriadis et al., 2003). For instance, McKinsey Global Institute study has suggested that up to 375 million employees around the world may be forced to change their occupations by 2030 through their pro-automation. This results to conflict in organizations because members are worried with their ability to keep on being employed within the respective organization. To address those concerns, organizational leadership have to explain how technology in the context of human enhancement will complement the human capital. If the persons, who have concerns, are communicated concerns and demonstrate how these innovations may lead to the employment of more people and a rise in skills level of many employees, the effect of insecurity may be minimized (Coenen, 2024).

Inadequate Knowledge: One of the main challenges that affect the necessity level of recognizing and adopting technology the higher is the rather weak recognition and information about the capabilities of human enhancement technology. Most employees remain ignorant of how these technologies function and how they can benefit the organization. Such ignorance can breed doubt and opposition due to the fact that individuals exhibit resistance to novel concepts which they lack understanding. Education is again key in ensuring that these problems are corrected; so training sessions and other educational activities should be employed. Organizations should embark on extensive training with the purpose of not only elucidating the gist of novelties in organizational technologies, but also the pragmatic uses and value thereof. It makes people naturalized in the certain tools and/or applications, thus decreasing the level of threat and resistance in organizations in the long run (Harel et al., 2021).

Cultural and Generational Differences: Other factors include cultural and generation gaps which are present in the employees. Experts found that the older employees who usually have low IT literacy are likely to show higher level of resistance as compared to the young employees. In particular, conflicts are often observed between generations, which makes it difficult for organizations to implement new technologies seamlessly. To fill this gap, there should be efficient measures right from the organizations enforcing diversity in the working force. This composes of initiating the special training for working with the patients of the unique age groups and with their attention to details and learning approach. Promoting cross- generational cooperation and.This means that employing older employees can also be used to encourage inter generation/ mentor ship, which would ensure that the older personnel do not feel alone while embracing new technologies (Kimiagari & Baei, 2022).

¬**Ethical Concerns**: This has acted as another major factor that has led to the emergence of resistance with regard to the concept of ethics inherent in the human enhancement technologies. The people at the workforce may have different issues to embrace regarding the privacy, freedom, and independence violations by these technologies. For example, the following such concepts as the Biometric data for the security or Performance Management by AI can be looked at as surveillance and data misuse. Ethical issues can be managed they require to be addressed in a way that is cleansing and preventive (Lee et al., 2013). Therefore, the organizations should set broad codes of ethics and regulation concerns the use of human enhancement technologies. Educating the employees to let them know that one can talk about ethical issues and involving them in decision making process can also help in eradicating the fears. Also, the augmentation of data protection and the recognition of the principles of an individual right to privacy can assist in the eradication of the ethical issues (Fisher, 2021).

¬**Organizational Inertia:** The other factor is organizational resistance; because of champions' preference for these technologies, the current organizational status quo becomes a barrier to altering the current leadership and organizations. Primarily, the issue of status quo is most fitting as this pertains to the unwillingness of organizations and leaders to change because the current systems and processes have been beneficial to them in the past. Such a mechanism of resistance originating from the upper echelon of the organization can easily spread to other levels of the organization thus constraining the possibility of innovation at the organizational level (Mao & Palvia, 2006). On this basis, it is suggested that the organizational leadership commits to the change by transforming organizational culture and developing the organizational need for innovation as well as acceptance of the technology. The proposed strategies for managers include advocating post-humanist ontology and enhancing change preparedness at the organizational level. It may be noted that delivering a culturally propelled agenda that aligns with the firm's strategic course can help to attain the twin objectives of speed as well as acceptance across the system (Lee et al., 2011).

Psychological Resistance to Change: One category is the changing of psychological inertia which is widely described in the literature and has a major influence on acceptance of technologies. In general, many employees are likely to exhibit discomfort or even fear considering new technologies because such changes can be viewed as a threat to their normal working schedule. This resistance is usually due to their need to create work environment that is more stable and predictable than it has been in the past (Marangunić & Granić, 2015). Regarding the last type of psychological resistance, it is critical for organizations to work on specific educational campaigns to involve and convince all stakeholders about the reasons for introducing certain technologies and the advantages that can be gained from such changes. Engaging the employees thus by asking for their opinion and trying to considering their opinion as to why change has to be made will go a lengthy way in minimizing feelings of such feelings of uncertainty. This situation means that constant support and encouragement as well as provision of other resources in the phase of transition can make the change process easier when it comes to adopting new technologies by the personnel (Mao & Palvia, 2006).

Resource Constraints: This is even as lack of financial resources, and lack of technical know how, among other factors may limit the implementation of human enhancement technologies. The actual use of sophisticated technologies frequently demands substantial capital in facilities, human resources development, and sustaining recourse. Challenges include; Lack of resources, where some organizations may not have adequately appointed IM systems or the funds to do so (Moffett et al., 2002). It should also be noted that organizations have to look for the opportunities to reduce the costs and to concentrate more on the technologies, which can be efficiently used to increase the efficiency of work for achieving the organizational objectives. Technology partners can also help, or else look for more funding outside the classroom or apply for a grant. Also, active collaboration with the academic community and professionals can contribute to the elimination of technical knowledge deficiencies and obtain relevant sources for effective implementation (Marangunić & Granić, 2015).

¬**Anticipation and Resistance to Technological Values:** Technology also incurs resistance for it puts into use objectified performance indicators when applied. This is part of the intrinsic nature of these types of systems because the majority of the employee population is going to view it as either arrogant or prejudicial, in that they are being evaluated by this automatic decision-making tool while they expect the opposite. This results in discontentment and, consequently, mistrust in the smart technology. To rule out resistance to the application of the above figures of performance due to new technologies, organizations should be very clear and fair on the use of the systems. Thus, by being open to the objectives that are set on the performance appraisal and involving subordinates in the formulation and the development of the existing systems, trust can be established. One is training, education or communication which creates awareness on issues related to change and other is feedback which can also assist in addressing such issues that users or employees may have that is likely to increase acceptance while reducing resistance to change (Molino et al., 2020).

¬**Concerns About Skill Obsolescence:** Some of the problems that may be precipitated by a fast Gulf stream of technological disturbances include the reduction of competence of the employees. It is also important to emphasize that there will always be resistance to change; more so when employing new technologies introduces new forms of work that cast the previous knowledge and skill taught as irrelevant. The employees may be aware of threats that as they emerge, they will fail to manage with new technologies hence losing their position within the labour market Mun et al., 2006). To ensure that such fears are not realized, organizations can consider the following strategies of eliminating skill obsolescence Emphasis on skills acquisition. Providing ordinary training sessions, courses geared toward increasing one's level of expertise, or showing the employees how they can advance within the company can assist in keeping the workers aware of particular technologies and assist them in feeling more certain about practical applications. The two areas of focus, emphasis on growth mentality and learning throughout one's lifetime can also contribute to promoting agility and tenacity (Kaur & Arora, 2023).

It will be shown that the age of human enhancement offers many possibilities to optimize human performances and thus foster the success of organizations. However, the major success factor that has received less attention is obviously the resistance to change with regards to technological acceptance. Thus, organizations must have a broad perspective of all the aspects, such as job insecurity, knowledge limit, cultural differences, ethical issues, the organizational culture that resists change, psycho-biological resistance against change, lack of resources, resistance against the technology and performance enshrinement, and threat of skills becoming redundant, in order to evolve a strategy towards changing the attitude towards technology (Blok,2024).

Mitigating these factors interdependently calls for an appreciable overall approach such as; emphasizing on effective communication, effective training, ethically sound approach, consideration of diversity, leadership's full endorsement, and the provision of consistent support. Through proper assessment of the resistance and action in preparation to foster the climate of acceptable innovations, human enhancement technologies can be optimally infused into organizations as a resource lever toward productivity, creativity, and competitiveness in today's global economy (Santini et al., 2019). Therefore, it is suggested that organizations may limit the extent to which their skills become out of date by incorporating learning. Annual training sessions, a move to advance into higher categories of training and staff development could help the employees to address them and bring them to the level of the modern technological world, and boost their self efficacy. Thus, the learning culture and characteristics of learning organizations can also foster and support resilience and adaptability (Kaur & Arora, 2023).

Thus, thinking of enhancement of human beings seems to point at extensive opportunities for enhancing work performance, which is critical, when it comes to organizations. However, one of the issues that always prevail is the resistance to change and in this regard, the acceptance of technology is not exempted from it. Some of the reasons that are linked to resistance are; job-loss, ignorance, cultural and/or, age differences, ethical issues, bureaucratic structure, psychological resistance to change, lack of capital, ethical issue with performance metrics including technology and fear of becoming outdated (Siegel, 2008). Therefore, solutions for those factors relate to communication, teaching/training, adherence to ethical cultures, accessibility emphasis, leadership, and support. The authors establish that organizations which could anticipate and mitigate technological resistance are in a good position to assure that high added value derives from human enhancement technologies and results in increased productivity, creativity, and competitiveness of organizations in the modern economy (Sun & Zhang, 2006).

ROLE OF HUMAN RESOURCE DEPARTMENT IN PROMOTING TECHNOLOGICAL ACCEPTANCE: APPROACHES, CULTURES FOR GROWTH, AND ADMINISTRATOR POSITIONS

In the constantly evolving world of technology, thus being able to initiate the process of popularizing new technologies is very vital in order to enhance their effectiveness. However, culture remains an issue that hampers the adoption of technology at present as a major challenge in many organizations and throughout different tiers of society. To overcome such a resistance, what is needed is an adoption of more approaches, creation of support culture and active assumption of leadership roles in change management. This study discusses how human resource department can assist in dealing with the issue of resistance to change, key actions that contribute to the development of acceptance, and leaders' roles in influencing the introduction of technology (Van Offenbeek, Boonstra, & Seo, 2013).

STRATEGIES FOR OVERCOMING RESISTANCE

1. **Education and Awareness:** Education, trying to clear up misunderstandings and raising the visibility of the advantages offered by new tools are the first seeds of combating the issue. Stakeholders in educational programs should ensure that they present visionary benefits and ensure that beliefs on threats such as loss of jobs due to technology advancement are demystified, and where technology advancement are ensured to bring out efficiency and efficacy in discharging organizational goals efficiently (Venkatesh & Bala, 2008).
2. **User-Centred Design:** Getting feedback from the users and making certain that the distinct technologies are relevant to the user requirements and needs help in boosting acceptance. The also make use of user centred design principles which ensure that networks are easy to use and accessible hence more user friendly to make sure the users encounter less steep learning curves (Chau & Hu, 2002).
3. **Pilot Projects and Incremental Implementation:** Pilot implementations when it is used to introduce new technologies allows for the identification of functionalities that may turn out to be useless, the issues that are always associated with the new implementation and how to tackle them before whole scale implementation. It also does not create conflict and slowly builds up confidence among the stakeholders (Zhang & Gutierrez, 2007).
4. **Change Management Strategies:** The presented types of strategic change processes presuppose the need for preparation in advance like identifying the needs to be concerned and addressing the intentions of management to introduce technologies. Reducing the fears, providing the training and making them feel

that it is their own change are some factors of change management (Demetriadis et al., 2003).

5. **Incentives and Rewards:** Other organizational rewards that include the use of appreciation schemes or extra working allowances that are dependent on the willingness of use of the innovative technology can also build up the pressure on people and units on their use. Thanks and recognition to the pioneers; the act of 'toasting success' could have an impact on others to do the similar (Chau & Hu, 2002).
6. **Leadership Support:** Cultural resistance to change should be fought out at different level of the organization, especially, leaders' backing. The uncertainties which affect the information technology implementation must be handled by the heads of organization through promotion of ideas, preparing the budget to purchase the technologies, and training and educating the staffs (Zhang et al., 2019).

DEVELOPING ACCEPTANCE AS A CULTURE

Acceptance technological culture means the states of mind that predominate in organization practices, scientific advance, calculated risk-taking, and perpetual improvement. Key elements of fostering such a culture include the following aspects (figure 3):

Figure 3. Ways to overcome resistance to technological acceptance in organizations

Strategies for Overcoming Resistance to Adoption

1. Education and Awareness
   - Clearing Misunderstandings
   - Highlighting Advantages

2. User-Centered Design
   - Gathering User Feedback
   - Aligning Technologies with User Needs

3. Pilot Projects and Incremental Implementation
   - Identifying Issues Early
   - Building Stakeholder Confidence

4. Change Management Strategies
   - Preparation and Planning
   - Addressing Stakeholder Concerns

5. Incentives and Rewards
   - Recognition and Appreciation
   - Performance-Based Incentives

6. Leadership Support
   - Promoting Cultural Change
   - Providing Resources and Training

1. **Leadership by Example:** Competition wise, people get to practically engage in supporting organizational culture whereby leaders are involved in portraying the culture. In this way, together with the support of the technology projects, stimulating measures of risk-taking, and demonstration of openness to alters the leaders link subordinates to the ultra-modernist outlook of the organization (Demetriadis et al., 2003).
2. **Encouraging Collaboration:** Integrated working and adoption of cross-functional teams as this these enhances sharing of knowledge among the employees. Facilitate debates and ensure the creation of channels for exchange of ideas and information concerning the application of technology in business (Kaur & Arora, 2023) .

3. **Promoting Learning and Development**: Support awareness training and ongoing education of many programs and workshops in this case, which will bring the employee to the required competence levels in managing new technologies. These make the workers to be more assertive and this helps them feel that they are capable of handling issues in relation to technological developments (Kimiagari & Baei, 2022).
4. **Celebrating Innovation:** Promote those people and organizations which participate in the social construction of technology, as well as those who can apply the existing technology in creative way. Initiate public activities in overcoming the above challenges and arriving at positive results through the use of Information Technology implementations (Kaur & Arora, 2023) .
5. **Establishing Clear Goals and Metrics:** There should always be clear purposes for the implementation of technologies and such purposes should always be formulated in a way that they contain factors that can be used to measure the effectiveness of the projects. It must also be noted that these goals and metrics should be frequently repeated to the stakeholders in order to constantly remind everyone about them (Demetriadis et al., 2003) .

LEADERSHIP ROLES IN THE MANAGEMENT OF ACCEPTANCE OF TECHNOLOGIES

The study shows that it is the leaders' responsibility to champion technological acceptance both at the workplace and society. Effective leadership in this context involves: Leadership has the following definitions in this case:

1. **Visionary Leadership:** Suggest how technology have to be used for the change to start in the organization or in the entire society. Make sure this vision is very clear and that technological activities match up with the firm's other strategic objectives in order to encourage the players (Kimiagari & Baei, 2022) .
2. **Empowering Teams:** However, centralization should be done to some extend with the delegation of decision making in regards to the management of the technology to the teams. Being quite open and managing to support the culture of freedom as well as the offer of independence and creativity to come up with new ideas while at the same time establishing a framework that can absorb risks and losses (Demetriadis et al., 2003).

3. **Building Coalitions:** Develop and sustain proper working relations with each department in the company, every outside supplier, and the suppliers in the vicinity. In matters touching on technology work on the common cause, share information and advocate on issues surrounding technology.
4. **Cultivating Resilience:** Most commonly known risks should be managed that are in the foreground in turning the adoption of technology. Promote problem solution agility and persistence of the teams for the negative change by teaching them how to cope successfully and implying that it is a learning process (Demetriadis et al., 2003). .
5. **Promoting Ethical Use of Technology:** One can also suppose that the more exactly the organizational involvement sources are countered with the help of initiatives supported by the organizational culture, and leadership the technological acceptance can be achieved and established. Thus, applying specific measures, nurturing culture for idea acceptance and implementation, constructing and managing the organization with regard to all these aspects, the organizations and in general the communities can to some extent follow the technological change wave and deliver further sustainable development of competitiveness and quality of life. Such attitude is not a question of failed attempts to adopt instruments, but an attitude concerning the acceptance of a process which frames the technological advance and prepares the stakeholders for what is to come (Kimiagari & Baei, 2022).

THE FUTURE OF TECHNOLOGICAL INTEGRATION

Expecting the continuation of the development of technologies as human enhancement, people are now stepping over the edge of the next age of people's abilities and cultural metamorphosis. The advancement in biological sciences and information technology, along with the growing fields of study pertaining to the brain, compels man to begin to critically establish human identity. Neurally operated limb prostheses provide better control and copying of the motion of man's body and the relations between man and a mechanism. Neural implants or BCI are the exciting technologies in which one can technically update memory, or learning rate, or mood as well. Techniques like CRISPR-Cas 9 is enabling the modifying of our genome: for example, there are realistic solutions to cure inherited diseases or achieve the desires of an extraordinary body (Kaur & Arora, 2023).

However, at the same time such opportunities raise the unsolved ethical dilemmas and socially significant problems on the further development of the humanity. Questions of justice, especially in relation to social justice and equal opportunities, arise here as such technologies become available only to those who could afford them and this only aggravates social inequalities (Fisher, 2021). Enhancements give reasons to doubt what it means to be a human being and questions about the nature of what is normal/ enhanced appear. Besides, the use of such technologies should also be accompanied by other important components that are the legal consideration and using the said technology as an instrument of inflicting harm on any particular individual or organization. The existing laws can only be characterized as passive which implies that the policies must be much more active as pertains to the spectrum of innovation, yet, at the same time, society's best interest must remain paramount (Blok, 2024).

As for the prospects, one can expect that integration of technologies into human enhancement will have enormous influence on the society. It will not only change health care and industries but also culture and ethics all around the world. It hardly came into existence and as we tread the realm, there is a need to involve all the key stakeholders that include scientist, ethicist, policy makers, the society and others to fashion out a world where these enhancement technologies will be in a way that will promote the best interest of human beings as they enhance them, uphold, respect the universally acceptable ethical standards and most daunting of all; provide equitable access to such enhancement to all human beings. It is in the present that we make decisions that will determine the future that will be one of human enhancement done in a proper manner and with ethically sound technologies that will greatly enhance the quality of life of people through organizing a more efficient society.

STRATEGIES TO ALWAYS KEEP AN EYE ON RESISTANCE

1. **Education and Awareness:** As for the sources of resistance, they observed that the stakeholders are either ignorant of, or do not place adequate value on, the new technologies. In different case-studies, presentations and demonstrations of best-performance, hearings should encompass positive ramification which will remove misunderstanding and relate ideal schooling (Mao & Palvia, 2006) .
2. **Regulatory and Policy Frameworks:** Ethical implementation of the issues, privacy policies and safety measures, the formal rules and legislation which control new technologies has to be adopted because once individuals strengthen confidence in the technologies they will use them to the maximum. This is the role that government, industry and academic stakeholders have to build kinetic regulations – meaning that innovation as well as the spirit of a society is possible to be fostered (Fisher, 2021).

3. **Collaborative Innovation Ecosystems:** It is for this reason that since the technology was made people aware of the possibilities and the importance of of harnessing an open collaborative environment for the different stakeholders in the advancement of technology purposes as such, encompasses the different stakeholders such as the researchers, policy makers, the industrial players and members of the civil societies. The proposals for the ecological fund and the pilot schemes, for the establishing of the knowledge sharing, enable the members of the stakeholders, to work in a manner of building solutions, for the issues of implementation.
4. **User-Centred Design:** The incorporation of user-centered design and development as a strategy in the development of new technologies useful in the future is beneficial as regards use. Involving the end users with how these technologies will be implemented and where they will utilized in the usability testing addressing the successive changes in usage does ensure that these technologies fit the end users' consumption models (Mao & Palvia, 2006) .
5. **Continuous Evaluation and Adaptation:** Technology is a cycle that will need at one point or another testing and adjustments as in order to become efficient. Individuals highlighted the following as vital in the improvement of the use of technologies and their returns: The strict cyclical observation of the application results, the accumulation of feed-backs to be received from the stakeholders on the strategies utilized and the information gathered from the statistics (Blok, 2024).

Hence, one should be prepared and come up with strategies to meet the future acceptance challenges pertaining to technological advancements. Based on importance of education and awareness, creating flexible and adaptive regulations, collaboration between the ecosystems, user centrism, and continuous evaluations, the stakeholders can-oppose resistance and unleash new technologies' opportunities. Responsible approach to adoption of the new technologies means receptive approach that will ensure that stakeholder's ethic is followed in order to form a future that is profitable for the economy and healthy for the society.

CONCLUSION

The study defines the characteristics of technological phases as organizations move from the phase where there is no acceptance to the phase whereby there is acceptance of the technologies introduced and as this unfolds, it becomes crucial for any organization willing to compete effectively in the modern world to ensure that it sets the right culture that will enhance technological experience. This change requires something more than the transformation from the technical viewpoint; there is a need to incorporate people into the system of the organization. Therefore, learning about the emergence of resistance and the dynamics of its evolution, organizations can develop strategies to reduce the possible consequences and provide clarity to misconception regarding innovations.

In this process, leadership is crucial, which implies inspiring people and being a role model that might illustrate why there is a demand for technological progress and what benefits it can bring. The management and other key personnel should demonstrate how the new technologies are advantageous to the new employees and those already in the workplace as well as how those new technologies will assist in the realization of the set organizational goals. Furthermore, they should guarantee that there is policy and psychological safety, in which an employee can discuss his or her concern and be given assistance out of the organization. Some of the possible antecedents of technology that can be employed in order to enhance the level of acceptance include education and training. Based on the identification of the mode of learning of an individual, extensive training interventions boost assurance and effectiveness when the employees are training in new tools. Existential training and knowledge allow employees in such organizations to remain up to date and, thus, effectively harness technology to enhance their efficiency and creativity.

Also, the practice of cultivating positive attitudes towards the use of feedback and continuous improvement mechanisms will further improve the adoption process. It is effective to discuss the case in groups and hear from the employees regarding their experiences and ideas as this not only assist in the improvement of the implementation strategies but also it aids in the employees feeling involved. By appreciating the early starters and change agents, a similar attitude can be observed from other employees, which can enhance change. Summing up, it can be noted that the process of calculating a proper culture of accepting technologies is a complex task which implies the integration of leadership, communication, education, and feedback functions. Realising that human factor is the key to the technological change, organisations can easily turn those who are against such changes into people who are for such changes thus creating a foundation for sustainable innovation and achievements of organisational goals and objectives in the world that has gone digital. Technology is not a phenomenon that you follow; it is the concept of an organization that is geared for constant progression fostering flexibility in organizational development.

REFERENCES

Abbasi, M. S., Tarhini, A., Hassouna, M., & Shah, F. (2015). Social, organizational, demography and individuals' technology acceptance behaviour: a conceptual model. European Scientific Journal, 11(9), 48-76.

Anandarajan, M., Igbaria, M., & Anakwe, U. P. (2000). Technology acceptance in the banking industry: A perspective from a less developed country. *Information Technology & People*, 13(4), 298–312. 10.1108/09593840010359491

Blok, V. (2024). Materiality versus Metabolism in the hybrid world: Towards a dualist concept of materialism as limit of post-humanism in the technical era. *Philosophy & Technology*, 37(2), 1–22. 10.1007/s13347-024-00751-x

Carmen, C., & María José, G. (2008). The role of technological and organizational innovation in the relation between market orientation and performance in cultural organizations. *European Journal of Innovation Management*, 11(3), 413–434. 10.1108/14601060810889035

Chau, P. Y., & Hu, P. J. (2002). Examining a model of information technology acceptance by individual professionals: An exploratory study. *Journal of Management Information Systems*, 18(4), 191–229. 10.1080/07421222.2002.11045699

Chen, M. Y. C., Lin, C. Y. Y., Lin, H. E., & McDonough, E. F.III. (2012). Does transformational leadership facilitate technological innovation? The moderating roles of innovative culture and incentive compensation. *Asia Pacific Journal of Management*, 29(2), 239–264. 10.1007/s10490-012-9285-9

Ciganke, A. P., Mao, E., & Srite, M. (2008). Organizational culture for knowledge management systems: A study of corporate users. [IJKM]. *International Journal of Knowledge Management*, 4(1), 1–16. 10.4018/jkm.2008010101

Coenen, C. (2024). Spotlights on the history of human enhancement discourse. In *The Routledge Handbook of the Ethics of Human Enhancement* (pp. 18–29). Routledge.

Demetriadis, S., Barbas, A., Molohides, A., Palaigeorgiou, G., Psillos, D., Vlahavas, I., Tsoukalas, I., & Pombortsis, A. (2003). "Cultures in negotiation": Teachers' acceptance/resistance attitudes considering the infusion of technology into schools. *Computers & Education*, 41(1), 19–37. 10.1016/S0360-1315(03)00012-5

Fisher, J. A. (2021). *Enhancing 'Human Nature': The Human Enhancement Debate in US Bioethics*. Columbia University.

Harel, R., Schwartz, D., & Kaufmann, D. (2021). Organizational culture processes for promoting innovation in small businesses. *EuroMed Journal of Business*, 16(2), 218–240. 10.1108/EMJB-03-2020-0027

Kaur, J., & Arora, R. (2023a, June). Exploring the dimensionality of employee silence in healthcare sector. In AIP conference proceedings (Vol. 2782, No. 1). AIP Publishing. 10.1063/5.0154178

Kaur, J., & Arora, R. (2023b, June). Exploring the impact of employee silence in private hospitals-a structural equation modeling approach. In AIP Conference Proceedings (Vol. 2782, No. 1). AIP Publishing. 10.1063/5.0154173

Kimiagari, S., & Baei, F. (2022). Promoting e-banking actual usage: Mix of technology acceptance model and technology-organisation-environment framework. *Enterprise Information Systems*, 16(8-9), 1894356. 10.1080/17517575.2021.1894356

Lee, Y. H., Hsieh, Y. C., & Chen, Y. H. (2013). An investigation of employees' use of e-learning systems: Applying the technology acceptance model. *Behaviour & Information Technology*, 32(2), 173–189. 10.1080/0144929X.2011.577190

Lee, Y. H., Hsieh, Y. C., & Ma, C. Y. (2011). A model of organizational employees'e-learning systems acceptance. *Knowledge-Based Systems*, 24(3), 355–366. 10.1016/j.knosys.2010.09.005

Mao, E., & Palvia, P. (2006). Testing an extended model of IT acceptance in the Chinese cultural context. *The Data Base for Advances in Information Systems*, 37(2-3), 20–32. 10.1145/1161345.1161351

Marangunić, N., & Granić, A. (2015). Technology acceptance model: A literature review from 1986 to 2013. *Universal Access in the Information Society*, 14(1), 81–95. 10.1007/s10209-014-0348-1

Moffett, S., McAdam, R., & Parkinson, S. (2002). Developing a model for technology and cultural factors in knowledge management: A factor analysis. *Knowledge and Process Management*, 9(4), 237–255. 10.1002/kpm.152

Molino, M., Cortese, C. G., & Ghislieri, C. (2020). The promotion of technology acceptance and work engagement in industry 4.0: From personal resources to information and training. *International Journal of Environmental Research and Public Health*, 17(7), 2438. 10.3390/ijerph1707243832260142

Mun, Y. Y., Jackson, J. D., Park, J. S., & Probst, J. C. (2006). Understanding information technology acceptance by individual professionals: Toward an integrative view. *Information & Management*, 43(3), 350–363. 10.1016/j.im.2005.08.006

Santini, F. D. O., Ladeira, W. J., Sampaio, C. H., Perin, M. G., & Dolci, P. C. (2019). A meta-analytical study of technological acceptance in banking contexts. *International Journal of Bank Marketing*, 37(3), 755–774. 10.1108/IJBM-04-2018-0110

Siegel, D. M. (2008). *Accepting technology and overcoming resistance to change using the motivation and acceptance model*. University of Central Florida.

Sun, H., & Zhang, P. (2006). The role of moderating factors in user technology acceptance. *International Journal of Human-Computer Studies*, 64(2), 53–78. 10.1016/j.ijhcs.2005.04.013

Van Offenbeek, M., Boonstra, A., & Seo, D. (2013). Towards integrating acceptance and resistance research: Evidence from a telecare case study. *European Journal of Information Systems*, 22(4), 434–454. 10.1057/ejis.2012.29

Venkatesh, V., & Bala, H. (2008). Technology acceptance model 3 and a research agenda on interventions. *Decision Sciences*, 39(2), 273–315. 10.1111/j.1540-5915.2008.00192.x

Zhang, W., & Gutierrez, O. (2007). Information technology acceptance in the social services sector context: An exploration. *Social Work*, 52(3), 221–231. 10.1093/sw/52.3.22117850030

Zhang, Y., Luo, Y., Zhang, X., & Zhao, J. (2019). How green human resource management can promote green employee behavior in China: A technology acceptance model perspective. *Sustainability (Basel)*, 11(19), 5408. 10.3390/su11195408

Chapter 4
AI Integration in HR Processes

Satya Subrahmanyam
Holy Spirit University of Kaslik, Lebanon

ABSTRACT

This chapter explores the integration of artificial intelligence (AI) in human resources (HR) processes, examining its evolution, current applications, and future implications. Beginning with a definition of AI in HR, including machine learning and predictive analytics, it highlights AI's role in enhancing efficiency, decision-making, and employee experiences. The chapter delves into specific applications across recruitment, onboarding, training, performance management, and employee engagement, showcasing AI's transformative impact. Ethical considerations such as bias mitigation, data privacy, and human-AI collaboration are also discussed. Ultimately, it emphasizes AI's potential to revolutionize HR practices while advocating for responsible adoption and ongoing professional development in HR professionals.

1. INTRODUCTION

1.1 Defining AI in HR

Artificial Intelligence (AI) refers to the simulation of human intelligence in machines designed to think and act like humans. In the context of Human Resources (HR), AI encompasses a variety of technologies such as machine learning, natural language processing, and robotic process automation that can enhance and streamline HR functions (Bersin, 2019). AI applications in HR include:

DOI: 10.4018/979-8-3693-7668-3.ch004

- **Recruitment and Talent Acquisition**: Automated resume screening, chatbots for initial candidate interaction, and predictive analytics for candidate selection.
- **Employee Onboarding**: Virtual assistants guiding new hires through onboarding processes.
- **Performance Management**: AI-driven analytics for evaluating employee performance and providing personalized feedback.
- **Learning and Development**: Customized training programs and AI-driven skill assessments.
- **Employee Engagement**: Sentiment analysis and pulse surveys to gauge employee morale and engagement.

The importance of AI in modern HR management is underscored by its ability to enhance efficiency, reduce bias, improve decision-making, and provide personalized experiences for employees. By leveraging AI, HR professionals can focus on strategic tasks rather than routine administrative duties, ultimately contributing to a more dynamic and responsive HR function (Bersin, 2019).

1.2 Historical Context

Evolution of HR Processes and the Introduction of Technology

Human Resources has undergone significant transformations over the decades. Traditionally, HR functions were manual and paper-based, encompassing activities such as record-keeping, payroll processing, and basic employee management. The introduction of technology in HR began with the advent of computer systems and basic HR software in the late 20th century, which digitized many routine processes and improved efficiency (Kavanagh & Thite, 2009).

Key developments include:

- **1980s-1990s**: Introduction of HR Information Systems (HRIS), which digitized personnel records and payroll (Kavanagh & Thite, 2009).
- **2000s**: Emergence of cloud-based HR systems, enabling better data management and accessibility (Bondarouk & Ruël, 2009).
- **2010s**: Rise of big data and analytics in HR, allowing for more informed decision-making and strategic planning (Falletta, 2014).

Milestones in the Integration of AI in HR

The integration of AI into HR processes represents a significant leap forward. Notable milestones include:

- **2010s**: Early adoption of AI in recruitment through automated resume screening and applicant tracking systems (ATS).
- **2014**: Introduction of chatbots for candidate interactions, improving the candidate experience and freeing up HR professionals from routine queries.
- **2016**: AI-driven performance management systems begin to gain traction, offering real-time analytics and personalized feedback.
- **2018**: AI in learning and development sees widespread adoption, with platforms offering customized training programs and continuous learning opportunities.
- **2020 onwards**: Increased focus on AI for employee engagement and retention, utilizing predictive analytics to foresee turnover risks and sentiment analysis tools to gauge employee morale.

These advancements highlight how AI has transformed HR from a purely administrative function to a strategic partner within organizations, driving better business outcomes and enhancing employee experiences (Bersin, 2019).

2. RECRUITMENT AND TALENT ACQUISITION

2.1 Automated Resume Screening

How AI Algorithms Streamline Resume Screening and Candidate Shortlisting

AI algorithms in automated resume screening utilize natural language processing (NLP) and machine learning techniques to scan and evaluate resumes quickly. These systems can parse large volumes of resumes, extracting key information such as skills, experiences, and qualifications. AI models are trained on historical hiring data to identify patterns and determine the most relevant candidates for a given job description (Upadhyay & Khandelwal, 2018).

Benefits and Challenges of Automated Screening Systems

Benefits:

- **Efficiency:** AI systems can process thousands of resumes in a fraction of the time it would take human recruiters, significantly reducing the time-to-hire.
- **Consistency:** AI ensures uniform criteria are applied to all resumes, reducing human bias in the initial screening phase.
- **Cost-effectiveness:** By automating the screening process, organizations can save on recruitment costs and allocate human resources to more strategic tasks.

Challenges:

- **Bias:** If the training data contains biases, the AI system may perpetuate or even exacerbate these biases in candidate selection.
- **Transparency:** AI decision-making processes can be opaque, making it difficult for recruiters to understand why certain candidates are selected or rejected.
- **Adaptability:** Automated systems may struggle with non-standard resume formats or unconventional career paths, potentially overlooking qualified candidates.

2.2 Chatbots and Virtual Assistants

Role of AI-Powered Chatbots in Candidate Interaction and Support

AI-powered chatbots and virtual assistants enhance candidate interaction by providing instant responses to common queries, scheduling interviews, and guiding applicants through the application process. These tools are available 24/7, ensuring candidates receive timely support and a positive experience (Akkiraju et al., 2020).

Key Functions:

- **Answering FAQs:** Chatbots can handle frequently asked questions about the job, application process, and company policies.
- **Interview Scheduling:** Automating the scheduling of interviews reduces the administrative burden on HR teams and minimizes scheduling conflicts.
- **Pre-screening:** Chatbots can conduct initial assessments by asking candidates about their qualifications and experience, further streamlining the recruitment process.

Case Studies on Effectiveness and User Experience

Case Study 1: Unilever

Unilever implemented an AI-driven recruitment process using chatbots and automated screening tools. The company reported a reduction in hiring time by 75% and a 100% increase in applicant satisfaction due to the swift and responsive nature of the system (Unilever, 2017).

Case Study 2: Mya Systems

Mya, an AI recruitment assistant, has been used by several Fortune 500 companies. It has shown significant improvements in candidate engagement and recruiter productivity. Mya's ability to understand and process natural language queries has resulted in a more personalized and efficient candidate experience (Mya Systems, 2019).

2.3 Predictive Analytics in Hiring

Use of AI to Predict Candidate Success and Cultural Fit

Predictive analytics in hiring involves using AI to analyze data from various sources, such as resumes, social media profiles, and performance metrics, to forecast a candidate's potential success and cultural fit within an organization. These models assess various factors, including skills, personality traits, and past performance, to make informed predictions (Chamorro-Premuzic et al., 2016).

Key Applications:

- **Performance Prediction:** AI models can predict how well a candidate will perform in a role based on historical data and identified success factors.
- **Cultural Fit:** By analyzing personality traits and values, AI can estimate how well a candidate will integrate with the company culture.

Examples of Predictive Models and Their Impact on Hiring Decisions

Example 1: IBM Watson Talent

IBM Watson Talent uses AI and machine learning to predict candidate success by analyzing past hiring data and performance metrics. Companies using this tool have reported improved hiring outcomes and better alignment between candidates and job requirements (IBM, 2017).

Example 2: HireVue

HireVue's AI-driven platform assesses video interviews to evaluate candidate competencies and predict job performance. This tool has helped companies reduce hiring bias and increase the accuracy of their hiring decisions (HireVue, 2018).

Impact on Hiring Decisions:

- **Enhanced Accuracy:** Predictive analytics improves the accuracy of hiring decisions by identifying the best-fit candidates based on data-driven insights.
- **Reduced Turnover:** By selecting candidates who are likely to succeed and fit well within the company culture, organizations can reduce employee turnover rates.
- **Strategic Hiring:** Predictive models allow HR teams to make more strategic hiring decisions, focusing on long-term organizational needs and goals.

3. ONBOARDING AND TRAINING

3.1 Personalized Onboarding Programs

AI-Driven Customization of Onboarding Processes

AI-driven onboarding programs leverage machine learning and data analytics to tailor the onboarding experience to each new hire's specific role, background, and learning style. By analyzing data from various sources such as pre-employment assessments, resumes, and personal preferences, AI systems can create customized onboarding plans that better prepare new employees for their roles (Jiang, 2019).

Key Features:

- **Automated Paperwork:** AI tools automate administrative tasks, ensuring new hires complete necessary documentation efficiently.
- **Role-Specific Content:** AI systems deliver relevant information and resources based on the new hire's role and department (Society for Human Resource Management, 2020).
- **Progress Tracking:** Continuous monitoring of the onboarding process helps identify areas where new hires may need additional support.

Enhancing New Hire Experience Through Tailored Training Modules

AI enhances the onboarding experience by providing personalized training modules that cater to the individual needs of each employee. These modules can include interactive simulations, video tutorials, and quizzes that adapt to the user's progress and comprehension level (IBM, 2017).

Benefits:

- **Increased Engagement:** Personalized content keeps new hires more engaged and motivated during the onboarding process.
- **Faster Ramp-Up Time:** Tailored training ensures new employees acquire the necessary skills and knowledge more quickly, leading to faster productivity.
- **Improved Retention:** A customized onboarding experience helps new hires feel valued and supported, reducing turnover rates.

Case Study:

- **Cognizant:** Cognizant uses an AI-powered onboarding platform that personalizes the onboarding journey for each new hire. The system uses data from initial interviews and assessments to create a tailored onboarding plan, resulting in higher engagement and faster integration into the company (Cognizant, 2020).

3.2 Continuous Learning and Development

AI in Identifying Skill Gaps and Recommending Personalized Learning Paths

AI technologies can analyze employee performance data, feedback, and career progression to identify skill gaps. By understanding these gaps, AI systems can recommend personalized learning paths that are aligned with the employee's career goals and the organization's needs (Bersin, 2019).

Key Applications:

- **Skill Assessment:** AI tools continuously assess employees' skills through quizzes, performance data, and peer reviews.
- **Learning Recommendations:** Based on the assessments, AI systems suggest relevant courses, workshops, and training programs.
- **Career Development:** AI-driven insights help employees understand the skills they need to advance their careers, creating a more strategic approach to professional development.

Benefits:

- **Personalized Learning:** Employees receive training that is specifically relevant to their roles and career aspirations.
- **Efficient Resource Allocation:** Organizations can focus training resources on areas that will have the most significant impact.
- **Enhanced Employee Growth:** Continuous learning opportunities ensure employees keep their skills updated, fostering a culture of continuous improvement.

Case Study:

- **IBM:** IBM's AI platform, Watson, helps employees identify skill gaps and provides personalized learning recommendations. This system has enabled IBM to create targeted development plans that enhance employee skills and career growth (IBM, 2017).

Integration of AI With Learning Management Systems (LMS)

Integrating AI with Learning Management Systems (LMS) transforms traditional training programs by providing adaptive learning experiences. AI-enhanced LMS platforms can analyze learners' interactions and adjust content in real-time to optimize learning outcomes (Deloitte, 2019).

Features:

- **Adaptive Learning Paths:** AI algorithms adjust the learning path based on the learner's progress, ensuring that content is neither too easy nor too difficult.
- **Data-Driven Insights:** AI provides detailed analytics on learner engagement, performance, and content effectiveness.
- **Interactive Content:** AI tools can generate interactive and immersive content, such as virtual reality (VR) simulations, to enhance the learning experience.

Benefits:

- **Improved Learning Outcomes:** Personalized and adaptive learning paths ensure better comprehension and retention of information.
- **Enhanced Engagement:** Interactive and relevant content keeps learners engaged and motivated.
- **Scalability:** AI-powered LMS can cater to a large number of employees, providing personalized experiences at scale.

Case Study:

- **Deloitte:** Deloitte's AI-integrated LMS uses machine learning algorithms to personalize learning paths for employees. The system analyzes individual learning behaviors and performance to recommend relevant courses and resources, leading to improved learning outcomes and employee satisfaction (Deloitte, 2019).

4. PERFORMANCE MANAGEMENT

4.1 Real-Time Performance Tracking

AI Tools for Continuous Performance Monitoring and Feedback

AI-driven performance management tools offer continuous monitoring and real-time feedback, transforming how organizations track and enhance employee performance. These tools utilize data from various sources, such as project management systems, communication platforms, and work outputs, to provide a comprehensive view of employee performance (Aguinis, 2019).

Key Features:

- **Continuous Monitoring:** AI tools can track performance metrics in real-time, providing instant insights into employee productivity and engagement.
- **Automated Feedback:** AI systems can deliver immediate feedback based on performance data, helping employees to make timely improvements.
- **Behavioral Analytics:** AI analyzes communication patterns and collaboration behaviors to identify high and low-performing employees.

Benefits of Real-Time Analytics Over Traditional Performance Reviews

Benefits:

- **Immediate Improvements:** Real-time feedback allows employees to correct their course immediately, enhancing productivity and performance.
- **Reduced Bias:** Continuous monitoring minimizes the influence of recency bias present in traditional annual reviews.
- **Enhanced Engagement:** Regular feedback keeps employees engaged and motivated, fostering a culture of continuous improvement.
- **Data-Driven Insights:** Real-time analytics provide actionable insights that help managers make informed decisions about promotions, training needs, and performance interventions.

Case Study:

- **General Electric (GE):** GE transitioned from annual performance reviews to real-time feedback using an AI-driven performance management system.

This change led to a more agile workforce and improved overall productivity and employee satisfaction (GE, 2020).

4.2 Objective Setting and Evaluation

AI in Setting and Aligning Individual and Organizational Goals

AI can assist in setting and aligning individual and organizational goals by analyzing historical data, market trends, and individual performance metrics. This ensures that goals are not only realistic and achievable but also aligned with the strategic objectives of the organization (Upadhyay & Khandelwal, 2018).

Key Applications:

- **Goal Alignment:** AI tools ensure that individual goals are aligned with broader organizational objectives, enhancing coherence and strategic focus.
- **Personalized Goal Setting:** AI systems can personalize goals based on an employee's past performance, skills, and career aspirations.
- **Predictive Goal Setting:** By analyzing trends and historical data, AI can predict future performance and set goals accordingly.

Benefits:

- **Clarity and Focus:** Clear alignment of goals helps employees understand their contributions to the organizational strategy.
- **Motivation and Engagement:** Personalized and achievable goals enhance employee motivation and engagement.
- **Strategic Agility:** AI-driven goal setting allows organizations to quickly adapt to changing market conditions and internal dynamics.

Case Study:

- **Adobe:** Adobe replaced traditional performance reviews with a continuous feedback system, supported by AI tools for goal setting and tracking. This approach has resulted in higher employee engagement and better alignment with corporate goals (Adobe, 2018).

Data-Driven Evaluation of Employee Performance

AI-powered performance evaluation systems leverage data from multiple sources to provide a holistic and objective assessment of employee performance. These systems can analyze quantitative metrics, such as sales figures or project completions, alongside qualitative data, such as peer reviews and customer feedback (Guenole et al., 2017).

Key Features:

- **Comprehensive Data Analysis:** AI systems aggregate data from various sources to provide a well-rounded view of performance.
- **Objective Assessment:** Data-driven evaluations reduce human biases and ensure fair and accurate assessments.
- **Actionable Insights:** AI provides detailed reports and insights that help managers make informed decisions regarding promotions, bonuses, and development plans.

Benefits:

- **Fairness and Transparency:** Objective evaluations based on data enhance fairness and transparency in performance assessments.
- **Improved Decision Making:** Data-driven insights support better decision-making regarding employee development and organizational strategy.
- **Enhanced Accountability:** Continuous and objective performance tracking fosters a culture of accountability and continuous improvement.

Case Study:

- **IBM:** IBM uses an AI-driven performance management system that analyzes a wide range of performance data to provide objective evaluations. This system has improved the accuracy and fairness of performance assessments and has contributed to a more motivated and productive workforce (IBM, 2017).

5. EMPLOYEE ENGAGEMENT AND RETENTION

5.1 Sentiment Analysis

Using AI to Gauge Employee Sentiment Through Surveys and Feedback

AI-driven sentiment analysis involves using machine learning algorithms to analyze employee feedback from surveys, emails, and other communication channels to understand their emotions and attitudes. By evaluating the language used in these communications, AI can provide insights into overall employee sentiment, helping organizations to identify areas of concern and improve workplace culture (Giles, 2020).

Key Applications:

- **Employee Surveys:** AI analyzes responses from employee engagement surveys to detect sentiments and trends.
- **Feedback Analysis:** Continuous analysis of feedback from performance reviews, exit interviews, and suggestion boxes.
- **Real-Time Sentiment Monitoring:** AI tools monitor real-time communications, such as emails and chat messages, to gauge ongoing sentiment.

Benefits:

- **Proactive Management:** Early detection of negative sentiments allows for timely interventions.
- **Enhanced Engagement:** Understanding employee sentiment helps in creating initiatives that enhance engagement and satisfaction.
- **Data-Driven Decisions:** Sentiment analysis provides objective data, aiding in informed decision-making.

Case Study:

- **Microsoft:** Microsoft utilizes AI-driven sentiment analysis to evaluate employee feedback from surveys and other communication channels. This approach has helped them improve employee satisfaction and address issues promptly (Jain et al., 2019).

Applications of Natural Language Processing (NLP) in Understanding Employee Morale

NLP is a branch of AI that enables machines to understand, interpret, and respond to human language. In the context of HR, NLP is used to analyze text data from various sources to understand employee morale and engagement (Miner et al., 2020).

Key Features:

- **Text Mining:** NLP algorithms analyze large volumes of text data from employee communications to extract meaningful insights.
- **Emotion Detection:** NLP models detect and classify emotions expressed in written communication.
- **Trend Analysis:** Identifying trends in employee sentiment over time to understand morale dynamics.

Benefits:

- **Comprehensive Analysis:** NLP can process vast amounts of data quickly, providing a comprehensive view of employee sentiment.
- **Actionable Insights:** Detailed analysis of employee morale helps in designing targeted engagement strategies.
- **Reduced Bias:** Automated sentiment analysis reduces the risk of bias compared to manual interpretation.

Case Study:

- **IBM Watson:** IBM Watson uses NLP to analyze employee feedback and monitor sentiment. This has enabled IBM to maintain a positive work environment by addressing issues before they escalate (Miner et al., 2020).

5.2 Predictive Retention Models

AI in Predicting Employee Turnover and Identifying Retention Risks

Predictive retention models use AI to analyze employee data and identify patterns that indicate a risk of turnover. By examining factors such as job satisfaction, engagement levels, performance metrics, and external market conditions, AI can predict which employees are at risk of leaving (De Mauro et al., 2018).

Key Applications:

- **Turnover Prediction:** AI models analyze historical data to predict which employees are likely to leave.
- **Risk Identification:** Identifying high-risk groups based on factors such as role, department, and tenure.
- **Customized Interventions:** Developing tailored retention strategies for at-risk employees.

Benefits:

- **Reduced Turnover:** Early identification of retention risks allows for proactive measures to retain employees.
- **Cost Savings:** Reducing turnover helps in saving costs associated with recruitment and training of new employees.
- **Enhanced Engagement:** Addressing factors that contribute to turnover improves overall employee engagement.

Case Study:

- **SAP:** SAP uses AI to predict employee turnover by analyzing various data points such as job satisfaction, engagement scores, and external job market data. This predictive approach has significantly reduced their employee turnover rates (De Mauro et al., 2018).

Strategies for Using AI Insights to Enhance Employee Retention

Utilizing AI insights to enhance employee retention involves implementing targeted strategies based on predictive analytics and sentiment analysis.

Key Strategies:

- **Personalized Retention Plans:** Developing individualized retention plans based on AI insights into employee satisfaction and risk factors.
- **Career Development Programs:** Offering targeted career development opportunities to at-risk employees.
- **Improved Work Environment:** Addressing workplace issues identified through sentiment analysis to enhance job satisfaction.
- **Recognition and Rewards:** Implementing recognition and reward programs tailored to the preferences and contributions of employees.

Benefits:

- **Higher Retention Rates:** Targeted strategies help in retaining top talent.
- **Increased Engagement:** Personalized interventions improve employee engagement and satisfaction.
- **Better Workforce Planning:** Predictive analytics support strategic workforce planning by identifying future retention risks.

Case Study:

- **Google:** Google employs AI-driven retention strategies that include personalized career development plans and tailored rewards programs. These initiatives have helped Google maintain high employee retention and satisfaction levels (Hassan et al., 2020).

6. ETHICAL CONSIDERATIONS AND CHALLENGES

6.1 Bias and Fairness

Addressing the Risk of AI-Induced Bias in HR Processes

AI systems in HR can inadvertently perpetuate existing biases if not carefully designed and monitored. Bias can originate from biased training data or from the algorithms themselves, leading to unfair hiring, promotion, or performance evaluation practices (Raghavan et al., 2020).

Key Risks:

- **Historical Bias:** AI models trained on historical data may inherit biases present in past hiring decisions.
- **Algorithmic Bias:** Inherent biases in algorithmic design can lead to discriminatory outcomes.

Mitigation Strategies:

- **Diverse Training Data:** Ensuring training data is representative and free from historical biases.
- **Bias Audits:** Regular audits of AI algorithms to detect and rectify biases.
- **Transparent Models:** Using transparent and interpretable AI models to understand decision-making processes.

Case Study:

- **Amazon:** Amazon had to discontinue its AI recruiting tool after it was found to be biased against women. The tool favored resumes that used male-dominated language, highlighting the importance of bias mitigation in AI systems (Dastin, 2018).

Approaches to Ensure Fairness and Transparency in AI Algorithms

Ensuring fairness and transparency involves implementing best practices in the design, deployment, and monitoring of AI systems.

Best Practices:

- **Fairness Metrics:** Developing and using metrics to evaluate the fairness of AI decisions.
- **Explainability:** Designing algorithms that provide clear explanations for their decisions.
- **Stakeholder Involvement:** Involving diverse stakeholders in the development and evaluation of AI systems.

Regulatory Compliance:

- **GDPR:** Compliance with regulations such as the General Data Protection Regulation (GDPR) which mandates transparency and fairness in automated decision-making (Voigt & Von dem Bussche, 2017).
- **EEOC Guidelines:** Adhering to guidelines set by bodies like the Equal Employment Opportunity Commission (EEOC) to prevent discrimination (Raghavan et al., 2020).

Case Study:

- **Microsoft:** Microsoft has implemented fairness dashboards and rigorous testing protocols to ensure its AI systems operate fairly and transparently (Williams, 2018).

6.2 Privacy and Data Security

Safeguarding Employee Data in AI-Driven HR Systems

The use of AI in HR requires handling vast amounts of personal data, making data security a paramount concern. Ensuring the privacy and security of employee data involves robust data protection measures (Tankard, 2016).

Key Measures:

- **Data Encryption:** Encrypting data both in transit and at rest to prevent unauthorized access.
- **Access Controls:** Implementing strict access controls to ensure only authorized personnel can access sensitive data.
- **Anonymization:** Anonymizing data wherever possible to protect employee identities.

Challenges:

- **Data Breaches:** Risk of data breaches due to cyberattacks.
- **Compliance:** Meeting various regulatory requirements related to data protection.

Case Study:

- **Salesforce:** Salesforce employs comprehensive data security measures, including encryption and rigorous access controls, to protect the data within its AI-driven HR systems (Tankard, 2016).

Regulatory Considerations and Best Practices

Compliance with data protection regulations and adoption of best practices are critical for maintaining trust and ensuring legal compliance.

Key Regulations:

- **GDPR:** Imposes strict requirements on data protection and the use of personal data in automated decision-making.
- **CCPA:** The California Consumer Privacy Act (CCPA) provides similar protections for residents of California.

Best Practices:

- **Privacy by Design:** Incorporating data protection principles from the start of the system design.
- **Regular Audits:** Conducting regular audits to ensure compliance with data protection regulations.
- **Employee Training:** Educating employees about data protection practices and their importance.

Case Study:

- **IBM:** IBM has developed stringent data protection protocols and regularly audits its AI systems to ensure compliance with global data privacy regulations.

6.3 Human-AI Collaboration

Balancing Automation With the Human Element in HR

While AI can automate many HR processes, it is essential to maintain a balance between automation and the human touch to ensure empathy and personal connection (Brynjolfsson & McAfee, 2014).

Key Strategies:

- **Human Oversight:** Ensuring human oversight in decision-making processes to provide empathy and context.
- **Complementary Roles:** Using AI to handle routine tasks, freeing up HR professionals to focus on more strategic and personal aspects of their roles.
- **Feedback Mechanisms:** Implementing feedback loops where AI decisions are regularly reviewed and validated by humans.

Benefits:

- **Efficiency:** Automation of repetitive tasks increases efficiency and allows HR professionals to focus on high-value activities.
- **Personalization:** Human oversight ensures personalized and empathetic handling of employee concerns.

Case Study:

- **Cognizant:** Cognizant employs AI to automate administrative HR tasks while HR professionals focus on strategic initiatives and employee engagement, ensuring a balanced approach (Brynjolfsson & McAfee, 2014).

Preparing HR Professionals for an AI-Integrated Workplace

As AI becomes more prevalent in HR, it is crucial to prepare HR professionals to work effectively in an AI-integrated environment.

Training and Development:

- **AI Literacy:** Training HR professionals in basic AI concepts and applications.
- **Continuous Learning:** Encouraging continuous learning to keep up with evolving AI technologies.
- **Change Management:** Implementing change management strategies to ease the transition to AI-integrated processes.

Cultural Shift:

- **Collaboration:** Promoting a culture of collaboration between humans and AI systems.
- **Innovation:** Encouraging an innovative mindset to leverage AI capabilities fully.

Case Study:

- **PwC:** PwC provides extensive training programs for its HR professionals to ensure they are well-versed in AI technologies and can effectively integrate them into their workflows (Huang & Rust, 2018).

7. FUTURE TRENDS AND INNOVATIONS

7.1 AI and the Future of Work

Emerging Trends in AI and Their Potential Impact on HR

As AI continues to evolve, several emerging trends are set to transform HR practices, making processes more efficient, data-driven, and personalized (Chui et al., 2016).

Key Trends:

- **Hyper-Personalization:** AI-driven tools are expected to offer hyper-personalized employee experiences, from recruitment to retirement, tailoring interactions and opportunities based on individual preferences and behaviors.
- **AI-Driven Decision Making:** Enhanced AI capabilities will enable more accurate and data-driven decision-making in areas such as talent acquisition, performance management, and employee engagement.
- **Augmented Workforce:** AI will augment human capabilities, allowing HR professionals to focus on strategic initiatives by automating routine and administrative tasks.
- **Ethical AI:** There will be an increasing focus on developing ethical AI frameworks to ensure fairness, transparency, and accountability in AI applications.

Impact on HR:

- **Efficiency and Productivity:** AI will streamline HR operations, leading to significant improvements in efficiency and productivity.
- **Employee Experience:** AI will enhance the employee experience by providing personalized and responsive HR services.
- **Talent Management:** Predictive analytics will improve talent management, from identifying high-potential employees to forecasting workforce needs.

Case Study:

- **Unilever:** Unilever uses AI to analyze video interviews and evaluate candidates based on various attributes, significantly reducing the time and bias involved in the recruitment process (Harvard Business Review, 2019).

Predictions on the Evolving Role of AI in Shaping the Future Workplace

AI is poised to play a critical role in shaping the future workplace, driving innovation, and transforming the way work is conducted.

Predictions:

- **Remote and Hybrid Work:** AI will support remote and hybrid work models by providing tools for virtual collaboration, performance tracking, and employee engagement.

- **Wellness and Well-being:** AI will play a pivotal role in monitoring and enhancing employee well-being through tools that assess stress levels, suggest interventions, and provide support.
- **Reskilling and Upskilling:** AI will identify skill gaps and recommend personalized learning paths, facilitating continuous learning and adaptation to new job roles.

Impact:

- **Workplace Transformation:** AI will transform the physical and virtual workplace, making it more adaptive and responsive to employee needs.
- **Workforce Dynamics:** The integration of AI will change workforce dynamics, requiring new skills and leading to the creation of new job roles focused on managing and interacting with AI systems.

Case Study:

- **PwC:** PwC leverages AI to support remote work, using AI-driven tools for performance management, virtual collaboration, and employee engagement, fostering a more flexible and productive work environment (Snyder, 2020).

7.2 Integration With Genomics

Exploring the Intersection of AI, Genomics, and HR

The integration of AI with genomics is an emerging field with the potential to revolutionize HR practices, particularly in areas such as personalized health and wellness programs (Topol, 2019).

Key Applications:

- **Personalized Health Plans:** AI can analyze genomic data to create personalized health and wellness plans for employees, enhancing overall well-being and productivity.
- **Predictive Health Analytics:** AI-driven genomic analysis can predict potential health issues, allowing for early interventions and proactive health management.
- **Talent Optimization:** Understanding genetic predispositions can help in tailoring job roles and work environments to optimize employee performance and satisfaction.

Benefits:

- **Improved Health Outcomes:** Personalized health plans can lead to better health outcomes and reduced healthcare costs.
- **Enhanced Employee Well-being:** Proactive health management can improve employee well-being and reduce absenteeism.
- **Data-Driven Decisions:** AI and genomics provide data-driven insights that can inform HR decisions related to employee health and productivity.

Case Study:

- **Health Nucleus:** Health Nucleus uses AI and genomic data to provide personalized health insights and proactive wellness recommendations, demonstrating the potential of integrating AI and genomics in employee health management (Topol, 2019).

Potential Applications and Ethical Implications

While the integration of AI and genomics holds great promise, it also raises significant ethical and privacy concerns that must be addressed (Rothstein, 2015).

Potential Applications:

- **Workplace Wellness Programs:** Using genomic data to design tailored wellness programs.
- **Predictive Maintenance of Health:** Identifying employees at risk of certain conditions and providing preventive care.
- **Customized Job Roles:** Aligning job roles with employees' genetic predispositions to enhance performance and job satisfaction.

Ethical Implications:

- **Privacy Concerns:** Ensuring the confidentiality and security of genetic data to protect employee privacy.
- **Discrimination Risks:** Preventing genetic discrimination in hiring, promotions, and other HR decisions.
- **Informed Consent:** Obtaining informed consent from employees before collecting and using their genomic data.

Case Study:

- **23andMe:** 23andMe provides genomic testing and health insights, highlighting the importance of privacy and ethical considerations in the use of genomic data in any context (Goddard et al., 2013).

8. CASE STUDIES

8.1 Success Stories

Detailed Examination of Organizations that Have Successfully Integrated AI Into Their HR Processes

IBM

IBM has integrated AI across various HR functions, including recruitment, employee engagement, and performance management. IBM's AI tool, Watson, analyzes resumes, assesses job fit, and provides insights into candidate suitability, significantly reducing the time and effort involved in hiring (IBM, 2019).

Key Achievements:

- **Efficiency in Recruitment:** Watson has reduced the time-to-hire by 30% by automating initial resume screening and candidate matching.
- **Employee Retention:** AI-driven predictive analytics identify employees at risk of leaving, allowing IBM to implement targeted retention strategies, resulting in a 20% decrease in turnover.

Best Practices:

- **Holistic Integration:** Incorporating AI across multiple HR functions to leverage comprehensive data insights.
- **Continuous Monitoring:** Regularly updating AI algorithms based on feedback and new data to maintain accuracy and relevance.

Unilever

Unilever has adopted AI to revolutionize its recruitment process. By using AI-powered video interviews and gamified assessments, Unilever has enhanced the candidate experience and improved the efficiency of its hiring process (Harvard Business Review, 2019).

Key Achievements:

- **Reduced Bias:** AI tools have helped reduce unconscious bias in hiring decisions, promoting a more diverse workforce.
- **Enhanced Candidate Experience:** The use of AI-driven assessments has made the application process more engaging and efficient, reducing the time from application to offer by 75%.

Best Practices:

- **Transparent Communication:** Clearly communicating the role of AI in the recruitment process to candidates to build trust and transparency.
- **Bias Mitigation:** Regularly auditing AI tools to ensure they remain unbiased and inclusive.

PwC

PwC has implemented AI in their talent management systems, focusing on personalized learning and development. AI algorithms analyze employee skills and recommend personalized learning paths, leading to higher employee satisfaction and improved skill development (Snyder, 2020).

Key Achievements:

- **Skill Development:** AI-driven personalized learning has increased employee engagement with training programs by 40%.
- **Performance Improvement:** Targeted skill development has led to a 15% improvement in employee performance metrics.

Best Practices:

- **Personalization:** Using AI to deliver personalized learning experiences tailored to individual needs and career goals.
- **Feedback Integration:** Continuously integrating employee feedback to refine AI recommendations and improve outcomes.

8.2 Lessons From Failures

Analysis of Challenges and Failures in AI Implementation in HR

Amazon

Amazon faced significant challenges with its AI-powered recruitment tool, which was found to be biased against female candidates. The AI system favored resumes that used male-dominated language, reflecting historical biases present in the training data (Dastin, 2018).

Key Issues:

- **Bias in Training Data:** The AI system was trained on resumes submitted over a decade, most of which came from male candidates, leading to biased outcomes.
- **Lack of Diversity:** The AI tool failed to promote diversity and inclusivity, reinforcing existing biases.

Lessons Learned:

- **Data Quality:** Ensuring that training data is diverse and representative to prevent bias.
- **Regular Audits:** Conducting regular audits of AI systems to identify and rectify biases early.

HR Chatbot Implementation at a Tech Company

A tech company faced challenges when implementing an AI-powered HR chatbot intended to handle employee queries. The chatbot struggled with understanding complex queries and providing accurate responses, leading to employee dissatisfaction.

Key Issues:

- **Inaccurate Responses:** The chatbot frequently provided incorrect or irrelevant answers, frustrating employees.
- **Limited Understanding:** The AI struggled with understanding the context of complex HR queries, limiting its effectiveness.

Lessons Learned:

- **Continuous Training:** Regularly updating and training the chatbot to improve its understanding and accuracy.

- **Human Oversight:** Maintaining human oversight to handle complex queries that the AI cannot address effectively.

Strategies for Overcoming Common Pitfalls

- **Comprehensive Testing:** Thoroughly testing AI systems before deployment to identify potential issues and ensure they function as intended.
- **Stakeholder Involvement:** Involving diverse stakeholders in the development and implementation process to provide multiple perspectives and ensure inclusivity.
- **Transparent Communication:** Clearly communicating the role and limitations of AI systems to employees to manage expectations and build trust.
- **Regular Monitoring and Updates:** Continuously monitoring AI systems and updating them based on new data and feedback to maintain their effectiveness and relevance.
- **Ethical Considerations:** Ensuring ethical considerations are prioritized in the design and deployment of AI systems, including fairness, transparency, and accountability.

9. CONCLUSION

9.1 Summary of Key Points

This comprehensive exploration of AI in HR highlights its transformative impact on various aspects of human resource management. AI has proven instrumental in optimizing recruitment processes, fostering employee development through personalized learning paths, and facilitating data-driven decision-making that enhances organizational effectiveness. Despite these advancements, challenges such as bias mitigation, ethical considerations, and the balance between automation and human touch remain critical areas for ongoing refinement.

AI-driven innovations in HR have introduced efficiencies that were previously unimaginable, allowing organizations to adapt quickly to changing market dynamics and employee expectations. The ability to harness vast amounts of data for predictive analytics has revolutionized talent management strategies, enabling proactive interventions to boost employee engagement and retention.

9.2 Final Thoughts

Reflecting on the transformative power of AI in HR underscores its role as a catalyst for organizational agility and competitiveness in the digital age. The ongoing evolution of AI technologies presents HR professionals with unprecedented opportunities to reshape workplace practices and enhance the employee experience. Embracing AI not only streamlines administrative tasks but also empowers HR teams to focus on strategic initiatives that drive innovation and sustainable growth.

Looking to the future, it is clear that AI will continue to redefine the HR landscape, offering new possibilities for talent acquisition, performance management, and employee well-being. HR leaders are encouraged to embrace AI-driven solutions proactively, fostering a culture of continuous learning and adaptation to maximize the potential of these technologies. By staying ahead of technological advancements and leveraging AI responsibly, organizations can position themselves as leaders in attracting, developing, and retaining top talent in an increasingly competitive global marketplace.

In conclusion, the integration of AI in HR represents not just a technological advancement, but a paradigm shift in how organizations nurture their most valuable asset – their people. By harnessing AI's capabilities effectively and ethically, HR professionals can pave the way for a future where human potential is maximized, organizational goals are achieved with greater precision, and workplaces thrive in an era of unprecedented digital transformation.

REFERENCES

Adobe. (2018). Adobe's continuous performance management system. Retrieved from https://www.adobe.com/

Aguinis, H. (2019). *Performance management for dummies*. John Wiley & Sons.

Akkiraju, R., Sinha, V., Xu, A., Mahmud, J., & Liu, Z. (2020). Characterizing Machine Learning Processes for AI Explainability: A Process Model Inventory Approach. *IBM Journal of Research and Development*, 64(4/5), 1–10.

Bersin, J. (2019). HR technology market 2019: Disruption ahead. *Josh Bersin Report*. Retrieved from https://joshbersin.com

Bondarouk, T., & Ruël, H. (2009). *Electronic HRM in Theory and Practice*. Emerald Group Publishing.

Brynjolfsson, E., & McAfee, A. (2014). *The second machine age: Work, progress, and prosperity in a time of brilliant technologies*. W.W. Norton & Company.

Chamorro-Premuzic, T., Akhtar, R., Winsborough, D., & Sherman, R. A. (2016). The datafication of talent: How technology is advancing the science of human potential. *Harvard Business Review*.

Chui, M., Manyika, J., & Miremadi, M. (2016). Where machines could replace humans—and where they can't (yet). *McKinsey Quarterly*. Retrieved from https://www.mckinsey.com

Cognizant. (2020). Cognizant's AI-Powered Onboarding Platform. Retrieved from https://www.cognizant.com/

Dastin, J. (2018). Amazon scraps secret AI recruiting tool that showed bias against women. *Reuters*. Retrieved from https://www.reuters.com

De Mauro, A., Greco, M., & Grimaldi, M. (2018). A formal definition of big data based on its essential features. *Library Review*, 65(3), 122–135. 10.1108/LR-06-2015-0061

Deloitte. (2019). Global Human Capital Trends 2019: Leading the social enterprise. Retrieved from https://www2.deloitte.com/content/dam/Deloitte/at/Documents/human-capital/hc-trends-2019.pdf

Falletta, S. V. (2014). In search of HR intelligence: Evidence-based HR analytics practices in high performing companies. *People & Strategy*, 36(4), 28–37.

GE. (2020). GE's transition to real-time performance management. Retrieved from https://www.ge.com/

Giles, S. (2020). AI in HR: Using machine learning to enhance employee engagement and retention. *HR Technology Review*, 34(1), 22–28.

Goddard, K. A., Robitaille, J., Dowling, N. F., Parrado, A. R., Fishman, J., Bradley, L. A., & Korf, B. R. (2013). Health-related direct-to-consumer genetic tests: A public health assessment and analysis of practices related to internet-based tests for risk of thrombosis. *Public Health Genomics*, 16(5), 223–234.23886854

Guenole, N., Ferrar, J., & Feinzig, S. (2017). *The Power of People: Learn How Successful Organizations Use Workforce Analytics to Improve Business Performance*. Pearson.

Harvard Business Review. (2019). How Unilever uses AI to hire. *Harvard Business Review*. Retrieved from https://hbr.org

Hassan, S., Jehan, S., & Khan, A. (2020). Predictive analytics in human resource management: A comprehensive review of machine learning applications. *International Journal of HR Studies*, 10(1), 1–15.

HireVue. (2018). HireVue's AI-driven platform. Retrieved from https://www.hirevue.com/

Huang, M. H., & Rust, R. T. (2018). Artificial intelligence in service. *Journal of Service Research*, 21(2), 155–172. 10.1177/1094670517752459

IBM. (2017). The enterprise guide to artificial intelligence. Retrieved from https://www.ibm.com/watson/enterprise-ai

IBM. (2019). IBM Watson talent management solutions. *IBM*. Retrieved from https://www.ibm.com

Jain, A., Singhal, P., & Singh, R. (2019). Sentiment analysis and its application in human resources. *International Journal of Advanced Research in Computer Science and Software Engineering*, 9(5), 1–8.

Jiang, F. (2019). Leveraging AI for employee onboarding and training. *Journal of Human Resources Management*, 58(3), 42–51.

Kavanagh, M. J., & Thite, M. (2009). *Human Resource Information Systems: Basics, Applications, and Future Directions*. SAGE Publications.

Miner, L., Petry, L., & De Meo, J. (2020). Understanding the impact of NLP on HR functions: A case study of IBM Watson. *Journal of Artificial Intelligence Research*, 67(3), 621–634.

Mya Systems. (2019). Mya: The AI Recruitment Assistant. Retrieved from https://www.mya.com/

Raghavan, M., Barocas, S., Kleinberg, J., & Levy, K. (2020). Mitigating bias in algorithmic hiring: Evaluating claims and practices. In *Proceedings of the 2020 Conference on Fairness, Accountability, and Transparency* (pp. 469-481). 10.1145/3351095.3372828

Rothstein, M. A. (2015). Ethical issues in big data health research. *The Journal of Law, Medicine & Ethics*, 43(2), 425–429. 10.1111/jlme.1225826242964

Snyder, B. (2020). How PwC is using AI to reshape the future of work. *Forbes*. Retrieved from https://www.forbes.com

Society for Human Resource Management. (2020). The Benefits of AI in Onboarding. Retrieved from https://www.shrm.org/

Tankard, C. (2016). What the GDPR means for businesses. *Network Security*, 2016(6), 5–8. 10.1016/S1353-4858(16)30056-3

Topol, E. (2019). *Deep medicine: How artificial intelligence can make healthcare human again*. Basic Books.

Unilever. (2017). How Unilever brought AI into its recruitment process. Retrieved from https://www.unilever.com/

Upadhyay, A. K., & Khandelwal, K. (2018). Applying artificial intelligence: Implications for recruitment. *Strategic HR Review*, 17(5), 255–258. 10.1108/SHR-07-2018-0051

Voigt, P., & Von dem Bussche, A. (2017). *The EU General Data Protection Regulation (GDPR): A practical guide*. Springer International Publishing. 10.1007/978-3-319-57959-7

Williams, S. (2018, March 7). Microsoft's AI bias efforts focus on accountability and transparency. CIO Dive. Retrieved from https://www.ciodive.com/news/microsofts-ai-bias-efforts-focus-on-accountability-and-transparency/518784/

Chapter 5
Genomics in HR: A New Frontier for Employee Wellness and Organizational Growth

Dixit Kalra
https://orcid.org/0000-0002-1744-6717
Parul University, Vadodara, India

Medha Gupta
https://orcid.org/0000-0003-2789-9784
Chandigarh University, India

Swati Gupta
https://orcid.org/0009-0005-5795-8353
Chitkara University, Punjab, India

Harpreet Singh
Punjabi University, India

ABSTRACT

Genomics advancements have shaken the human resources (HR) field, giving insights into employee and organizational health and an understanding of the individualized approach to medicine. This study seeks to explore the evolving trends within human resource management (HRM) and genomics by reviewing past research. It also aims to identify key themes and outline potential directions for future investigations in these fields. The current research utilized electronic Scopus databases, which include all major prior studies. This bibliometric analysis utilizes content analysis to explore the methodology, annual contribution trends, and leading countries in the research area. VOS viewer software is also used for bibliometric analysis, incorpo-

DOI: 10.4018/979-8-3693-7668-3.ch005

rating bibliographic coupling, co-occurrence, and co-authorship analyses, which provide various graphical representations of the research network. Furthermore, thematic mapping is conducted to identify different clusters within the topic and highlight recent contributions, where new topics emerge, and new methodologies are employed to study consumer behavior. The current study aims to compile ideas for future research and explore how genomics can be pivotal and beneficial in comprehending HRM in the modern age.

INTRODUCTION

Genomics and Technological development have intensified over the years and this has impacted all the fields among them being human resource management. There is a broad understanding of the term Technology in HR popularly known as HR Tech. These are sophisticated tools and systems that improve the working of HR. This transformation is leading to the formation of new prospects in enhancing the health status of the employees besides fostering organizational development.

Most technological enhancements like Artificial Intelligence (AI), Machine Learning (ML), Big data analytics, and Cloud computing have affected HR functions. They enable organizations to delegate some of the simple but time-consuming tasks which if handled by the HR personnel would take a lot of their time while handling crucial issues related to employees' satisfaction and well-being (Stone et al., 2015). For example, in the case of recruitment, there are AI tools that can look for the best candidates rapidly, or in the case of an organization, there are other tools with analytic means to have a better perception of workforce planning and talent management (Davenport, Harris, & Shapiro, 2010).

The wellness of its employees is therefore an important aspect when it comes to Human Resource Technology. Digital wellness and mobile solutions are the ways to provide organizations with the opportunity to deliver efficient wellness programs that are targeted at the specific needs of the employee (Mulcahy & Betts, 2018). These platforms can track physical activity, counsellors can be provided for mental health care nutritional information can be given, and so on, making the employees and thus the organization healthier. Additionally, wearable technology which includes health trackers like fitness trackers enables this due to a real-time report of health status which can be used to enhance the adoption of healthy lifestyles and mitigate diseases (Zhu, Azar, Meng, et al., 2020).

Also, technology creates a more connected and actively participating population in the workforce. Tools like Slack and Microsoft Teams improve communication throughout the teams and keep the workers linked, especially in a distant or blended workforce (van Zoonen & Sivunen, 2020). These tools enhance efficiency and also foster camaraderie and assuage the stress for employees which is imperative for their stability. The adoption of HR Tech also brings challenges most of which are rooted in data privacy and security issues (Gupta & Arora, 2024). Employers must safeguard and use employee information appropriately and follow the provisions of such laws as the General Data Protection Regulation (Moore 2017). However, there is a challenge of training needs to facilitate the enhancement of the proper use of such technologies in their practice among HR professionals (Bondarouk et al., 2017).

The technology implementation in the HRM function presents a strategic opportunity to bring improvement to employees' well-being and organizational development. Therefore, by utilizing these tools and platforms, one can reach the level where the organization and, as a result, the workforce is healthier, and people are more engaged (Soonthodu et. al.,2021). However, it is indicated that there are ethical and security issues that must be solved to avoid the negative use of HR Tech.

RESEARCH METHODOLOGY

This research paper aims to investigate the recent advancements in genomics concentrating mostly on their management implications. A bibliometric analysis which is one way of visually analyzing a certain area of study was used by researchers (Kalra et al., 2024). We started this research by using the SLR technique to collect the necessary literature. This process involves choosing relevant keywords, conducting searches with these keywords, and then choosing the most important literature by employing a set of criteria (Gupta et al., 2023). Furthermore, the study utilized bibliometric methods in its examination of literature while it disclosed significant insights into changing patterns of influential publications, key writers, crucial articles, and important areas for future investigation (Kaur et al., 2024).

SEARCH CRITERIA

Utilizing the wide-ranging Scopus database renowned for its all-inclusive accumulations of articles was done. Publishing research results serves as a database of established information. In these publications, citations underscored the key variables examined and various relationships found (Kalra et al., 2023). Here is Figure 1 that shows our approach to gather the total set of 191 articles. We initiated this exercise by reviewing 914 papers in the database related to Genomics and HRM. To narrow down our selection we implemented rigorous screening procedures that included evaluating subject areas, document types, language, content relevance, and removing incorrect records, among other criteria. This thorough sifting ensured only pertinent and high-quality articles were included in our study thereby enhancing the reliability of our findings through being robust. Leveraging the Scopus database plus running meticulous screenings enabled us to curate a focused and relevant collection of literature essential to understanding genomics' management functions within HR.

Table 1. Search Criteria Table

Particulars	Reject	Accept
Search Term- "Genomics" AND "human relations" OR "HR" OR "human relation management" OR "HRM"		914
Subject Area- Medicine, Biochemistry, Genetics and Molecular Biology, Multidisciplinary, Business, Management and Accounting	258	656
Document Type- Article	68	588
Language- English only	6	582
Keywords- Related to genomics and hr	121	461
Source Type- Journals only	1	460
Publication Stage- Final	2	458
Open Access articles are selected for research	91	367
Total Articles for review		367

Source – Author's Compilations from Scopus Database

RESULTS

This section of the results examines the research questions outlined in this document, concentrating on publication trends and identifying the leading authors with the highest citation counts, country-wise citing, and noting journals are found significant in covering Genomics in HR: A New Frontier for Employee Wellness and Organizational Growth.

The section maps out the publishing landscape through an examination of key figures in academic discourse and shows how these interactions transpire across academic boundaries. Consequently, this chapter explores impact factors and influential publications within Genomics in the HR framework. Further, it assesses how these articles influence management practices, thus demonstrating how scholarly works relate to practical business decisions. This thesis provides a detailed analysis of research trends, authorship patterns, geographical distribution of citations, and overall scholarly impacts that offer insights into the changing nature of genomics HRM worldwide.

PATTERNS IN PUBLICATIONS

The article under consideration highlights the application of genomics in HR and managerial activities as depicted in Figure 1. It depicts the publishing of the first article related to this subject in the year, 2015. From 2020 to 2021, the number of articles related to genomics in HR was relatively low, but in 2022, this trend grew more interesting, and published 60 articles. This number was 52 in 2023 and again rose to 20 in 2024 proving that the number of articles is on the rise. The data reveals an increasing academic interest in genomics within human resources, with a marked rise beginning in 2020. This trend highlights the growing significance of genomics in HR, employee wellness, and organizational development, indicating a burgeoning field of study. The emphasis on genomics reflects its expanding relevance and importance in both academic and professional circles. As interest continues to grow, the integration of genomics into HR practices is becoming more prominent, demonstrating the evolving landscape of human resources and its focus on innovative approaches to employee management and organizational growth.

Figure 1. Sequential Publishing Trend.

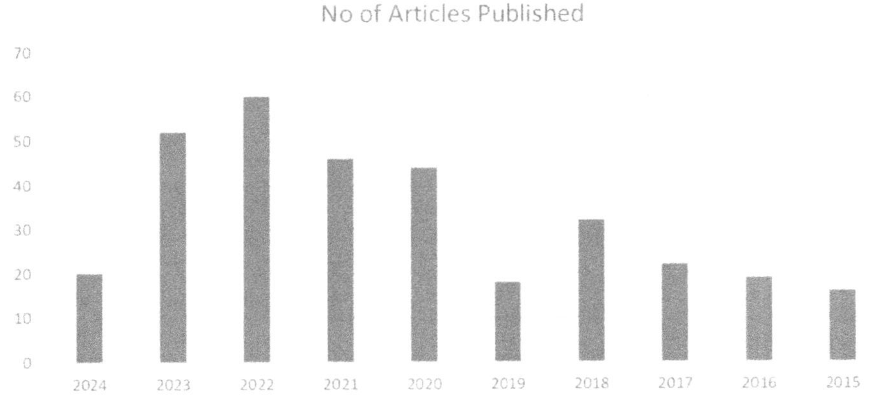

Source – Author's Compilations from Scopus Database.

TOP-CITED AUTHORS

The Scopus database was used to pinpoint the authors with the most citations, and the results are shown in Table 2. This table lists the top 10 authors by their citation counts, along with their respective citation numbers. Leading the list are Ceccaldi R., Amunugama R., Liu J.C., Hajdu I., Petalcorin M.I.R., Primack B., and O'Connor K.W., each with 623 citations. They are followed by Inouye M., Wood A.M., Abraham G., Sweeting M.J., Nelson C.P., Dudbridge F., Kaptoge S., Lai F.Y., Wang T., Brozynska M., and Ye S., all with 481 citations. Additionally, Yang D., Sun Y., Khan S., Hess K., Shmulevich I., Zhang W., and Sood A.K. have each accumulated 470 citations, marking them as highly referenced authors. This table offers an in-depth look at the leading authors in the dataset, showcasing their scholarly influence as reflected in citation counts. It underscores the importance of their contributions to their fields, demonstrated by the high frequency with which their work is cited in academic literature.

Table 2. Top-Cited Authors

Author	Title	No. of Citations
Ceccaldi R.; Amunugama R.; Liu J.C.; Hajdu I.; Petalcorin M.I.R.; Primack B.; O'Connor K.W.	Homologous-recombination-deficient tumors are dependent on Polθ -mediated repair	623
Inouye M.; Wood A.M.; Abraham G.; Sweeting M.J.; Nelson C.P.; Dudbridge F.; Kaptoge S.; Lai F.Y.; Wang T.; Brozynska M.; Ye S.;	Genomic Risk Prediction of Coronary Artery Disease in 480,000 Adults: Implications for Primary Prevention	481
Yang D.; Sun Y.; Khan S.; Hess K.; Shmulevich I.; Zhang W.; Sood A.K.;	Association of BRCA1 and BRCA2 mutations with survival, chemotherapy sensitivity, and gene mutator phenotype in patients with ovarian cancer	470
Davenport E.E.; Burnham K.L.; Radhakrishnan J.; Humburg P.; Hutton P.; Mills T.C.; Rautanen A.; Gordon A.C.; Garrard C.	Genomic landscape of the individual host response and outcomes in sepsis: A prospective cohort study	466
Scicluna B.P.; van Vught L.A.; Zwinderman A.H.; Wiewel M.A.; Davenport E.E.; Burnham K.L.;	Classification of patients with sepsis according to blood genomic endotype: a prospective cohort study	335
Vleeshouwers V.G.A.A.; Rietman H.; Krenek P.; Champouret N.; Young C.; Oh S.-K.; Wang M.	Effector genomics accelerates discovery and functional profiling of potato disease resistance and Phytophthora Infestans avirulence genes	314
Paynter N.P.; Chasman D.I.; Paré G.; Buring J.E.; Cook N.R.; Miletich J.P.; Ridker P.M.	Association between a literature-based genetic risk score and cardiovascular events in women	305

continued on following page

Table 2. Continued

Author	Title	No. of Citations
Goldstein L.J.; Gray R.; Badve S.; Childs B.H.; Yoshizawa C.; Rowley S.; Shak S.; Baehner F.L.; Ravdin P.M.; Davidson N.E.	Prognostic utility of the 21-gene assay in hormone receptor-positive operable breast cancer compared with classical clinicopathologic features	298
Lalonde E.; Ishkanian A.S.; Sykes J.; Fraser M.; Ross-Adams H.; Erho N.; Dunning M.J.; Halim S.; Lamb A.D.; Moon N.C.; Zafarana G.; Warren A.Y.; Meng X.; Thoms J.; Grzadkowski M.R.; Berlin A.	Tumour genomic and microenvironmental heterogeneity for integrated prediction of 5-year biochemical recurrence of prostate cancer: A retrospective cohort study	266
Cullen J.; Rosner I.L.; Brand T.C.; Zhang N.; Tsiatis A.C.; Moncur J.; Ali A.; Chen Y.; Knezevic D.; Maddala T.; Lawrence H.J.; Febbo P.G.; Srivastava S.; Sesterhenn I.A.; McLeod D.G.	A biopsy-based 17-gene genomic prostate score predicts recurrence after radical prostatectomy and adverse surgical pathology in a racially diverse population of men with clinically low- and intermediate-risk prostate cancer	263

Source – Author's Compilations from Scopus Database.

PUBLISHING TRENDS BY COUNTRY

Figure 2 depicts the distribution of citations by country, showcasing the academic contributions from various nations. It emphasizes the United States' dominant position with 215 publications, followed by the United Kingdom with 53 documents and China closely behind with 52. Canada and Italy occupy the next positions with 39 and 38 documents, respectively. This analysis shows that the United States is the leading producer of literature on this topic, with the United Kingdom and China following closely. While the U.S., U.K., and China dominate in academic publications, the data points to an emerging opportunity for more research contributions from India in this field.

Figure 2. Publishing Trends by Country

Source – Author's Calculations from Scopus Database.

TOP-RATED ACADEMIC JOURNALS WITH THE MOST CITATIONS

The leading journals in the field of Genomics in HR, specifically JAMA and The Lancet Respiratory Medicine, have received 935 and 801 citations respectively. These figures underscore the importance and impact of these journals within the domain. An important trend that can be drawn from the table provided is that most of the papers, nearly all starting from 2020 up to 2023, have been published in these leading journals; therefore, they confirm the key status of the platforms of scholarly discourse. The high citation counts and general trends in publications extracted from these prominent journals show their influence on the development of knowledge related to Genomics in HR and their management. This preference for reputable journals thus represents, in part, a scholarly community that recognizes the value and impact of publishing in an esteemed outlet.

Table 3. Top-rated academic journals with the most citations

Journal	Total Citations
JAMA	935

continued on following page

Table 3. Continued

Journal	Total Citations
The Lancet Respiratory Medicine	801
Journal of Clinical Oncology	781
Nature	623
Journal of the American College of Cardiology	606
PLoS ONE	536
The Lancet Oncology	551
European Urology	544
Blood	478
Journal of the National Cancer Institute	347

Source – Author's Calculations from Scopus Database.

BIBLIOGRAPHIC LINKAGES BETWEEN NATIONS

Figure 3 illustrates the countries that have produced the highest number of collaborative papers with international authors. The United States tops the list with numerous partnerships, followed by the United Kingdom, China, Canada, Italy, Germany, France, Spain, the Netherlands, Australia, Belgium, Sweden, South Korea, Switzerland, and Japan. It is evident from these data that these countries participate in international research collaborations in a major way, signifying the combined effort globally to merge shared expertise in some projects that are perceived to have mutual benefits for each other. As such, the international collaboration of these publications by the above-mentioned countries reflects the interwoven fabric making up the global scientific community and is seen as the most powerful force behind the strengthening of knowledge and innovation. Through collaborative efforts, nations can pool their variegated vantage, and resources, and address the complex issues that go beyond national borders in efforts to bring a better vigor to the global scientific landscape.

Figure 3. Bibliographic Linkages between Nations

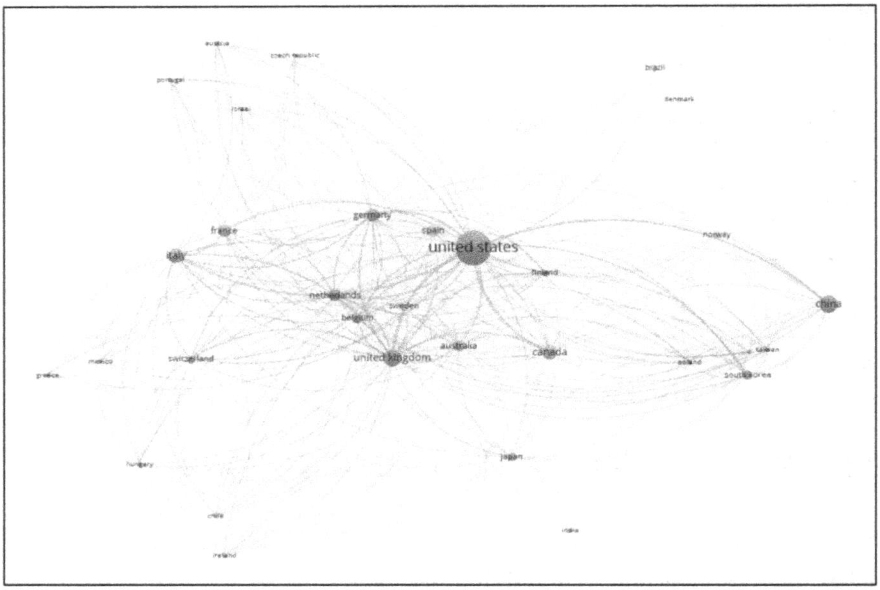

Source – Author's Compilations from Scopus Database

Figure 4. Groups of authors collaborating internationally

Cluster	Countries
Cluster 1	China, Finland, Netherlands, Poland, Singapore, UK, US, South Korea
Cluster 2	Belgium, Chile, France, Greece, Mexico, Switzerland
Cluster 3	Austria, Germany, Israel, Italy, Portugal
Cluster 4	Brazil, Denmark, Norway, Spain
Cluster 5	Australia, India, Canada, Japan

Source – Author's Compilations from Scopus Database.

BIBLIOGRAPHIC LINKAGES BETWEEN KEYWORDS

Next, the bibliometric data is analyzed using VoS Viewer, focusing on the most frequently used author keywords. Figure 5 displays the resulting visualization. To be included in the analysis, documents must have a minimum of ten citations. The graphical representation clearly shows that the journal's articles encompass various topics within genomics, human resource management, and human studies.

The authors suggest categorizing approved submissions into four groups using author-supplied keywords such as Genomics in HR, Employee Wellness, and Organizational Growth, as illustrated in figure 6. The analysis indicates a significant unmet need for journals to expand their scope in the areas of genomics in HR, employee wellness, organizational growth, and related fields.

Figure 5. Bibliographic Linkages of Keywords

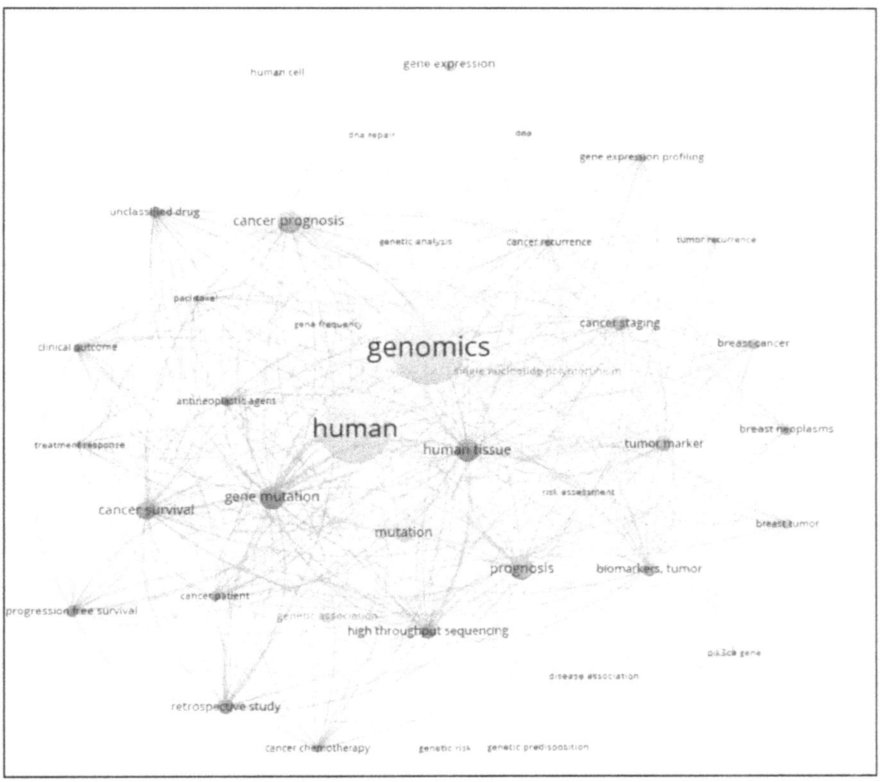

Source – Author's Calculations from Scopus Database

Figure 6. Groups of Keywords

Cluster	Keywords
Cluster 1	Gene mutation, human tissue, cancer chemotherapy, treatment response, antineoplastic agent, cancer survival
Cluster 2	Cancer prognosis, cancer recurrence, cancer staging, prognosis, tumor maker
Cluster 3	Genomics, human, genetic analysis, genetic association, genetic risk, gene frequency
Cluster 4	DNA, DNA repair, gene expression, human cell

Source – Author's Compilation from Scopus Database.

LIMITATIONS

While this study adopts a comprehensive approach, there are several limitations to consider. First, data collection is exclusively based on the Scopus database. Although Scopus is vast, it may not encompass all pertinent studies, particularly those published in non-indexed journals or grey literature. Additionally, the inclusion criteria were limited to English-language publications, possibly omitting significant research in other languages. There is also a possibility that the study doesn't fully reflect the latest advancements and trends in genomics and HRM, due to the time lag between research completion, publication, and indexing. Although the employed VOSviewer software is very powerful and efficient, there can exist flaws in the data analysis and information graphics reviewed that may hurt the depth of results presented. Finally, the thematic mapping might distort the relations between various clusters of research, and therefore give a rather simplified view of the issue.

FUTURE RESEARCH

It is recommended that future studies be done to address the mentioned limitations and expand from the established results. The way it is now it could be broader with the inclusion of other databases such as Web of Science, PubMed, and Google Scholar. This is because the filters set while conducting the systematic review and meta-analysis would complement the goal of addressing language bias by including other studies published in other languages. Exploratory studies could also be proposed aiming at examining changes that have taken place over time in genomics and the effects of those changes on the area of HRM. Using tools and methods of bibliometrics, in particular, the more modern ones could also contribute to the deepening and quantitative sophistication of the study. Another factor that can be recommended is the increase in interdisciplinary research focus that will include genomics as well as other fields like psychology, sociology, and business in understanding its impact on HRM. Further, the integration of both the bibliometric analysis and the qualitative research approaches can just possibly provide an in-depth into the contextual factors that may affect the genomics integration in the HRM.

CONCLUSION

This chapter presents a comprehensive bibliometric review of the crossroads of genomics and Human Resource Management (HRM). This meta-analysis defines emerging topics, essential topics, and leading scholars in this emerging field. Therefore, based on the content analysis and using VOSviewer software, the study outlines the main clusters and recent publications, which can act as the roadmap for further research in the field. However, there are some limitations which are listed below: These are the possibilities for further studies to elaborate on the current results. The integration of genomics into HRM practices has the potential to transform the arena by enabling personalized employee health strategies and enhancing organizational well-being. As this area of study continues to develop, future research should aim to address these limitations, explore cross-disciplinary methodologies, and employ sophisticated analytical tools to fully unlock the potential of genomics within HRM.

REFERENCES

Bondarouk, T., Parry, E., & Furtmueller, E. (2017). Electronic HRM: Four decades of research on adoption and consequences. The International Journal of Human Resource Management, 28(1), 98-131.

Davenport, T. H., Harris, J., & Shapiro, J. (2010). Competing on talent analytics. Harvard Business Review, 88(10), 52-58.

Gupta, M., & Arora, R. (2024). Thriving in the Hybrid Era: Investigating the Relationships between Leadership, High-Performance Work Systems, Employee Resilience, and Engagement in the IT Industry. Pacific Business Review International, 16(11), 117-128.

Gupta, M., Arora, R., & Taneja, S. (2023). Bibliometric analysis on employee engagement and human resource management. In *Enhancing Customer Engagement Through Location-Based Marketing* (pp. 185–205). IGI Global. 10.4018/978-1-6684-8177-6.ch013

Kalra, D., Gupta, M., Gupta, S., Singh, H., & Agrawal, P. (2023). Exploring the impact of green marketing on consumer behavior in the FMCG industry: A sustainable path to consumer engagement. Green Management- A New Paradigm in the World of Business, 81-93.

Kalra, D., Gupta, M., Gupta, S., Singh, H., & Agrawal, P. (2024). Quantifying Influence: A Bibliometric Analysis of Robo-Advisors in Management. Robo-Advisors in Management, 175-186.

Kaur, G., Kalra, D., Kumari, I., Gupta, S., & Saluja, S. (2024). Exploring the Fusion of Neuroscience and Leadership Through Bibliometric Analysis. In Building Organizational Resilience With Neuroleadership (pp. 280-295). IGI Global. 10.4018/979-8-3693-1785-3.ch019

Moore, A. D. (2017). *Privacy rights: Moral and legal foundations*. Penn State Press.

Mulcahy, D., & Betts, T. (2018). The evolution of wellness in the workplace. American Journal of Health Promotion, 32(8), 1689-1695.

Soonthodu, S., & Shetty, S. (2021). Innovative Technology and Human Resource Management. .10.4018/978-1-7998-4787-8.ch016

Stone, D. L., Deadrick, D. L., Lukaszewski, K. M., & Johnson, K. R. (2015). The influence of technology on the future of human resource management. Human Resource Management Review, 25(2), 216-231.

van Zoonen, W., & Sivunen, A. (2020). The impact of remote work and mediated communication frequency on isolation and organizational commitment. Technology in Society, 63, 101377.

Zhu, Y., Azar, K. M. J., & Meng, H. (2020). Incorporating wearable device data into electronic health records: Analysis of the Health eHeart Study. Journal of Medical Internet Research, 22(3), e15021.

Chapter 6
Transforming People Management:
Harnessing HR Analytics in the Transhuman Era

Kapil Kumar Aggarwal
https://orcid.org/0000-0002-2752-9495
Chandigarh University, India

Ravneet Kaur
https://orcid.org/0009-0008-3101-5738
Chandigarh University, India

ABSTRACT

The transhuman era has seen a significant shift in the public sector, with machine learning and artificial intelligence transforming HR analytics. This field has seen rapid growth, but its application in the public sector remains underexplored. This review explores the concepts and practices of HR analytics, proposing a five-step methodology for effective adoption. The study provides insights into workforce planning, recruitment, HR development, and performance management in the context of the transhuman era. It addresses critical issues such as privacy, data integrity, algorithmic bias, and public awareness. Key conditions for successful HR analytics implementation include robust data management, enhanced staff competencies, and broad acceptance of these technologies.

DOI: 10.4018/979-8-3693-7668-3.ch006

INTRODUCTION

The transhuman era, marked by the convergence of human capabilities and advanced technologies, is revolutionizing how we live, work, and interact. As we venture deeper into this era, the implications for human resources (HR) are profound and multifaceted. "Harnessing HR Analytics in the Transhuman Era" aims to discover the transformative capability of HR analytics in managing a workforce that increasingly integrates human talent with technological augmentation. This comprehensive introduction delves into the essence of transhumanism, the evolving role of HR, and pivotal role of analytics in navigating this complex landscape.

Understanding Transhumanism

Transhumanism is a philosophical and scientific movement that advocates for the use of technology to enhance the human condition. This can include cognitive enhancements through AI, physical enhancements through wearable technologies and biomedical implants, and even genetic modifications to improve health and longevity. Transhumanism aims to transcend the limitations of the human body and mind, creating a future where humans can achieve unprecedented levels of performance and well-being.

As these technologies become more integrated into everyday life, the concept of what it means to be human is expanding. In the workplace, this means employees may increasingly be augmented by technologies that enhance their capabilities, creating a workforce that blends human and machine in ways previously unimaginable. This hybrid workforce presents unique challenges and opportunities for HR professionals tasked with managing, developing, and optimizing this new breed of worker.

The Evolving Role of HR in the Transhuman Era

The traditional role of HR is evolving rapidly in response to these technological advancements. HR professionals are no longer just responsible for hiring, training, and managing employees. They now play a critical role in integrating new technologies into the workforce, ensuring that both human and technological resources are utilized effectively. This requires a deep understanding of both human behavior and technological capabilities, as well as the capability to navigate the ethical and practical implications of these changes.

One of the key challenges in the transhuman era is managing the interface between human employees and the technologies that augment them. This includes ensuring that employees are comfortable and competent in using new technologies, as well as addressing any concerns or resistance they may have. It also involves managing the data generated by these technologies, which can provide valuable insights into employee performance and well-being but also raises important privacy and security concerns.

The Power of HR Analytics

HR analytics involves the use of data analysis techniques to improve HR practices and outcomes. By collecting and analyzing data on employee behavior, performance, and engagement. HR professionals can gain valuable insights that inform decision-making and strategy. In the transhuman era, HR analytics takes on even greater significance, as the integration of advanced technologies generates vast amounts of data that can be harnessed to optimize workforce management.

HR analytics is its ability to provide a holistic view of the workforce. By integrating data from multiple sources, including traditional HR systems, wearable devices, and other technological enhancements. This enables them to identify patterns and trends, predict future outcomes, and make data-driven decisions that improve organizational performance.

HR analytics can be used to identify high-performing employees and understand the factors that contribute to their success. HR analytics can be used to identify potential issues, such as declining employee engagement or rising turnover rates, and implement proactive measures to address them before they become significant problems.

Ethical Considerations

As with any technological advancement, the use of HR analytics in the transhuman era raises important ethical considerations. One of the primary concerns is privacy. The integration of advanced technologies into the workplace generates vast amounts of data, much of which is personal and sensitive. Another key ethical consideration is the potential for bias. HR professionals must be vigilant in identifying and addressing any potential biases in their data and analytics processes to ensure that decisions are fair and equitable.

Employees have a right to know how their data is being collected, used, and stored, and how it will impact their employment. HR professionals must communicate clearly and transparently with employees about the use of HR analytics, addressing any concerns and ensuring that employees are comfortable with the process.

Human Resource Analytics

Human Resource Analytics is a methodology that helps firms effectively and efficiently manage their human resource data, leading to organizational success. Anticipating critical qualities is crucial for making good decisions regarding a vast amount of employee data. HR Analytics is a collection of statistical techniques used to acquire, analyze, measure, and predict data. By employing analytics, human resource professionals may gain a deeper understanding of current trends, future prospects of workers, and the most efficient strategies to enhance productivity. When firms are compelled to establish a distinct market niche, they face a shortage of trained people and a growing need for effective workforce planning. Human resource managers have always had the crucial duty of ensuring that employees can effectively showcase their skills and competencies. To assess the characteristics and performance of employees, it is essential to use workforce analytics, which will assist in transforming data into measurable outcomes. This may be accomplished by employing workforce analytics.

Planning the workforce for the top managers will be easier as a result of this, and strategic decisions will be easier to make. It is possible to determine whether or not the outcomes of HR analytics are accurate through the utilization of human resource measurements. The efficacy and efficiency of the company's policies on human resources may be evaluated with the use of this. What is happening, why it happened, and what the result of that is are all things that may be reported by analytics. If there is a managerial-level view on the parameters of the employees in a firm, data is collected to actualize the opinion. This is done by analysing the associated metric, which may pertain to the cost of lost productivity due to absenteeism, the cost per hire, or any other pertinent metric. Through the use of analytics, one may determine the reasons and determine how they impact the business.

Furthermore, this information aids in identifying possible issues in evaluating HR metrics, so offering managers valuable insights that can be used to rectify or normalize the measurements. As a result, the impact of weaker metric data is reduced. Human resources metrics, commonly referred to as HR metrics, are crucial statistics in the corporate realm. They aid firms in assessing their human capital and deciding the most effective ways to employ their workforce.

This complicates the process of creating employee metrics, but HR analytics facilitates the effective connection between performance measurements and enables the tracking of the influence of metrics on HR and business outcomes. In order for businesses to remain competitive in a world driven by technology, they need to foster innovation. It is beneficial to have an understanding of how to achieve corporate objectives by means of providing quantified sustained outcomes, doing analysis, and making predictions. In the past, human resource analytics concentrated on the data that was currently available, such as the cost of hiring an employee. However, in today's businesses, they have come to realize that it is essential to have a consistent and general perspective in order to carry out effective management and optimization.

To achieve this, it is crucial to recruit HR analysts who have the requisite skills and expertise to examine ways to enhance the company's ROI by strategically managing human resources. The HR Analytics implementation approach is characterized by its elegance and deserves your attention. HR analytics is the systematic gathering, analyzing, converting, and overseeing of data pertaining to human resources. This approach is further used to analytical data models. Analytics is a process that aims to align human resources practices with strategic business strategies. To reach desired goals, it utilizes descriptive, predictive, and prescriptive methodologies, each with its own distinct advantages:

Descriptive analytics, the initial level of analysis, necessitates a thorough understanding of past data and patterns. This level of analysis specifically delineates the link between the variables in question. By employing descriptive analysis, it is possible to monitor personnel turnover rates, cost per hire, and absence rates.

Optimization Analytics, the next level of analysis, aims to achieve optimal outcomes by maximizing resource efficiency. This approach employs techniques like as linear programming, simulations, and the creation of mathematical models.

Using Human Resource Analytics to Control Key Aspects of the Payroll System

There are several areas in which HR analytics is essential, and these areas show to be advantageous for the better management of the personnel inside an organization:

It offers corrective measures that are supported by evidence, with the aim of reducing employee turnover that is not desired. This, in turn, enables an organization to make appropriate recruiting decisions for crucial roles and accurately anticipate top talent and potential successors.

- By conducting an analysis of the absence rate and the causes that contribute to excessive absenteeism, it becomes possible to take steps that will aid in maintaining a check on the overall welfare of the business.
- By displaying diagnostic data on the revenue per employee (RPE), which is a measurement of the amount of contribution that each employee makes to the organization, it provides valuable insights. It assists in the diagnosis of the employee productivity index, which in turn assists in focusing on the personnel. discover those who are performing well and those who are not performing well.

- It proves to be beneficial in assessing the degree of morale and motivation among the workforce.
- The system offers logical and quantitative information on the validations of the cost of hiring each individual employee.
- As a result of its ability to create results, it is able to maintain a check on the cost of training an employee, so ensuring that excessive expenditures are avoided. In addition, this will be of use to management in revaluating the success of their training rating system.
- It plays a crucial function in retaining personnel and reducing the talent turnover rate, and it also assists in establishing connections between aspects that are similar.
- HR metrics, which are also known as key performance indicators (KPIs), are a useful tool for determining how well an organization is achieving its goals and implementing its policies. It is possible to measure the costs and analyze the effect of each human resource policy practice with the assistance of the HR analytics data, which in turn provides areas that require development.

In addition to providing vital data string sufficient for a comprehensive strategy reformation based on real facts, it enables managers to leave opportunity for modification in order to promote the growth of the organization.

LITERATURE REVIEW

The convergence of human resources (HR) analytics and transhumanism represents a cutting-edge frontier in organizational management. HR analytics involves the application of data analysis techniques to human resource management to improve decision-making, enhance workforce performance, and drive strategic business outcomes. Transhumanism, on the other hand, advocates for the use of technology to enhance human physical and cognitive capabilities. This literature review explores the intersection of these two fields, examining the potential benefits, challenges, and ethical considerations of harnessing HR analytics in the transhuman era.

HR analytics has evolved significantly over the past few decades. Initially focused on basic metrics and reporting, it has now expanded to encompass advanced data analysis techniques, including predictive analytics and machine learning.

Recent studies highlight the importance of data-driven decision-making in HR. For instance, Bassi (2011) emphasizes that organizations leveraging HR analytics are better positioned to identify talent gaps, predict turnover, and develop targeted employee development programs. Moreover, Fitz-enz (2010) underscores the role of analytics in developing workforce productivity and reducing operational costs.

Transhumanism envisions a future where human capabilities are significantly augmented through technology. This includes the use of artificial intelligence (AI), biotechnology, and cybernetic enhancements to improve physical and cognitive functions (More & Vita-More, 2013). In the workplace, these advancements could lead to unprecedented levels of productivity, creativity, and innovation.

Scholars argue that transhuman technologies have the potential to revolutionize various aspects of work. Kurzweil (2005) predicts that AI and machine learning will augment human decision-making, enabling employees to solve complex problems more efficiently. Similarly, Bostrom (2014) suggests that cognitive enhancements, such as brain-computer interfaces, could enhance learning and memory, leading to a more skilled and adaptable workforce.

The integration of HR analytics with transhuman technologies promises to transform talent management and organizational performance. By analyzing data from enhanced employees, organizations can gain insights into the effectiveness of cognitive and physical augmentations, ultimately refining their HR strategies.

For example, AI-driven analytics can assess the impact of cognitive enhancements on employee performance, identifying which technologies yield the greatest productivity gains (Davenport & Harris, 2017). Additionally, wearable technologies that monitor physiological data can provide real-time insights into employee health and well-being, enabling proactive interventions to prevent burnout and improve overall wellness (Bersin, 2020).

Despite the potential benefits, the integration of HR analytics and transhuman technologies raises significant challenges and ethical concerns. Privacy is a major issue, as the collection and analysis of personal data from enhanced employees could lead to surveillance and discrimination (Zuboff, 2019). Organizations must navigate the balance between leveraging data for performance improvement and respecting employee privacy rights.

The ethical implications of cognitive and physical enhancements must be considered. There is a risk that such technologies could exacerbate existing inequalities, creating a divide between enhanced and non-enhanced employees (Hughes, 2004). Ensuring equitable access to transhuman technologies and addressing potential biases in HR analytics are critical to fostering an inclusive workplace.

The combination of quantitative and qualitative data and information, as described by Jain and Nagar (2015), provides valuable insights that management may use to inform their decision-making process.

According to Kapoor and Sherif (2012), human resource analytics is keeping track of important documents and data pertaining to human resources in order to analyze it with business analytics models and then share the results with decision makers so they may make informed choices.

Decisions in evidence-based human resource management (EBHR) are based on a combination of critical thinking and the best available scientific evidences and business facts (Reddy and Lakshmikeerthi, 2017). Profitability, customer happiness, and quality are some of the business outcomes that may be understood via the use of data, analysis, and research in this area.

The field of human resource analytics encompasses more than just research design and statistics (Boudreau and Ramstad, 2004; Levenson, 2005). It also involves developing and implementing standards for rigor and relevance, improving HR analytical competencies company-wide, and determining and answering relevant questions.

According to Bhattacharyya (2017) - Using analytical reasoning in human resource management.

The harnessing of HR analytics in the transhuman era presents a promising yet complex landscape. While the integration of advanced data analytics and transhuman technologies can drive significant improvements in workforce performance and organizational outcomes, it also necessitates careful consideration of privacy, ethical, and equity issues. Future research should focus on developing frameworks and guidelines to navigate these challenges, ensuring that the benefits of this convergence are realized in a responsible and inclusive manner.

Obstacles and Potential Solutions in HR Analytics

The one and only goal of human resources has always been to create better ways to save employee information, whether it be performance reviews, remuneration, training and development, or printed copies, and this has been true even before the advent of the internet(Aggarwal & Taneja, 2024). The shift from making judgments based on gut feelings to using facts has garnered a lot of attention with the explosion of technical advancements. In order to arrive at strategic decisions for the workforce, it is necessary to mine this collective data and fix the pattern of decision-making. Having predictive analytics expertise is crucial for organizations to have a competitive edge and thrive in the long run.

Possibilities for Applying HR Analytics

The typical HR processes display intricate and ever-changing responsibility due to the constant flow of digital communication. The human resources department must prioritize making critical decisions about the organization's staff in 2021, when the digital disruption age reaches its pinnacle. With that said, here are a few ways in which HR analytics may improve workforce management:

- Improved HR metric trade-offs: By utilizing HR analytics diagnostic methodologies, organizations may analyze past and present patterns, gather performance data to support policy changes, and effectively manage trade-offs between workforce metrics with higher and lower impact.
- Gain a competitive advantage: A company may achieve its intended goals by making data-driven and strategic personnel decisions, which give them a leg up in their everyday operations.
- Investment in analytical tools will enable the organization to discover genuine HR data, ensuring transparency to stakeholders.
- Enhanced efficiency: By automating several personnel subsystems using analytical tools, HR executives will be able to escape their mundane tasks and focus on other matters.
- Enhanced quality standards: The organization's data is consistently monitored through accurate analysis of employee data, leading to improved quality standards in HR metrics.

Obstacles to Human Resource Analytics

HR diagnostic technologies that analyze data are urgently needed. Human resources leaders should think about the requirement of looking at things from many angles. For this reason, among many others, the following may explain why certain companies have been slow to use HR analytics:

- HR executives' reluctance to participate in sophisticated analytics that benefit the workforce is an example of a lack of collaboration.
- The novel technology is not being used exactly due to a lack of qualified human resources experts. The ability to analyze big data requires a competent staff with specialized training.
- Capital investment burden: The organization fails to employ modern HR analytics technologies that incorporate machine learning because it has the false idea about how much capital is needed.
- The absence of real data: The absence of true employee data, which is fundamental to effective governance, due to a lack of legality and ethics. Decisions driven by data, when the data is very relevant, are analytics' bread and butter.
- Top management's reluctance to use analytical tools makes them ill-equipped to deal with cognitive intelligence software's associated costs and potential data integrity issues.

HR executives and organizations are increasingly worried about the possibility of hackers gaining unauthorized access to critical human data due to the prevalence of digital communication.

Uses of Human Resource Analytics in Government Personnel Administration

Modern HR methods improve workforce planning by using analytics for strategic staffing, an important part of corporate management. Analytics that make use of predictive modeling help businesses plan ahead for their personnel requirements by taking factors like projected growth, skill demands, and attrition rates into account. With the use of analytics, talent pipelines are effectively handled, allowing for the identification of high-potential individuals and the smooth flow of responsibilities during expansion or turnover. Insights into present skill sets are made possible by skill mapping, which in turn allows for focused training or recruitment efforts to coincide with developing company demands. By simulating various scenarios with the use of workforce planning analytics, firms may better understand how changes in market circumstances or company goals could affect their personnel needs.

Analytics have completely transformed recruitment, a fundamental HR activity, allowing for far more efficient applicant selection. Allocating resources to the most effective recruiting channels, such as job boards, social media, or referrals, is what sourcing optimization is all about. Using data analytics to uncover important success indicators and anticipate applicants' prospective success improves overall hiring quality and reduces subjectivity in selection procedures, which in turn benefits candidate evaluation processes. Analysis of time-to-hire data simplifies hiring procedures, shortening the time it takes to bring on board competent individuals. By monitoring and evaluating demographic data at various phases of the recruiting process, analytics also plays an important role in fostering inclusion and diversity.

Organizational growth relies on filling skill shortages, and HR development with analytics support allows for strategic goal-aligned, targeted training. Analytics tools provide a thorough picture of the current talent landscape by conducting thorough skills evaluations, which include both technical and soft skills. In order to guarantee that training is in line with present and future job requirements, analytics correlates skill evaluations with corporate goals. Analytics evaluate the influence on employee performance and skill improvement by measuring the success of training programs through specified KPIs. By implementing and optimizing continuous learning platforms with analytics, organizations may cultivate a culture of lifelong learning.

Analytics improve performance management, an essential HR function, by providing impartiality and insights derived from data. Beyond conventional key performance indicators (KPIs), analytics creates objective performance measurements that include project completion timelines, customer satisfaction scores, and other measurable success factors. An all-encompassing picture of an employee's performance may be obtained through the use of comprehensive 360-degree feedback analysis, which gathers comments from superiors, peers, and subordinates. To ensure that high-performing people are recognized and that collaborative efforts leading to team achievement are acknowledged, performance analytics evaluates both individual and team performance. Using data, HR is able to pinpoint where employees are falling short and provide them with targeted training to help them improve.

Concerns About Ethics Are Major Factors in the Use of HR Analytics

In order to responsibly and successfully implement HR analytics, firms must carefully manage the hurdles posed by ethical issues. With challenges about how to protect individual privacy arising from the gathering and analysis of employee data, privacy concerns are a major issue. To combat this, businesses should implement stringent data security measures, define data usage rules precisely, and check that they are in accordance with all applicable privacy laws. Preventing prejudice and bias requires decision-making procedures to be honest and fair at all times. A dedication to fair treatment, impartial assessments, and equal opportunity should permeate HR analytics. If you want to find and fix any unintended biases in your algorithms and decision models, you need to audit and evaluate them often. The possibility that analytics tools may reinforce preexisting prejudices in decision-making is known as algorithmic bias, and it presents a serious ethical dilemma. Businesses should think carefully about the data they utilize and use fairness-enhancing strategies to find, fix, and reduce bias in their algorithms. In order to gain the public's and workers' trust, transparency is essential. The kinds of data gathered, how it is used, and how it could affect hiring choices should all be transparent. Improved transparency and informed consent may be achieved by open and honest communication with workers about the aim and implications of HR analytics methods. Maintaining openness, justice, and responsible data use—backed by frequent ethical assessments and stakeholder communication—is essential for proactively addressing these ethical concerns.

Critical Analysis

In order to successfully deploy HR analytics, it is crucial to carefully analyze the essential requirements that lay the foundation for effectiveness. The accuracy of data plays a crucial role in the dependability of HR analytics, thus firms must actively assure it as part of data management. In order to get reliable insights, it is necessary to establish strong procedures for data validation and maintenance. At the same time, protecting confidential employee data must be a top priority. A thorough security architecture that stops unauthorized access or breaches must have encryption, access restrictions, and other safeguards. Organizations should also make sure HR specialists are accessible when they need to be. While keeping sensitive information safe, HR teams should have unrestricted access to data that is useful for analysis.

Organizations should prioritize educating HR personnel to successfully utilize analytics tools as it pertains to staff competence. Efficient data analysis, interpretation, and strategic application skills necessitate well-planned training programs. Human resources professionals must be proficient in the analytics tools used to draw useful insights and contribute to informed decision-making.

HR analytics are accepted by the organization's ecosystem. Getting buy-in from stakeholders and workers is crucial for HR analytics to be successful. In order to cultivate a favorable impression, it is crucial to communicate clearly about the advantages, goals, and ethical concerns of HR analytics. In order to foster an atmosphere where acceptance is common, it is critical to proactively address any worries or misunderstandings that workers may have regarding HR analytics through honest communication and openness. By giving these requirements the attention they deserve, firms may set the stage for HR analytics to be successfully implemented. This will result in better data management, more competent workers, and wider support from stakeholders.

CONCLUSION

"Harnessing HR Analytics in the Transhuman Era" explores the transformative potential of HR analytics in managing a hybrid workforce that blends human and technological capabilities. By understanding the implications of transhumanism, evolving the role of HR, leveraging the power of analytics, and addressing ethical considerations, organizations can navigate this complex landscape and unlock new levels of performance and success. As we venture deeper into the transhuman era, the insights gained from HR analytics will be crucial in shaping the future of work and ensuring that both human potential and technological advancements are maximized to their fullest extent.

The revolutionary technology known as HR analytics has shown to be effective in helping employees make well-informed and strategic decisions. Many large-scale firms have already used it. There is a widespread failure to use HR analytics in several industries, despite the obvious advantages. It is critical for firms to acknowledge and utilize analytical tools to optimize their workforce management procedures, since this omission highlights the necessity of doing so. Human resource professionals are beginning to focus more on analytical data integration in response to the pressing need to maintain an effective workforce across a variety of industries. The focus here isn't just on adoption; it's on using analytics to foresee and solve problems with things like talent management strategies, learning and development programs, and employee turnover. This study article sets out to investigate HR analytics using secondary data, gleaning insights and facts from a wide variety of sources like as websites, blogs, and journals. In order to better understand how analytics will impact workforce management in the future, this article will focus on HR practices and how they are changing.

The Future of HR in the Transhuman Era

As we look to the future, it is clear that the role of HR will continue to evolve in response to the ongoing advancements in technology and the increasing integration of these technologies into the workforce. HR professionals will need to develop new skills and competencies, including a deep understanding of data analysis and the ethical implications of technology use. They will also need to be agile and adaptable, able to navigate the rapidly changing landscape of the transhuman era.

The potential benefits of harnessing HR analytics in the transhuman era are immense. By leveraging the power of data, HR professionals can optimize workforce management, improve employee performance and well-being, and drive organizational success. However, realizing these benefits will require a thoughtful and strategic approach, one that balances the potential of technology with a deep commitment to ethical principles and respect for employee rights.

REFERENCES

Afzal, M. (2019). Hr Analytics: Challenges and Prospects of Indian It Sector. *International Journal of Management*, 9(7), 404–415.

Aggarwal, K. K., & Taneja, D. (2024). Impact of organizational culture on employee satisfaction, motivation, performance, commitment, and engagement in educational institutions. In *Building Sustainable Human Resources Management Practices for Businesses*. IGI Global. 10.4018/979-8-3693-1994-9.ch012

Barbar, K., Choughri, R., & Soubjaki, M. (2019). The Impact of HR Analytics on the Training and Development Strategy - Private Sector Case Study in Lebanon. *Journal of Management and Strategy*, 10(3), 27. 10.5430/jms.v10n3p27

Bassi, L. (2011). Raging debates in HR analytics. *People & Strategy*, 34(2), 14–18.

Ben-Gal, H. C. (2018). An ROI-based review of HR analytics: Practical implementation tools. *Personnel Review*.

Bersin, J. (2015). Becoming irresistible: A new model for employee engagement. Deloitte Review, 16.

Bersin, J. (2020). *HR technology 2020: What's hot in the new decade*. Deloitte Insights.

Bhatnagar, J. (2007). „Predictors of organizational commitment in India: Strategic HR roles, organizational learning capability and psychological empowerment". *International Journal of Human Resource Management*, 18(10), 1782–1811. 10.1080/09585190701570965

Bhatnagar, J. (2008). „Managing capabilities for talent engagement and pipeline development". *Industrial and Commercial Training*, 40(1), 19–28. 10.1108/00197850810841602

Bhatnagar, J., & Pandey, A. (2005). HR Metrics: HR Six Sigma in Indian Organizations. *Management and Labour Studies*, 30(4), 327–354. 10.1177/0258042X0503000403

Bhattacharyya, D. K. (2017). *HR Analytics: Understanding Theories and Applications*. SAGE Publications.

Bostrom, N. (2014). *Superintelligence: Paths, dangers, strategies*. Oxford University Press.

Chib, S. (2019). *Monograph on HR reporting using using HR dashboards*. International Journal of Scientific and Research Publications.

Davenport, T. H., & Harris, J. G. (2017). *Competing on analytics: Updated, with a new introduction: The new science of winning*. Harvard Business Review Press.

Fitz-enz, J. (2010). *The new HR analytics: Predicting the economic value of your company's human capital investments*. AMACOM.

Fred, M.O. & Kinange U.M. (2015). Overview of HR Analytics to maximize Human capital investment.

Gurusinghe, N., Arachchige, B. J. H., & Dayarathna, D. (2019), Identified Research Gaps in HR Analytics, Conference: Challenges to Humankind in the Face of New Technologies at Colombo, Sri Lanka.

Hughes, J. (2004). *Citizen Cyborg: Why democratic societies must respond to the redesigned human of the future*. Westview Press.

Jabir, B., Falih, N., & Rahmani, K. (2019). HR analytics a roadmap for decision making: Case study. *Indonesian Journal of Electrical Engineering and Computer Science*, 15(2), 979–990. 10.11591/ijeecs.v15.i2.pp979-990

Jain, A., & Nagar, N. (2015). An Emerging Trend in Human Resource Management. *International Journal of Advance & Innovative Research*, 5(3), 1–11.

Kaur, J. (2024a). Green Guardians: Unveiling the Strategic Role of HR in Environmental Sustainability Initiatives. In Building Sustainable Human Resources Management Practices for Businesses (pp. 125-143). IGI Global.

Kaur, J. (2024b). Tech Unleashed: The Influential Power of Artificial Intelligence on Venture Capital and Startups. In Fostering Innovation in Venture Capital and Startup Ecosystems (pp. 219-241). IGI Global.

Kirtane, A. (2015). corporate sustainable HR Analytical practices. *Journal of Management & Administration Tomorrow*, 4(1), 33–40.

Kurzweil, R. (2005). *The singularity is near: When humans transcend biology*. Penguin.

Levenson, A. R. (2005). Harnessing the power of HR analytics: Why building HR's analytics capability can help it add bottom-line value. *Center for Effective Organizations*, 4(3), 3–12.

Lochab, A., Kumar, S., & Tomar, H. (2018). Impact of Human Resource Analytics on Organizational Performance: A Review of Literature Using R-Software. *Management and Allied Sciences*, 5, 282–284.

Meri, M. M. (2016). The Spectacular Development of Human Resources Management: From Descriptive to Predictive in Modern Organizations. *Journal of Business Economics*, 7(11), 1932–1945. 10.15341/jbe(2155-7950)/11.07.2016/014

Mohammed, A. Q. (2019). HR analytics: A modern tool in HR for predictive decision making. *Journal of Management*, 6(3), 51–63. 10.34218/JOM.6.3.2019.007

Momin, W. Y. M., & Mishra, K. (2016). HR analytics: Re-inventing human resource management. *International Journal of Applied Research*, 2(5), 785–790.

Momin, Yancey, & Kushendra. (2015). HR Analytics as a Strategic Workforce Planning. International Journal of Advanced Research, 1(4), 258–60.

More, M., & Vita-More, N. (Eds.). (2013). *The transhumanist reader: Classical and contemporary essays on the science, technology, and philosophy of the human future*. Wiley-Blackwell. 10.1002/9781118555927

Opatha, H. (2009). *Human Resource Management: personnel.* University of Sri Jayewardenepura. 10.31357/bkc.fmsc.00001

Opatha, H. H. D. P. J., & Uresha, K. I. (2020). HRM and its impact on employee happiness: An empirical study on Sri Lankan employees. *Asian Journal of Social Sciences and Management Studies*, 7(2), 114–123. 10.20448/journal.500.2020.72.114.123

Reddy, P. R., & Lakshmikeerthi, P. (2017). 'HR Analytics' - An Effective Evidence Based HRM Tool. *International Journal of Business and Management Invention*, 6(7), 23–34.

Reena, R., Ansari, M. M. K., & Jayakrishnan, S. S. (2019). Emerging trends in human resource analytics in upcoming decade. *International Journal of Engineering Applied Sciences and Technology*, 4(8), 260–264. 10.33564/IJEAST.2019.v04i08.045

Singh, P., Upadhyay, R. K., & Srivastava, M. (2017). The role of HR analytics in higher education institution. *International Journal of Engineering Sciences & Research Technology*, 6(7), 92–100.

Telu, S., & Yoginder V. (2019). Human Resource Analytics : An Overview of Latest Development in Business Organizations. *International Journal of Research and Analytical Reviews*, 6(2), 274–281.

Tomar, S. (2020). HR Analytics in Business : Role, Opportunities, and Challenges of Using It. Journal of Xi'an University of Architecture & Technology, 12(7), 1299–1306. doi: 10.37896/JXAT12.07/2441

Verma, S. & Mehrotra, R. (2017) Research Paper on Role of Analytics in Renovating Human Resource Management. IPASJ International Journal of Management (IIJM), 5(5), 17–23.

Zuboff, S. (2019). *The age of surveillance capitalism: The fight for a human future at the new frontier of power*. PublicAffairs.

Chapter 7
Genomic Testing in Workplace Wellness: Employee Perspectives, HRM Approaches, and Insights

Harwinder Singh
York University, Canada

Aparna Saluja
Independent Researcher, Canada

ABSTRACT

Genomic testing in workplace wellness programmes provides personalised health information, but it also gives rise to substantial ethical and privacy issues among employees. This study examines the viewpoints of employees, uncovering a combination of positive and negative sentiments regarding the potential advantages in comparison to the hazards associated with data security. HRM techniques must effectively manage these problems by prioritising transparency and obtaining permission. Efficient incorporation of genomic testing necessitates unambiguous communication, ethical principles, and strong data security measures. The findings of this research emphasise the importance of developing customised HRM methods that promote acceptance and engagement, while still protecting employee privacy.

INTRODUCTION

The testing of genomics includes the examination of an individual's DNA in order to identify the differences that may be influencing their general health. These differences or these variations includes genetic markers. These genetic markers can

DOI: 10.4018/979-8-3693-7668-3.ch007

provide information related to a range of factors related to health and well-being of an individual (Ajunwa et al., 2016).

For example, an individual's susceptibility to disorders, the response of metabolism to range of foods, along with an individual's or a person's reaction to medicines and other medical interventions. This process traditionally includes collection of the sample of DNA, along with the saliva conducting a blood test or blood examination. The data is further examined in laboratory setup in order to detect, in order to detect the variations in the genomics profile of the individual. Moreover, the applicability of testing of genomics is very diverse.

For instance, well-less of employees at workplace or talking about the scenario of workplace wellness, the testing of genomics helps in providing important information or data related to risk of the employee or an individual related to health conditions for example, heart-related issues, diabetic conditions and cancerous conditions (Biswas et al., 2020).

The genomics testing can also reveal the risk of the employee or the tendency of the employee to acquire mental health conditions, including emotional exhaustion, anxiety, along with sensitivity to environmental factors. This information is very crucial for the organizations in order to plan and execute programs related to wellness of the employees, specially tailored, specially designed, according to the genetic profiles of the individuals, which can lead to better results and better health outcomes along with mental health outcomes also (Brown, 2016).

ADVANTAGES OF GENOMICS TESTING RELATED TO WORKPLACE WELLNESS PROGRAMS

The inclusion of testing of genomics for well-less programs at workplace offers variety of benefits (figure 1).

Figure 1. Advantages of genomics testing in workplace wellness programs

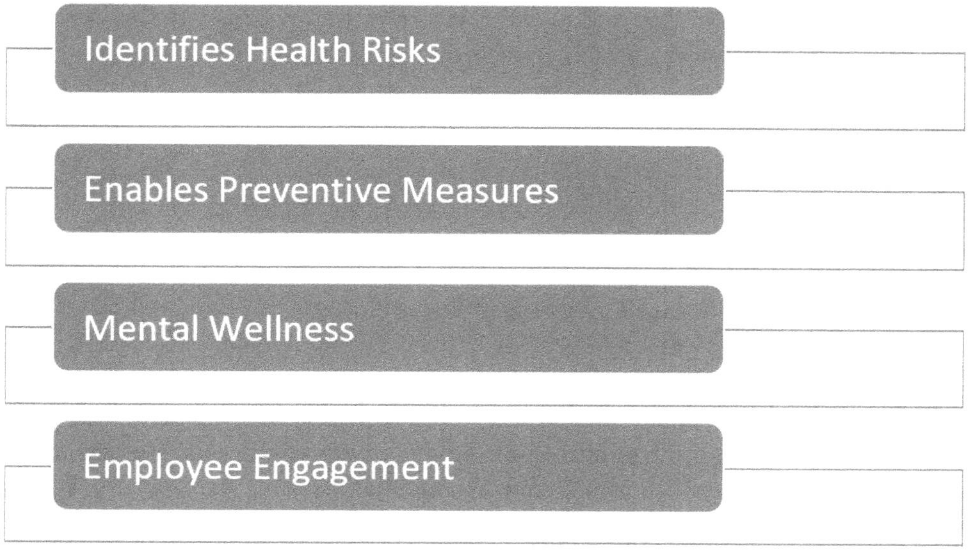

For example, genetic testing, it helps in shifting from traditional health management to proactive health management by identifying the tendency or the risk of the individual related to acquiring major or minor health issues diseases related to physical and mental health, further assisting in reducing the risk and helping in developing and planning of better wellness programs. Overall, if an individual after genomics testing may be identified of having higher level of risk related to diabetes, he or she might be then based on the results encouraged to adopt changes in the lifestyle, including exercises, including dietary changes in order to reduce the risk or in order to diminish the risk. In addition, genomics testing can help in enhancing in promoting better results of wellness initiatives (Brown, 2017).

In older times, the wellness programs recommended traditional health practices without differentiating or without taking the consideration of variation in the profile of the employees in terms of genetics. In today's time, the knowledge about the genomics of the individual helps in allowing in designing well-less programs according to the variations identified based on the testing. For example, the testing of genomics can help in identifying whether the individual has metabolism better in terms of certain nutrients so that he or she can be advised (Callier et al., 2016).

Certain dietary methods along with that genomics testing can help in identifying an individual mental well-less along with that susceptibility to certain psychological disorders. This identification can be used to plan better fitness, programs, in addition the information about the health of the employees, identified from genomics testing can help in improving the engagement of the employees and involvement of the employees in well-less programs. The individuals or the employees receiving specific focus based on their individual needs, which in term is based on genetic profile, are more likely to invest in their generally related to health. But this individualistic nature of the well-less programs can make them more successful and more beneficial in terms of results. Addressing the privacy concerns or privacy and ethical issues (Carnethon et al., 2009).

Genomics testing along with the benefits also raises concerns related to the privacy of the employees along with certain concerns related to ethics. Individuals or employees they may be apprehensive or they may not be willing to share their genetic profile. As a result of fear of unauthorized sharing of data and misuse of the data, employees having apprehension about discrimination or bias based on the genetic profiles can also leads to remaining them silent or not getting involved in the well-less programs. They may also have this fear of getting discriminated in terms of employee retention or opportunities related to promotions as well as discriminatory coverage in terms of health insurance (Cohn et al., 2023).

So, in order to resolve these concerns, the organizations need to take proactive actions to ensure strong protection of data including up to date encryption techniques to protect the genetic data of the employees along with restricting the excess of the data and giving the excess of the data to authorized personal only and making sure of conducting inspections so that this data is not used by unauthorized people. Along with this, legal compliance is also very essential. Certain laws are are required to provide protection to the employees against discrimination based on genetic data in terms of employment conditions and medical insurance. The companies need to ensure that the wellness programs are in line with the regulations or acts in order to safeguard the data of the patients' genomics data of the employees (Cummings, 2010).

IMPLEMENTATION OF GENOMICS TESTING IN TERMS OF WORKPLACE WELLNESS PROGRAMS

Implementation of the testing of genomics at workplace for wellness programs includes various considerations (figure 2).

Figure 2. Considerations of genomics testing in workplace wellness programs

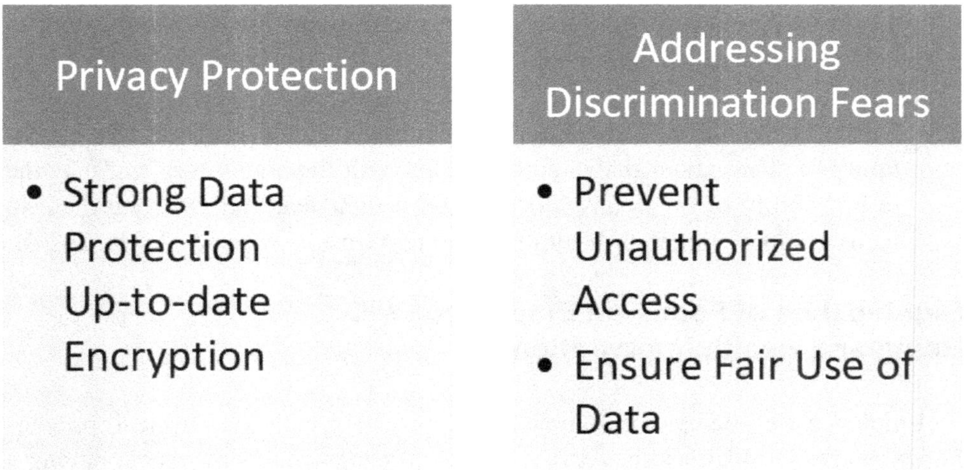

- First consideration is the association or partnership with testing providers who are reputable and provide reliable services. It is very crucial to select those providers who are adhering to superior standards or who are complying with proper guidelines regarding data protection and who also ensure ethical measures (Curnutte, 2012).
- Secondly, organizations should develop appropriate strategies for open communication in terms of dissemination of information regarding genetic testing at workplace. They also need to explain to the employees about the advantages of genomics testing, how the testing is conducted, the benefits of the testing in terms of mental and physical well-being and also the data security and privacy provided to the employees. Opportunities should also be given to the employees to share their concerns without any apprehension regarding their involved in such testing (Dashti et al., 2020).
- In addition, employee consent is very important. The inclusion of the employees in genetic testing must be voluntary and employees they must give their consent officially. Without the consent of the employees, no testing should be conducted and the consent should be taken on papers. The consent implies that the employees completely agree with terms and conditions of the testing and are aware of the likely implications of the tests and are also aware about their rights as well as responsibilities (David and Palaniappan, 2018).

- After the genomics testing is done, the management is required to clearly share and with the employees. Once the results are shared, employees have the right to access the opportunity to avail counseling services which can be very valuable in assisting the employees to fully understand the results and how to focus on planning about their physical and mental health for future (Delaney et al., 2016).
- During the counseling sessions, the employees should be given free opportunity to share about their apprehensions with the counselors and it is the responsibility of the counselors to remove their fear and motivate them to focus on improving their health for long term results (Kaur & Arora, 2023).

Significance of Focus on Employee Well-Being and Health Interventions

Employee well-being is a complex concept which includes physical, psychological and emotional health. Also, it has been recognized as a concept having a very productive role in terms of physical health as well as mental health (figure 3).

Figure 3. Significance of focusing on employee well-being and health interventions

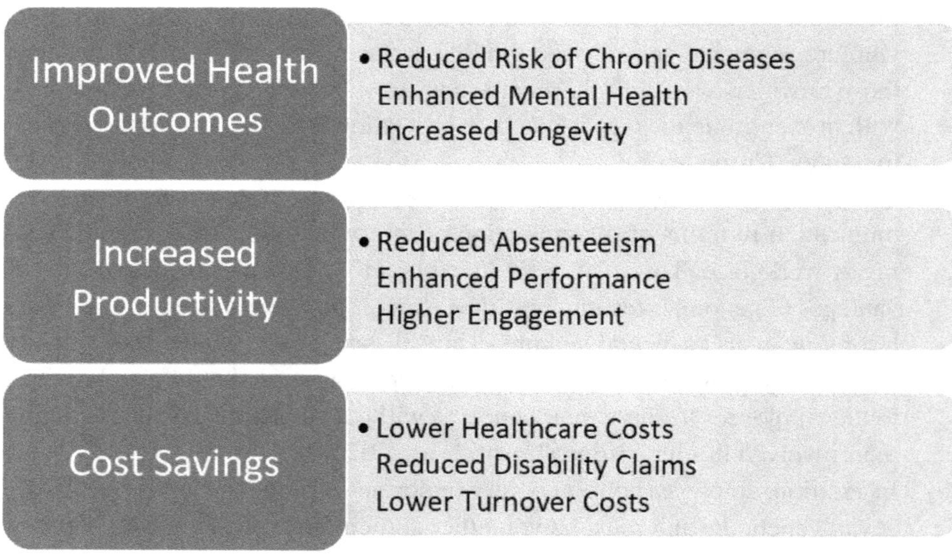

When the organizations promote well-being of the employees or they take initiatives in order to ensure better well-being of the employees, it helps in diminishing absenteeism, reducing medical costs and leading to a better open organizational climate. Overall, companies they are expected to invest in employee well-being programs in order to support the employees (Deverka et al., 2020). Earlier times, the employee wellness programs were designed as one-size-fits-all approach. This means that in earlier times the wellness programs were used to address common health issues and common psychological concerns. However, personalized interventions in today's time include specific focus on a person's genetic composition, personal lifestyle and external environment factors, thus allowing personalized interventions or initiatives to target the mental and physical well-being of the employees and developing personalized health strategies (Dohn, 2017). Hence, it is very crucial to ensure a personalized approach in terms of genomics testing. For instance, genomics testing can identify the health risks of the individuals like obesity, heart diseases, cancerous conditions. Organizations having such information about the employees can offer individualized or personalized health initiatives including special dietary programs, individualized exercise programs or stress reduction programs. Such personalized approach not only helps in improving the health of an individual but also assist in developing a feeling of trust and commitment among the employees towards the organization (Esmonde et al., 2023).

PURPOSE OF THE STUDY

This study aims to examine the usage of genomics testing in terms of workplace wellness programs, focusing on the advantages and the limitations of such programs.

- Very specifically, the first objective of the study is to provide a detailed knowledge of genomics testing and the advantages and related challenges of genomics testing related to workplace wellness.
- Second objective of the study is to investigate how the genomics data can help in framing and execution of wellness initiatives along with examining the likely impact of analysis of genomics data in terms of on individual's health and the well-being in terms of both physical and emotional well-being. The study also aims to highlight the advantages of genomics testing and well-being in enhancing the performance of the organizations.
- Third objective of the present research is to ensure compliance of genomics testing at workplace in terms of ethics and legal compliance along with examining the concerns of the employees in terms of their data privacy and the likely ethical challenges in terms of workplace discrimination. The research also focuses on practical challenges of implementing genomics testing at workplace.

- Lastly, the study aims to offer practical recommendations or suggestions to the organizations regarding the adoption of testing of genomics and incorporation of the testing in wellness programs. The study will provide guidance to the organizations on how the quality of wellness programs can be further enhanced by incorporating suitable strategies and by addressing the challenges or the concerns.

Employee Perception on Genomics Testing (Figure 4)

Figure 4. Employee perceptions on genomics testing

Positive Perceptions

- Personalized Health Insights
  Preventive Health Measures
  Risk Mitigation
  Enhanced Wellness Programs
  Customized Programs
  Improved Health Outcomes

Negative Perceptions

- Privacy Concerns
  Data Security
  Fear of Discrimination
  Job Security Concerns
  Ethical Concerns
  Data Misuse

Nowadays, the employers view the genomics testing as a novel concept which can actually enhance their well-being and health. Such employers think the initiatives taken by the organizations in terms of personalized programs for the employees is empowering and helps them in proactively managing their mental and physical health. The employees accept that the organization cares for them. Along with this, the employees who appreciate the personalized wellness programs also understand that the genomics testing can help in ensuring their fitness in terms of modifying their dietary habits according to their genetic makeup. Such individualized approach assist the employees in improving their productivity and being more engaged as a result of having better physical and mental health well-being. In addition, enthusiastic engaged employers who understand the importance of genomics testing recognizes the significance of such testing in identifying the medical risks before they actually turn into serious health conditions (Fonarow et al., 2015).

Hence, based on genomic testing, the employees can take the required preventive steps in order to address these risks. For example, proactive health screenings and adopting healthy dietary lifestyle leads to better health outcomes and also reduces the stress and anxiety among the individuals. In addition, when the employees receive individualized medical or health information from the organization based on genomics testing, it helps in fostering a feeling of trust within the organization and controlling their physical as well as psychological condition (Foss et al., 2022).

On the other hand, certain employees may have a concern in terms of the data privacy in genomics testing. Such concerns can be related to the possible misuse and sharing of the genetic information of the employers with unauthorized people. Employees may be very concerned who will be accessing the data, where the data will be stored, who will be having the authorized use of the data. This fear can lead to employees refusal to take part in the genomics testing. Employees may also have the apprehension that the organization may discriminate in terms of recruitment, training opportunities, career advancement and access to medical insurance. In addition, employees may be concerned about the constructiveness of the genomics testing, its validity as well as whether the tests are providing accurate data or not. This apprehension may be because of the lack of understanding of the concepts or scientific facts between behind genomics testing. Moreover, the discomfort of the employer regarding the sharing of the information collected from genomics testing may be shared with third party, who are not involved in the testing (Garcia et al., 2020).

Effect of the information derived from genomics testing in terms of psychology of the employees can have both positive effects and negative effects. In terms of information related to health, the employees may take the information as source of improving their health by taking preventive measures and to avoid the risk associated. For example, genetic knowledge can actually help in preventing serious health disorders like diabetes, cancer, heart diseases. However, the knowledge of such risks can also lead to anxiety, apprehension, stress and in severe case depression also (Goetzel and Ozminkowski, 2008).

Lastly, the effect of awareness of genetic information depends on the psychology of the employees and how their ability are on coping up with this. Some individuals are not able to cope up with the information properly and find it difficult to accept the risk associated with their health. This can lead to a strong feeling of helplessness, which can further have a diminished impact on the performance of the employees in terms of official activities as well as personal activities. Such employees having apprehension or fear may feel compelled to share this genetic information and health concerns with their loved ones which can further enhance their stress and strain (Gordon, 2022).

ORGANIZATIONAL INITIATIVES TO ADDRESS THE CONCERNS OF THE EMPLOYEES

As mentioned in the earlier sections, there is a wide range of concerns related to genomics testing of the employees. Hence, it becomes very crucial for the organizations to ensure adoption of the strategies in order to properly address the concerns of the employees effectively. Such initiatives includes open communication, developing trust among the employees for their organization, strong data protection activities and proper education of the employees about the potential advantages of genomics testing (Kaur & Arora, 2023).

1. **Clear communication:** Clear communication by the management with the employees is very important in terms of developing trust and removing the concerns and the apprehensions of the employees regarding genomics testing. The organizations need to ensure that clear information is timely provided to employees related to collection of genomics data and how it will be ensured that the data is protected and is not shared with unauthorized individuals. The employees must be provided information about the potential benefits of testing of genetics and how this data will be utilized to enhance the mental and physical well-being of the employees (Kaur & Arora, 2023).
2. **Informed consent:** The organizations also need to assure that employee consent is taken officially and they must be given the chance to withdraw from the test anytime without being forced to get involved. The consent consent process should be easily understandable by the employees and regular updates must be readily available to the employees so that whenever there is a concern or question that comes into the mind of the employees, it gets quickly addressed. Such initiatives will help in fostering an open climate where the employers think that the organization is honest with them, thus leading to active participation of the employees in genomics testing (Grifantini, 2017).

ROBUST DATA PROTECTION

In order to address the apprehensions of the employees with regard to the security of the genetic data, the organizations are expected to establish and implement adequate measures in order to ensure protection of the genetic information of the employees. These measures or initiatives involves higher standards of interpretation of data, secure storage of genetic data and strict controls in terms of accessibility of the data. The accessibility of the data should be given to authorized personal only. The organizations need to ensure that the security protocols are strictly adhered to and the employees must be assured that the company is fully committed to securing the data. In addition, the organizations are required to comply with the relevant rules and regulations and guidelines related to the genetic data (Hamilton et al., 2016).

Also, they are required to conduct regular inspections and check thoroughly the data repository and in case of any breach of data, suitable actions need to be taken. Such initiatives by the organization will demonstrate to the employees that the it is fully committed towards the protection of their data and is concerned about building confidence among the employees so that they feel more engaged and work better.In addition, proper education plays a very important role in addressing the concerns of the employees. With the help of proper education and information, the employees will understand the importance of the genomics testing. Organizations must provide educational resources to the employees, and also explain to them about the significance of genomics testing as well as its limitations (Harris and Wilkinson, 2013).

Genomics education can be provided with the help of workshops, seminars, sessions and access to experts who can answer all of their queries, concerns and if required, provide them personal guidance. In addition, the education needs to focus on interpretation of complex matters related to genomics to the employees and making them understand about the significance of the results of genomics testing. Also, after getting the results, the organizations need to provide counseling services to the employees in order to address their concerns (Horwitz et al., 2013).

These genetic counselors can offer the employees personalized advice, can address their concerns and help them in developing plans to manage any risk associated with their health. Such support can reduce the anxiety among the employees and can help the employees in making suitable decisions about their health-related matters. Overall, the organizations focusing on free communication, adequate data protection and proper education of the employees can effectively address employee concerns and develop a positive work climate where the employees accept proudly that their organization is taking initiatives for employee benefit (Hendricks-Sturrup et al., 2020).

HUMAN RESOURCE MANAGEMENT APPROACHES TO GENOMICS TESTING RELATED TO EMPLOYEE WELLNESS PROGRAMS

Ethical considerations related to genomics testing: The ethical considerations with regards to genomics testing in terms of workplace wellness programs is associated with complex combination of legal framework, ethical practices, informed consent or voluntary participation, appropriate data protection and prohibition of genetic discrimination. HRM practices should be designed in such a manner so as to prevent any unfairness against the employees and to support the employees in terms of mental health and protection of their rights (Hoffman, 2019).

- **Legal adherence:** Secondly, genomics testing at the workplace should be conducted in accordance with the legal norms or framework. In general, relevant authorities are expected to frame suitable legal frameworks in order to protect the rights of the employees in terms of genomics testing. The framework should be based on rule of autonomy, justice and trust (Ioan and Hanganu, 2023).
- **Autonomy and Trust:** Employee autonomy means that the rights of the individuals to make informed decisions and Justice means that employees are given complete information and no discrimination is done against them on the basis of genetic data. When organizations assure both employee autonomy and justice, this builds a sense of trust at workplace (Kaur and Arora, 2023).

Informed Consent or Participation of the Employees on Voluntary Basis

In relation to employee development, the employees have the right to not be forced to participate in genomics testing and their consent must be taken by the organizations before proceeding for the same. The employees should be clearly informed about the advantages, risks involved, purpose of genomics testing and how the data will be used and stored. The consent process should be designed in a clear manner, avoiding technical words that can mislead the employees. The employees before giving the consent must have opportunity to ask the management personnel, any doubts which they have in their mind. Their participation should be completely voluntary. They should not be forced to participate in genomics testing by giving any threat to their employment. In short, the employees should not have any fear of repercussions in case they refuse for genomics testing (Madison, 2016).

PROTECTION OF GENETIC DATA

Collection of genetic data is a sensitive matter in nature. Genetic data can reveal detailed information about a person's health risks and susceptibility to various diseases and health issues. Hence, it becomes very important that genetic data is delicately protected and the access should be given to only authorized people. Along with this, there should be stringent rules in terms of security and confidentiality (Mason, 2018).

The management needs to adopt strong advanced encryption technologies in order to secure the genetic data, preventing any unauthorized excess. The organizations also need to frame effective disciplinary policies which will be useful in case of any breach of data. Organizations should also conduct regular inspections of the genetic data in order to identify any breach or any vulnerability of getting shared with unauthorized people. There should also be transparency in terms of how the data will be used. The employees should be clearly informed about the purpose of such testing. Transparency will help in building trust and will reassure the employees that their data will be handled, handled responsibly (Maltby, 2009).

PREVENTION OF DISCRIMINATION BASED ON GENETICS

Although there are legal frameworks related to legal protection of genetic data, the fear of discrimination based on genetics is always there. Discrimination based on genetics can happen when the employees they are treated unfairly and hey are subjected to decisions based on their genetic information. So, in order to protect employees against any discrimination based on genetics, the management needs to establish strong non-discrimination policies. Such policies needs to clearly prohibit discrimination based on genetic data with regards to recruitment, selection, promotion, performance appraisal, opportunities of training and development. Hence, it's very important for the organizations to build a climate of trust and transparency among the employees so that they don't feel any fear of being discriminated on the basis of genetic data. When the employees they are clearly informed, they have trust in the organization, they don't have any fear or anxiety and their participation is clearly voluntary, such situations will help in fostering an environment where the employees will voluntary participate in genomics testing (Mazza and Fraker, 2010).

Integration of Ethics in Genomics Testing in Relation to Workplace Wellness Programs

In order to ensure integration of ethics efficiently, the organizations need to adopt and a holistic approach, which includes each and every aspect of genetic testing. In order to ensure this, the organizations need to develop appropriate policies and processes, in order to address each and every ethical aspect of genetic testing, including consent of employees, adequate protection of genetic data, and prevention of discrimination. Such policies should be regularly reviewed and proper upgradation should be done in order to ensure compliance with the best practices and regulatory frameworks (Mazza and Fraker, 2011).

Secondly, the organizations need to ensure that the employees they are provided, proper opportunities of education, training's, so that they are aware of the ethical regulatory and other practical aspects involved in genetic testing. Such type of training should include topic-like informed consent and prevention of discrimination of employees on the basis of genetic data. Thirdly, the organization should ensure that there are proper channels of communication, providing proper information to the employees about the significance and the limitations of genetic testing. Such open communication will help in developing trust and will also encourage participation of the employees in such testing programs (McDonald et al., 2020).

Employees should also be given proper opportunity to ask questions, addressing their concerns. Also, organizations should adopt latest technologies related to data protection in order to safeguard the genetic information of the employees, organizations should use proper encryption programs, latest storage programs and proper access of the controls of the genetic data in order to prevent unauthorized access and breach. They also need to conduct regular inspections and assessments in order to ensure security of genetic data. With the help of developing a work climate, that includes strong trust, complete transparency and organizational justice is very crucial for the success of genomics testing (Mess et al., 2024).

The management needs to demonstrate a strong commitment in terms of ethics by adhering to non-discrimination practices and by showcasing that each and every employee is treated with proper respect and dignity. The organizations by ensuring active engagement of the employees in genetic testing and wellness programs helps in addressing their concerns and developing confidence. By giving them proper opportunities for feedback, involving each and every employee in planning and implementation of wellness programs will help in making such programs successful. Overall, organizations ensuring ethical standards requires a committed approach that integrates legal ethical practices, proper protection of the data. In order to ensure that such testing programs, they are ethically sound and develop a positive work-climate of justice and fairness. Work-engagement enhancement initiatives by organizations is very crucial for success of genomics testing in relation to workplace wellness programs. To ensure employee engagement, open communication is the key. Open communication is the foundation of such wellness initiatives. This helps in ensuring that each and every employee understands the advantages as well as the limitations of genetic testing (Mujwara et al., 2023).

In regards to this, Human Resource Department plays a very important role in ensuring free communications channels with the help of informative programs, seminars, workshops. In order to provide detailed information to the employees related to genetic testing, HR department can also organize educational workshops and individualized counseling sessions of the employees. That can play a very significant role in developing work engagement among the employees. Individualized workshops and consultations can provide the employees with detailed understanding of the purpose of genetic testing, its potential benefits and how it can lead to overall positive, physical and mental well-being of the employees. Individualized counseling sessions with genetic counselors can help in addressing their concerns, can answer their questions queries and can provide them individualized advice based on their concerns and their genetic test results. This individualized approach can help in developing trust and will encourage the employees to participate in genetic testing wholeheartedly (Nissenbaum and Patterson, 2016).

During workshops, success stories can be shared with the employees about the significance of genetic testing and how it led to positive well-being or wellness of the employees by giving them practical examples of real world. The employees can understand the actual benefit of genomics testing, removing any apprehensions they have related to the outcome of such testing,by highlighting positive examples. The Human Resource Department can create a positive work climate where employees feel free to share their ideas and concerns and wholeheartedly participate in the genomics testing without remaining silent (Kaur & Arora, 2023). Designing workplace wellness programs along with other administrators, the Human Resource Department is responsible for designing or planning of wellness programs. Such wellness programs they need to be comprehensive and inclusive and such programs need to ensure that they address the requirements of the employees (Öncel, 2004).

A well-planned wellness program should include wide range of interventions. For example, individualized nutrition plans, physical fitness programs, mental fitness programs, according to the individualized need of the employees. Such programs need to address the health risks identified through genomics testing. HR Department by ensuring that such programs they are tailored according to the individualized needs of the employees. Example, age, gender, culture will help in ensuring a program which is flexible and according to the need of the employees. Along with proper planning and implementation of wellness programs, the HR Department also should ensure continuous evaluation of wellness programs (Olick, 2014).

HR Department needs to establish an evaluation criteria in order to assess the results of such programs on employee health and their work engagement and moral. Such metrics includes change in health data, example, heart rate, blood pressure, sugar level and satisfaction of the employees in terms of results of the program. Such evaluation can be with the help of regular surveys and one-to-one interaction along with health monitoring which can help in providing valuable insights in order to improve the future wellness programs (Kaur & Arora, 2023).

HR Department can implement written or oral surveys measuring the employee perception, satisfaction, moral and engagement levels about the program. The health monitoring can include regular health check-up of the employees measuring their blood pressure, heart rate, sugar level, cholesterol level etc. HR Department can also utilize focus groups to have one-to-one interaction with each and every member of the group. Such groups can provide a detailed information about their perception, experience and what can be done to further improve the programs. Overall, such feedback mechanisms, they are very vital for improvement of future wellness programs. Human Resource Department should establish free open channels for feedback, including suggestion boxes, open door policy, online feedback forms so that they get the idea about the perception of the employees related to genetic testing as well as the wellness programs (Patel et al., 2013).

MENTAL HEALTH SUPPORT

Along with physical health, mental health plays a very crucial role in employee wellness and it is the responsibility of both organizations as well as the employer to prioritize it. Employees they need to focus on both their physical and mental health along with the organizations taking initiatives in order to ensure the same. With the help of integration of genomics testing with workplace wellness programme, organizations can understand employee needs in terms of physical and mental health. For example, the genetic testing can help in identifying the genetic risks of the employees in terms of both physical and mental health conditions including depression, anxiety, heart disorders, diabetes, cancer (Penalvo et al., 2017).

When equipped with such information, the organizations can target the mental and physical health issues of the employees by ensuring interventions tailored to the individualized needs of the employees. This can be done by offering the employees mental health counsellor, psychologists who can give them individualized support and therapy based on their personalized genetic testing results. In addition to counselling sessions programmes targeting the physical and mental health of the employees, the organizations initiatives can also include meditation programmes, stress management programmes and by providing them a supporting work climate where they feel comfortable discussing about their mental and physical health issues without any fear of stigma (Pestka, 2022).

Ergonomic workspace is also a very crucial element of workplace wellness programmes. Ergonomics focuses on planning such work environment that enhances and employ productivity performance and reducing the risk of workplace injury. Improper ergonomics can lead to a variety of health problems including orthopaedics disorders, fatigue, mental stress that can negatively impact the well-being of the employees. The companies or organizations can regularly conduct ergonomics assessments in order to evaluate and examine the design of the workstations and identifying such areas of the work spaces that can lead to injury or discomfort for the employees. On the basis of such evaluation, the adjustment of the workplace can be done in terms of proper sitting chairs, desks, proper lightning and proper placement of the laptops and computers, improving the comfort of the employees and reducing the risk of workplace injury (Kaur & Arora, 2023).

Ergonomics also include providing the employees the proper tools and resources along with providing them proper education about the significance of ergonomics in enhancing workplace comfort and health. In addition, flexible work arrangements which includes work from home option, flexible work hours, compressed work weeks has been recognized as an important factor in enhancing employee moral well-being and working engagement. Such flexible arrangements can help and motivate the employees to manage their work better, reduction of work stress, improvement of job satisfaction. By giving them option of remote working and compressed work weeks, the physical and mental strength can be reduced (Pestka, 2022).

The organizations can also offer the employees flexible work options based on the individualized needs of the workforce. For example, employees can be given the option of part-time work or to work from any remote area which is feasible for them. Such flexible arrangements can help the employees in balancing their personal and professional lives and getting better control over their work schedules. Hence overall, it becomes very important for the organizations to clearly communicate to the employees regarding the availability of such flexible arrangements. Along with that HR department plays a very important role in ensuring that employees have equal accessibility so that they don't have any feeling of being discriminated in terms of flexible arrangement opportunities (Sanghavi et al., 2022).

DIGITAL PLATFORMS FOR HEALTH MONITORING

In today's time, technology plays a very important role in modern workplaces. Technology is required in each and every aspect of the work, along with playing a very crucial role in health monitoring and plan management. For example, digital platforms nowadays can help in proper storage and analysis of the health data of the employees, making it easier for the organizations and the employees to track the data and monitor the achievement of the goals (Schulte et al., 2012).

Digital platforms can provide opportunity for health monitoring. Such digital platforms can include variable devices, variable devices, health applications in order to have a detailed feedback about employees' health. These health monitoring platforms can offer the employees and organizations about individualized recommendations along with giving regular reminders for health screening and tracking the progress with time. They can make the tracking of the health more engaged for the employees (Kaur & Arora, 2023).

ROLE OF ARTIFICIAL INTELLIGENCE IN GENETIC DATA ANALYSIS

Artificial intelligence plays a very crucial role in revolutionizing the evaluation and examination of genetic data. It can provide a detailed insight and predictions about the health risks associated. Artificial intelligence algorithms can't investigate large sets of data in order to identify patterns and correlations that may not be identified or apparent by using traditional methods of analysis. For example, artificial intelligence can identify genetic defects which can be associated with specific medical conditions and on the basis of such information, it can provide suggestions regarding preventive measures or changes related to lifestyles that can help in reducing such health risks. Artificial intelligence can also provide recommendations on the basis of current trends so that the employees are up to date in terms of health advice (Seaver et al., 2022).

Early health, daily health consultations has become or has the potential of playing a very important role in workplace wellness programmes. It can offer employees convenient access to medical staff, reducing the requirement of being physical present at the workplace facility or health facility and costly in-person visits to the medical centre. Daily health consultations can also lead to addressing of health concerns of the employees promptly. Hence, integration of daily health consultations with genomics testing can lead to better effectiveness of workplace wellness programmes. For example, the employees receiving daily health counselling along with genetic counselling can lead to better individualized health results. Such combination can facilitate better health support, better follow-up care and ensures that the employees they are receiving guidance in terms of each and every aspect of their health (Seaver et al., 2022; Tutty et al., 2019).

CONCLUSION

Successful implementation of genetic testing, along with workplace wellness programs. requires a dedicated approach by the organizations, emphasizing on enhancing employee engagement, morale and their dedication towards the work, as well as ensuring their physical and mental well-being in general. Organizations should involve employers in the planning and development, as well as implementation of such wellness programs, in order to ensure that the individualistic requirements of the employees they are addressed properly according to their preferences. Such participatory approach can help in improving the commitment of the employees towards such programs.

In this regard, proper communication is the key to ensure success of workplace wellness programs. The employers should to make sure that multiple channels of communication are available in order to inform the employees about the significance, limitations, as well as the uses of the genetic data which can prove beneficial for their mental and physical well-being. Such communication channels will also help the employees in understanding how they can participate in the programs and what all resources are available for them to participate. In this regard, regular updates, sharing of success stories with the employees, as well as the testimonials of previous wellness programs will help in maintaining and the interest of the employees and will motivate them to participate in the programs. Also, the organizations they can provide the employees' incentives as well as rewards on the basis of their participation in order to enhance their engagement, offering such incentives for participation will encourage the employees to take active part in the programs. In addition to this, human resource department can organize recognition programs in order to highlight those employers who participated in the wellness programs wholeheartedly and such employers can share their experience and their point of views about the wellness programs, fostering a culture of understanding and health. In addition, ethical leadership is very important and fundamental for the success of successful combination of genomics testing and workplace wellness programs.

The management should demonstrate a strong commitment towards ethical practices, ensuring complete transparency and respecting the consent of the employees. This involves being clear about the use of genetic testing, how the data will be collected, how it will be used, where it will be stored and what all measures will be taken by the organizations in order to protect the privacy of the data. Transparency in terms of communication, of limitations and the potential risk involved in genetic testing should be clearly informed to the employees so that they have the opportunity to ask questions and the organization can address their concerns timely. The management should be very clear about the partnership with third party providers so that the employees they are not apprehensive about sharing their data with the organization. Moreover, the autonomy of the employees should be properly acknowledged and honored. They should be given proper opportunity to give consent related to collection of the data and of their health. Employees should not be pressurized to share or participate. The management should provide all necessary information to the employees so that they can make proper choices and their choice should be respected whether they decide to participate or not.

Along with this, the management should also ensure that the employees have access to their genetic data because non-transparency will not be beneficial for the success of wellness programs. Non-sharing of the genetic data with the participating employees will lead to a feeling of distrust in their organization and feeling of getting manipulated by the organization to participate in the genetic testing. Moreover, development of trust is very crucial for the successful adoption of wellness programs. Trust can be only built if the organization is realistic to ethical practices, complete transparency and showing commitment towards employee health and well-being. Employee sensing a supportive work climate will be safe and valued and they will willingly participate by giving their consent for genetic testing.

In order to ensure a positive participation of employees in wellness programs, management should take all the necessary measures in order to get the employees participate in genetic testing and wellness programs and by integrating genomics testing with wellness programs can lead to a new era of individualistic health management which offers a variety of benefits. With benefits comes challenges also but with the help of cooperation from the employees and by ensuring proper adherence to legal framework, ethical practices and proper trustful communication channels, organizations can help the employees in understanding the benefit of such practices and by aligning the wellness initiatives with the individualistic needs of the employees, the organizations can overall enhance the health outcomes reducing the cost as well as improving the engagement and productivity overall of the wellness programs. However, by adopting strategic human resource management practices, promoting a climate of health, utilizing advanced technologies, demonstrating ethical practices, organizations can promote effective and ethically sound wellness programs, contributing towards the success of the organization and ensuring that their goals are aligned with the individualistic needs of the employees.

REFERENCES

Ajunwa, I., Crawford, K., & Ford, J. S. (2016). Health and big data: An ethical framework for health information collection by corporate wellness programs. *The Journal of Law, Medicine & Ethics*, 44(3), 474–480. 10.1177/107311051666794327587451

Biswas, S., Medne, L., Devkota, B., Bedoukian, E., Berrodin, D., Izumi, K., Deardorff, M. A., Tarpinian, J., Leonard, J., Pyle, L., Gray, C., Montgomery, J., Williams, T., Fortunato, S., Weatherly, J., McEldrew, D., Kaur, M., Raible, S. E., Wilkens, A., & Krantz, I. D. (2020). A centralized approach for practicing genomic medicine. *Pediatrics*, 145(3), e20190855. 10.1542/peds.2019-085532102930

Brown, E. A. (2016). Are Wellness Programs Good for Business? A Legal and Ethical Checkup. *Business Law Review*, 49, 39.

Brown, E. A. (2017). Workplace wellness: Social injustice. *NYUJ Legis. & Pub. Pol'y*, 20, 191.

Callier, S. L., Abudu, R., Mehlman, M. J., Singer, M. E., Neuhauser, D., Caga-Anan, C., & Wiesner, G. L. (2016). Ethical, legal, and social implications of personalized genomic medicine research: Current literature and suggestions for the future. *Bioethics*, 30(9), 698–705. 10.1111/bioe.1228527767224

Carnethon, M., Whitsel, L. P., Franklin, B. A., Kris-Etherton, P., Milani, R., Pratt, C. A., & Wagner, G. R. (2009). Worksite wellness programs for cardiovascular disease prevention: A policy statement from the American Heart Association. *Circulation*, 120(17), 1725–1741. 10.1161/CIRCULATIONAHA.109.19265319794121

Cohn, B., Ryan, K. A., Hendy, K., Callahan, K., Roberts, J. S., Spector-Bagdady, K., & Lee, C. (2023). Genomic testing in voluntary workplace wellness programs in the US: Evidence and challenges. *Molecular Genetics & Genomic Medicine*, 11(11), e2245. 10.1002/mgg3.224537592452

Cummings, S. (2010). Genetic Counseling. In Genetic Diagnosis of Endocrine Disorders (pp. 293-302). Academic Press.

Curnutte, M. (2012). Consuming Genomes: The Coproduction of a New Scientific and Technological Order for Genetic Testing.

Dashti, H. S., Hivert, M. F., Levy, D. E., McCurley, J. L., Saxena, R., & Thorndike, A. N. (2020). Polygenic risk score for obesity and the quality, quantity, and timing of workplace food purchases: A secondary analysis from the ChooseWell 365 randomized trial. *PLoS Medicine*, 17(7), e1003219. 10.1371/journal.pmed.100321932692747

David, S. P., & Palaniappan, L. (2018). Clinical and personal utility of genetic risk testing. *American Family Physician*, 97(9), 600–602.29763257

Delaney, S. K., Hultner, M. L., Jacob, H. J., Ledbetter, D. H., McCarthy, J. J., Ball, M., Beckman, K. B., Belmont, J. W., Bloss, C. S., Christman, M. F., Cosgrove, A., Damiani, S. A., Danis, T., Delledonne, M., Dougherty, M. J., Dudley, J. T., Faucett, W. A., Friedman, J. R., Haase, D. H., & Green, R. C. (2016). Toward clinical genomics in everyday medicine: Perspectives and recommendations. *Expert Review of Molecular Diagnostics*, 16(5), 521–532. 10.1586/14737159.2016.114659326810587

Deverka, P. A., Goehringer, J., McDonald, W., Salvati, Z., Wagner, J., & Williams, M. S. (2020). *Stakeholders assessing genetics with employers (SAGE)*. NHGR Institute.

Dohn, M. R. (2017). Personal Genomics and Genetic Discrimination: Is Increased Access a Good Thing. *W. St. UL Rev.*, 45, 107.

Esmonde, K., Roth, S., & Walker, A. (2023). A social and ethical framework for providing health information obtained from combining genetics and fitness tracking data. *Technology in Society*, 74, 102297. 10.1016/j.techsoc.2023.10229737521714

Fonarow, G. C., Calitz, C., Arena, R., Baase, C., Isaac, F. W., Lloyd-Jones, D., Peterson, E. D., Pronk, N., Sanchez, E., Terry, P. E., Volpp, K. G., & Antman, E. M. (2015). Workplace wellness recognition for optimizing workplace health: A presidential advisory from the American Heart Association. *Circulation*, 131(20), e480–e497. 10.1161/CIR.000000000000020625869199

Foss, K. S., O'Daniel, J. M., Berg, J. S., Powell, S. N., Cadigan, R. J., Kuczynski, K. J., Milko, L., Saylor, K., Roberts, M., Weck, K., & Henderson, G. E. (2022). The rise of population genomic screening: Characteristics of current programs and the need for evidence regarding optimal implementation. *Journal of Personalized Medicine*, 12(5), 692. 10.3390/jpm1205069235629115

Garcia, E., Kundu, I., Kelly, M., Soles, R., Mulder, L., & Talmon, G. A. (2020). The American Society for Clinical Pathology's job satisfaction, well-being, and burnout survey of laboratory professionals. *American Journal of Clinical Pathology*, 153(4), 470–486. 10.1093/ajcp/aqaa00832080719

Goetzel, R. Z., & Ozminkowski, R. J. (2008). The health and cost benefits of work site health-promotion programs. *Annual Review of Public Health*, 29(1), 303–323. 10.1146/annurev.publhealth.29.020907.09093018173386

Gordon, T. M. (2022). Health, Wellness, and the Genome: Reconciling Biomedical Innovation with Market-Based Inequities after CRISPR. *The International Journal of Health, Wellness & Society*, 12(2), 11–24. 10.18848/2156-8960/CGP/v12i02/11-24

Grifantini, K. (2017). What? s Happening to Your DNA Data?: Genetic Testing Services Abound, but Consumers Opting to Use Them Should Be Aware of the Pitfalls. *IEEE Pulse*, 8(6), 10–14. 10.1109/MPUL.2017.275112129155371

Hamilton, J. G., Shuk, E., Arniella, G., González, C. J., Gold, G. S., Gany, F., Robson, M. E., & Hay, J. L. (2016). Genetic testing awareness and attitudes among Latinos: Exploring shared perceptions and gender-based differences. *Public Health Genomics*, 19(1), 34–46. 10.1159/00044155226555145

Harris, A., & Wilkinson, D. S. (2013). The Changing Healthcare Environment. Clinical Laboratory Management, 164-216.

Hendricks-Sturrup, R. M., Cerminara, K. L., & Lu, C. Y. (2020). A qualitative study to develop a privacy and nondiscrimination best practice framework for personalized wellness programs. *Journal of Personalized Medicine*, 10(4), 264. 10.3390/jpm1004026433287396

Hoffman, S. (2019). What Genetic Testing Teaches about Predictive Health Analysis Regulation. *North Carolina Law Review*, 98, 123.

Horwitz, J. R., Kelly, B. D., & DiNardo, J. E. (2013). Wellness incentives in the workplace: Cost savings through cost shifting to unhealthy workers. *Health Affairs*, 32(3), 468–476. 10.1377/hlthaff.2012.068323459725

Ioan, B. G., & Hanganu, B. (2023). Third-party sharing of genetic information. In *Clinical Ethics At the Crossroads of Genetic and Reproductive Technologies* (pp. 401–429). Academic Press. 10.1016/B978-0-443-19045-2.00005-2

Johnston, K. J., Hockenberry, J. M., Rask, K. J., Cunningham, L., Brigham, K. L., & Martin, G. S. (2015). Health care expenditures for university and academic medical center employees enrolled in a pilot workplace health partner intervention. *Journal of Occupational and Environmental Medicine*, 57(8), 897–903. 10.1097/JOM.000000000000048826247644

Kaur, J., & Arora, R. (2023a, June). Exploring the dimensionality of employee silence in healthcare sector. In AIP conference proceedings (Vol. 2782, No. 1). AIP Publishing. 10.1063/5.0154178

Kaur, J., & Arora, R. (2023b, June). Exploring the impact of employee silence in private hospitals-a structural equation modeling approach. In AIP Conference Proceedings (Vol. 2782, No. 1). AIP Publishing. 10.1063/5.0154173

Madison, K. M. (2016). The risks of using workplace wellness programs to foster a culture of health. *Health Affairs*, 35(11), 2068–2074. 10.1377/hlthaff.2016.072927834248

Maltby, L. (2009). Brave New Workplace: Genetic Discrimination. Illinois Public Employee Relations Report, 26(2).

Mason, S. (2018). Privacy of Information and DNA Testing Kits. *Cath. UJL & Tech*, 27, 161.

Mazza, A. M., & Fraker, M. (Eds.). (2010). *Direct-to-consumer genetic testing: summary of a workshop*. National Academies Press.

Mazza, A. M., & Fraker, M. (Eds.). (2011). *Direct-to-Consumer Genetic Testing: Summary of a Workshop*. National Academies Press.

McDonald, W. S., Wagner, J. K., Deverka, P. A., Woods, L. A., Peterson, J. F., & Williams, M. S. (2020). Genetic testing and employer-sponsored wellness programs: An overview of current vendors, products, and practices. *Molecular Genetics & Genomic Medicine*, 8(10), e1414. 10.1002/mgg3.141432715662

Mess, F., Blaschke, S., Schick, T. S., & Friedrich, J. (2024). Precision prevention in worksite health–A scoping review on research trends and gaps. *PLoS One*, 19(6), e0304951. 10.1371/journal.pone.030495138857277

Mujwara, D., Kintzle, J., Di Domenico, P., Busby, G. B., & Bottà, G. (2023). Cost-effectiveness analysis of implementing polygenic risk score in a workplace cardiovascular disease prevention program. *Frontiers in Public Health*, 11, 1139496. 10.3389/fpubh.2023.113949637497026

Nissenbaum, H., & Patterson, H. (2016). Biosensing in context: Health privacy in a connected world.

Olick, R. S. (2014). Genetic discrimination in the workplace after GINA. In *Genetic Discrimination* (pp. 128–140). Routledge.

Öncel, E. (2004). Effects of Human Genome Project on Organizational Behaviour (Master's thesis, Marmara Universitesi (Turkey)).

Patel, C. J., Sivadas, A., Tabassum, R., Preeprem, T., Zhao, J., Arafat, D., Chen, R., Morgan, A. A., Martin, G. S., Brigham, K. L., Butte, A. J., & Gibson, G. (2013). Whole genome sequencing in support of wellness and health maintenance. *Genome Medicine*, 5(6), 1–14. 10.1186/gm46223806097

Penalvo, J. L., Micha, R., Smith, J. D., Rehm, C. D., Bishop, E., Onopa, J. A., Uzhova, I., Wu, J. H., & Mozaffarian, D. (2017). Abstract MP005: Do Worksite Wellness Programs Improve Dietary Behaviors and Adiposity? A Systematic Review and Meta-analysis. *Circulation*, 135(suppl_1), AMP005–AMP005. 10.1161/circ.135.suppl_1.mp005

Pestka, E. L. (2022). Health Policy Implications of Genetics and Genomics on Advanced Practice Registered Nurses. Health Policy and Advanced Practice Nursing: Impact and Implications, 265.

Sanghavi, K., Cohn, B., Prince, A. E., Feero, W. G., Ryan, K. A., Spector-Bagdady, K., Uhlmann, W. R., Lee, C., Roberts, J. S., & Mathews, D. J. (2022). Voluntary workplace genomic testing: Wellness benefit or Pandora's box? *NPJ Genomic Medicine*, 7(1), 5. 10.1038/s41525-021-00276-835058451

Schulte, P. A., Pandalai, S., Wulsin, V., & Chun, H. (2012). Interaction of occupational and personal risk factors in workforce health and safety. *American Journal of Public Health*, 102(3), 434–448. 10.2105/AJPH.2011.30024922021293

Seaver, L. H., Khushf, G., King, N. M., Matalon, D. R., Sanghavi, K., Vatta, M., & Wees, K.Ethical and Legal Issues Committee. (2022). Points to consider to avoid unfair discrimination and the misuse of genetic information: A statement of the American College of Medical Genetics and Genomics (ACMG). *Genetics in Medicine*, 24(3), 512–520. 10.1016/j.gim.2021.11.00235253645

Tutty, E., Hickerton, C., Adamski, M. M., & Metcalfe, S. A. (2019). Personal genomic testing for nutrition and wellness in Australia: A content analysis of online information. *Nutrition & Dietetics: the Journal of the Dietitians Association of Australia*, 76(3), 263–270. 10.1111/1747-0080.1251630693631

Whitaker, J. (2019). From health care to self-care: workplace wellness programs and health discrimination (Master's thesis, Queen's University (Canada)).

Chapter 8

The Influence of Work From Home Policies on Employees' Well-Being and Moonlighting Behavior in the Transhuman Context

Diksha Taneja
Chandigarh University, India

ABSTRACT

When an employee is engaged in moonlighting, it means they are serving two different companies at the same time. There are different reasons behind the rise of this situation, such as the need to earn more to fulfill financial needs and dissatisfaction from existing pay. With the rise in work from home (WFH) culture, employees have access to multiple opportunities. With continued opportunities due to digitalization, work from home and moonlighting have made a significant impact on the overall well-being of employees. With that being said, this study is aimed to study the impact of moonlighting and work-from-home on employees' well-being in Delhi and Chandigarh and implications on moonlighting on their work-life balance. In order to fulfill the above objectives, this study is based on both primary and secondary data. Primary data is collected from an online survey conducted through self-structured questionnaire which is shared through Google Form. Survey data has been analyzed with Excel spreadsheet and statistical package for social sciences (SPSS) software. Secondary data has been collected from previous studies related to work from home

DOI: 10.4018/979-8-3693-7668-3.ch008

and moonlighting in various sectors. This study seeks to explore the impact of work-from-home and moonlighting on employees' wellbeing and their understanding of work commitment. The study was conducted on Delhi and Chandigarh regions where participants were chosen with random sampling. The dependent variable for this study was "work from home" and independent variable was "employees' well-being". This study found significant impact of work from home and moonlighting on employees' well-being ($p<0.05$) by conducting one-sample T-test using SPSS software. The study provides a valuable insight to corporate leaders and managers to manage the issues of employee turnover and declining job security.

1. INTRODUCTION

Frequency of constant financial stress and worries affecting work is known as financial insecurity (Kim & Garman, 2004). This financial insecurity has made the government provide short-term financial support to employees and businesses. There are several reasons why employees switch to moonlighting as Plan B of existing psychological and financial struggles and potential challenges in the labor market related to loss of jobs and layoffs (Asravor, 2021).

1.1 Background

Moonlighting is defined as an act of doing extra work, especially when doing it without informing the primary organization. It means if the employer catches the employees doing a second job, they may choose to fire the employees. There are different reasons for people to do moonlighting like a person looking to switch the career or join a new organization, using time to stay busy, making extra income, career growth, skill development, personal development, passion, social presence, having fun in another work, getting experience for new job, or job insecurity (Dickey et al., 2015; Böheim & Taylor, 2005; Kawakami, 2019; Heineck & Schwarze, 2004; Sakyi & Agomor, 2021).

Also known has "holding several jobs", moonlighting is a widespread concept among low-income groups who are primarily aimed to generate enough income and get their family and social status out of poverty (Hausken & Ncube, 2018; Urwick & Kisa, 2014). Especially those employees who are moonlighting are aware of the problem of financial stability in their lives.

2. LITERATURE REVIEWS

The process of performing a second job along with a primary job, either after working hours or during office hours, is called moonlighting (Yamini & Pushpa, 2016; Ashwini et al., 2017). Employees perform moonlighting usually with the approval of the main employer or without their knowledge. It may consist of working after regular working hours or on weekends, or employees may take some time away from the main employment to work for another employer. A lot of people are engaged in a second job to earn passive income, for a pastime, to follow their passion, or gain experience in another sector. If it affects job performance, raises conflict of interest, or violates company policy, it may lead to conflict with the primary employer. Employees must know about the regulations related to moonlighting in their organization before taking a second job.

Choudhary & Saini (2021) examined the aim of mid-level employees of some IT companies in south India. They looked into a lot of reasons behind moonlighting performed by the employees and concluded that organizational commitment was lost without proactive benefits for retaining expert, devoted, and loyal employees and they were engaged in secondary jobs to fulfill their own needs. There are several reasons why an employee should be devoted at work and rules that both employees and bosses must follow when working together as per the circumstances. There are several factors to be considered when it comes to finding out if moonlighting is allowed.

As long as it is evident that specific terms of bargaining agreement should always be considered to find out the appropriateness of any discipline, it is possible to form a proper policy for moonlighting. An employee may be penalized if they accept a position with a company which affects their capacity to accomplish their tasks. Even though there are events when the matter is not very serious, it doesn't mean an employee cannot be laid off if they constantly come late, miss a lot, or just performing poorly due to the exhaustion of their second job. George et al. (2022) compared the moonlighting rate in the IT sector and relation between secondary and primary jobs of employees.

According to Jaggarwal (2022), employees are constantly facing the pressure of moonlighting because of the state of the job market and global economy. This study has also discussed the importance of non-financial aspects to companies when it comes to developing the right HR policies and controlling the risk of moonlighting. The choice of moonlighting relies on personal needs and circumstances of employees. There are different benefits of moonlighting like career opportunities, personal growth, and financial rewards. But it also affects productivity, causes exhaustion, and affects interpersonal relations between employees and primary employers.

Before starting a side or second business. It is important to consider the pros and cons properly (Gaitonde et al., 2022).

Pros

Extra income – One can earn more money by doing a second job and reach their financial goals quickly, while paying off their debts.
Networking – It is possible to grow a professional network by working in a new industry or position, which can help in the long run.
Skill development – An employee can learn new skills or hone existing skills to advance in their profession.
Flexibility – An employee can have more control over their work hours and schedule if there is no flexibility in their primary job.

Cons

Time constraints – Moonlighting needs employees to commit their extra time they could spend with their family, friends, and things they love doing.
Exhaustion – Several jobs increase the risk of mental and emotional burnout as they can be physically and mentally draining.
Conflicts – It could be hard to manage first employment, which could lead to conflict of interest with the employer and employees may face consequences if the employer is not informed.

2.1 Research Gap

These days, a lot of organizations in India have allowed employees to work from home. Along with serving primary employers, majority of employees started moonlighting due to uncertainties in job market, fulfilling their financial obligations, and other reasons. Hence, this study fills the most important knowledge gap by discussing the impact of moonlighting and work from home on their overall well-being.

2.2 Research Question

- What is the impact of work from home and moonlighting on employees' wellbeing?
- What are the implications of moonlighting on the work-life balance of employees?

2.3 Research Objectives

- To find out the impact of work from home and moonlighting on wellbeing of employees
- To discuss the implications of moonlighting on work-life balance of employees

2.4 Hypothesis

H1 – There is a significant impact of work from home and moonlighting on wellbeing of employees

H0 – There is no significant impact of work from home and moonlighting on wellbeing of employees

3. RESEARCH METHODOLOGY

3.1 Research Method & Design

The research design for this study is exploratory. It consists of an online survey conducted on the sample of 153 employees in selected IT companies in Delhi and Gurgaon to find out their perception on working from home. A self-structured questionnaire is prepared to share using Google Form. The dependent and independent variables for this study are as follows –

- **Dependent Variable** – Work from Home and Moonlighting
- **Independent variable** – Employees' wellbeing

3.2 Research Approach

Research approach of this study consists of both primary and secondary data. Primary data is collected using an online survey and analyzed with Excel Spreadsheet and SPSS Software. In addition, secondary data is collected through literature search using Google Scholar.

4. ANALYSIS OF STUDY

4.1. Demographics

In this study, 91 (60%) participants are aged 28 to 35 years, 31 (20%) participants are 18 to 25 years old, 16 (11%) participants are 36 to 45 years old, and 15 (10%) participants are above 45 years old (Table 1) (Figure 1).

Table 1. Age

	Frequency	Percent	Valid Percent	Cumulative Percent
Valid 18 to 25 years	31	20.3	20.3	20.3
26 to 35 years	91	59.5	59.5	79.7
36 to 45 years	16	10.5	10.5	90.2
Above 45 years	15	9.8	9.8	100.0
Total	153	100.0	100.0	

Figure 1. Age of the participants

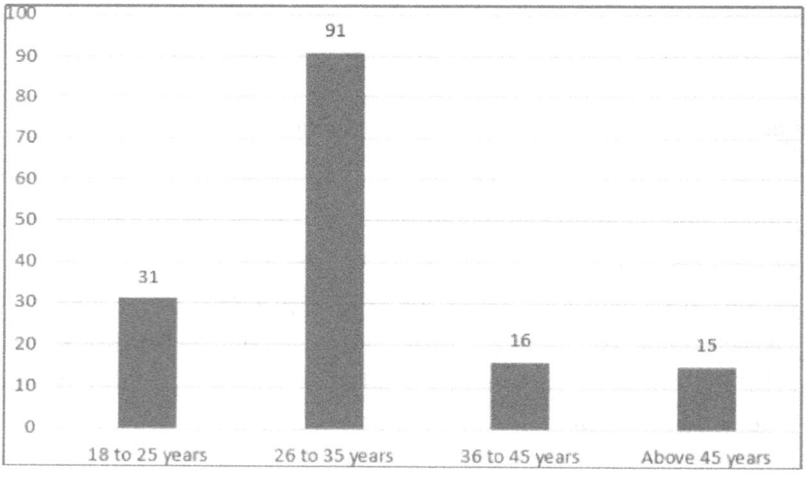

When it comes to gender of the participants, 100 (65%) participants are female and 53 (35%) participants are male (Table 2) (Figure 2).

Table 2. Gender

	Frequency	Percent	Valid Percent	Cumulative Percent
Valid Female	100	65.4	65.4	65.4
Male	53	34.6	34.6	100.0
Total	153	100.0	100.0	

Figure 2. Gender

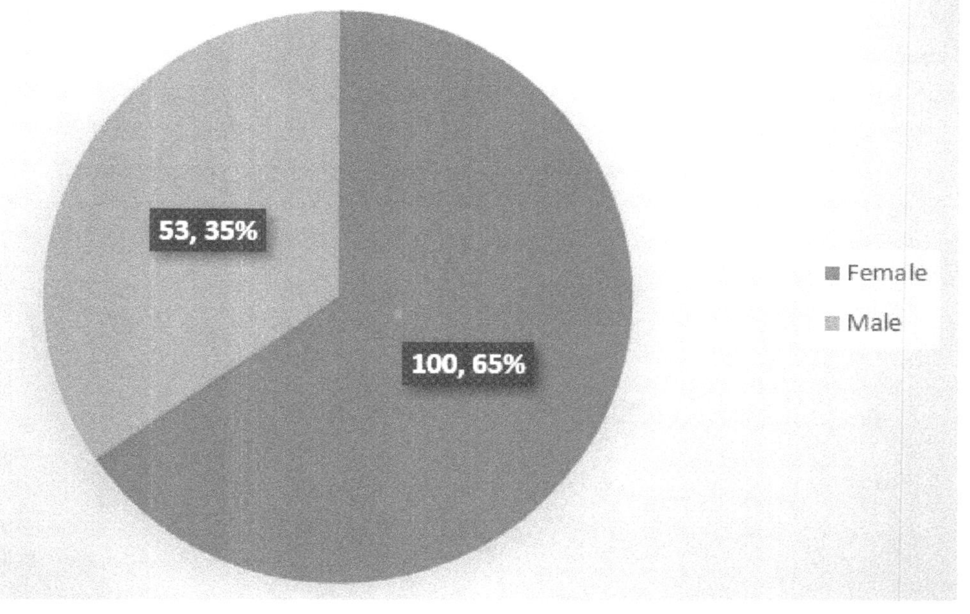

When it comes to size of the organization where employees were working in, there are 73 (48%) participants who were working in organization with over 500 employees, 48 (31%) participants were working in companies with 101 to 500 employees, 24 (16%) participants were working in companies with less than 50 employees, and 8 (5%) participants were working in companies with 50 to 100 employees (Table 3) (Figure 3).

Table 3. Size of organization

	Frequency	Percent	Valid Percent	Cumulative Percent
Valid 101 to 500 employees	48	31.4	31.4	31.4
50 to 100 employees	8	5.2	5.2	36.6
Less than 50 employees	24	15.7	15.7	52.3
More than 500 employees	73	47.7	47.7	100.0
Total	153	100.0	100.0	

Figure 3. Size of organization

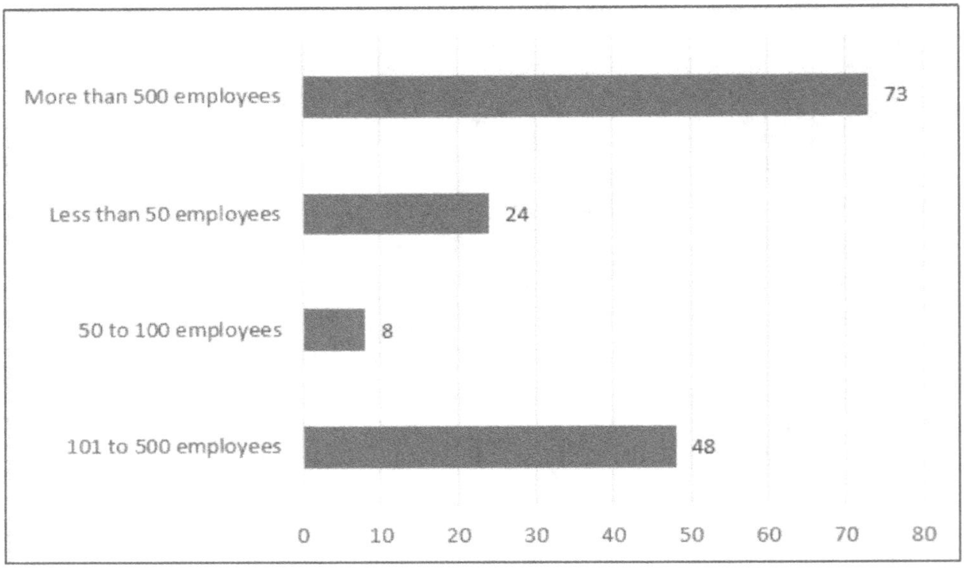

There are 100 (65%) participants working in private limited companies, 32 (21%) participants working in multinational companies, and 21 (14%) participants working in public limited companies (Table 4) (Figure 4).

Table 4. Type of organization

	Frequency	Percent	Valid Percent	Cumulative Percent
Valid Multination al	32	20.9	20.9	20.9

continued on following page

Table 4. Continued

	Frequency	Percent	Valid Percent	Cumulative Percent
Private Ltd	100	65.4	65.4	86.3
Public Ltd	21	13.7	13.7	100.0
Total	153	100.0	100.0	

Figure 4. Type of Organization

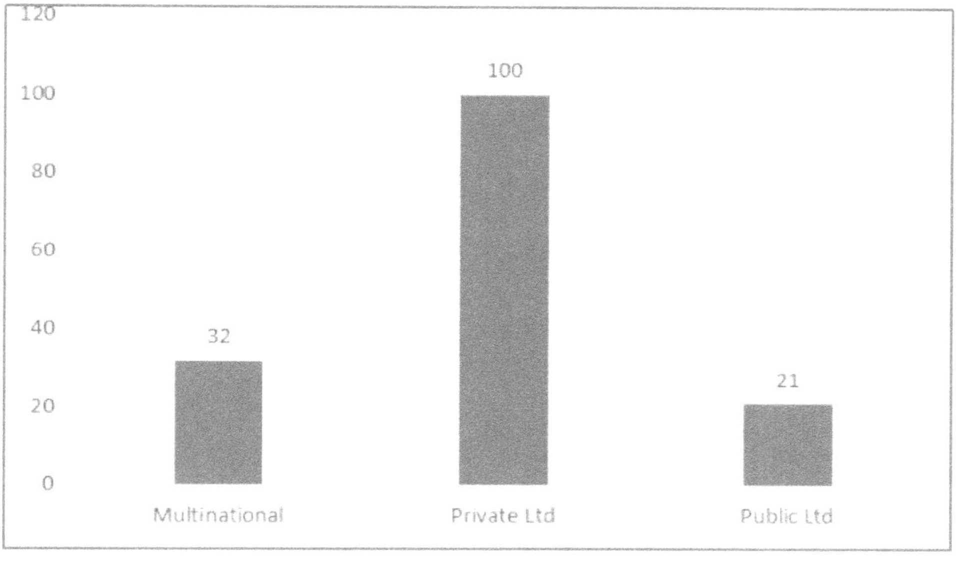

When it comes to work experience, there are 76 (50%) participants who have up to 5 years of experience, 46 (30%) employees who have 6 to 10 years of work experience, 24 (16%) participants who have 11 to 20 years of work experience, and 76 (50%) participants who have up to 5 years of experience (Table 5) (Figure 5).

Table 5. Work experience

	Frequency	Percent	Valid Percent	Cumulative Percent
Valid 11 to 20 years	24	15.7	15.7	15.7
6 to 10 years	46	30.1	30.1	45.8

continued on following page

Table 5. Continued

	Frequency	Percent	Valid Percent	Cumulative Percent
Above 20 years	7	4.6	4.6	50.3
Up to 5 years	76	49.7	49.7	100.0
Total	153	100.0	100.0	

Figure 5. Work experience of participants

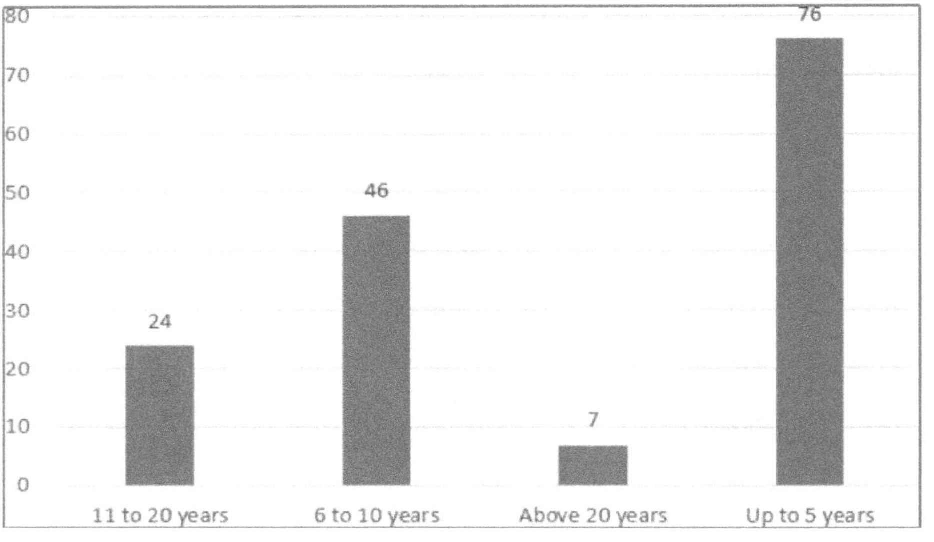

4.2. Employees' Perception on Work From Home and Moonlighting

There are 67 (44%) participants who strongly agree and 22 (14%) participants agree that their productivity has been improved by working from home, while 32 (21%) participants were neutral, 16 (11%) participants disagree, and 16 (11%) participants strongly disagree with that (Table 6) (Figure 6).

Table 6. Your productivity has been improved by working from home

	Frequency	Percent	Valid Percent	Cumulative Percent
Valid Strongly Disagree	16	10.5	10.5	10.5
Disagree	16	10.5	10.5	20.9
Neutral	32	20.9	20.9	41.8
Agree	22	14.4	14.4	56.2
Strongly Agree	67	43.8	43.8	100.0
Total	153	100.0	100.0	

Figure 6. Rise in productivity by working from home

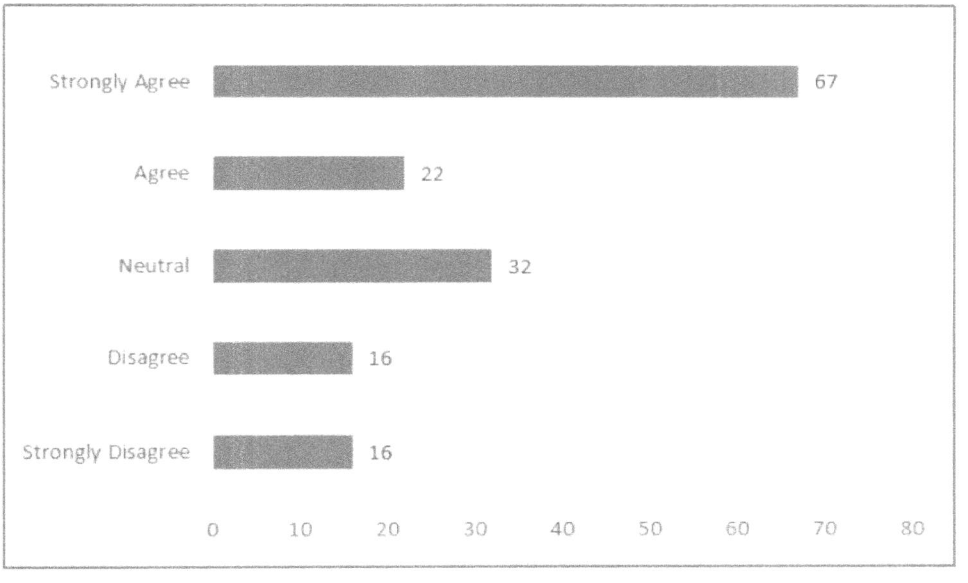

There are 47 (31%) employees who strongly disagree and 24 (16%) participants who disagree that their organization allows employees to moonlight after working hours, while 68 (44%) participants were not sure (Table 7) (Figure 7).

Table 7. Your organization allows employees to moonlight after working hours

	Frequency	Percent	Valid Percent	Cumulative Percent
Valid Strongly Disagree	47	30.7	30.7	30.7
Disagree	24	15.7	15.7	46.4
Neutral	68	44.4	44.4	90.8
Agree	7	4.6	4.6	95.4
Strongly Agree	7	4.6	4.6	100.0
Total	153	100.0	100.0	

Figure 7. Moonlighting after working hours

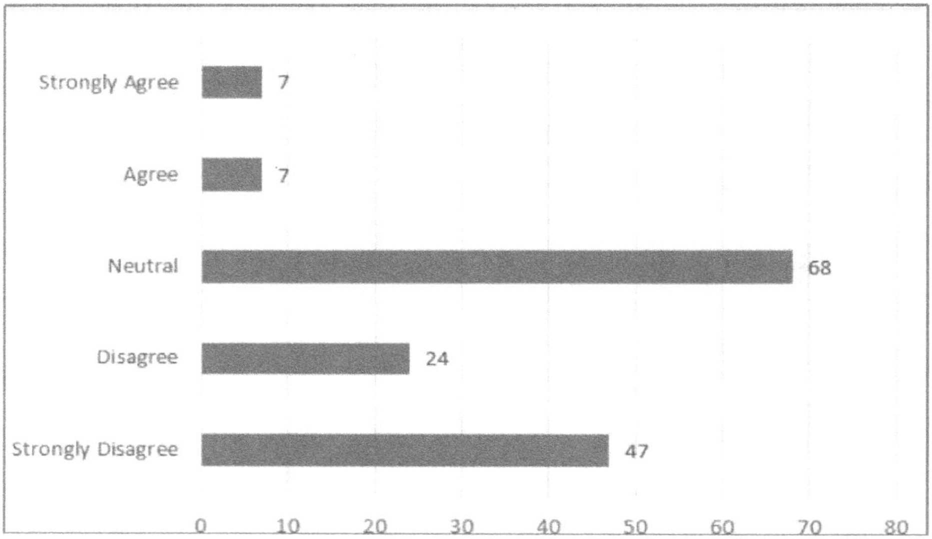

There are 37 (24%) participants who strongly agree and 15 (10%) participants who agree that they cannot fulfill their financial needs by working from home only for their company, while 30 (20%) participants were neutral, 33 (22%) participants disagree, and 38 (25%) participants strongly disagree with that (Table 8) (Figure 8).

Table 8. You are not able to fulfill your financial needs by working from home only for your company

	Frequency	Percent	Valid Percent	Cumulative Percent
Valid Strongly Disagree	38	24.8	24.8	24.8
Disagree	33	21.6	21.6	46.4
Neutral	30	19.6	19.6	66.0
Agree	15	9.8	9.8	75.8
Strongly Agree	37	24.2	24.2	100.0
Total	153	100.0	100.0	

Figure 8. Meeting financial needs only with one job

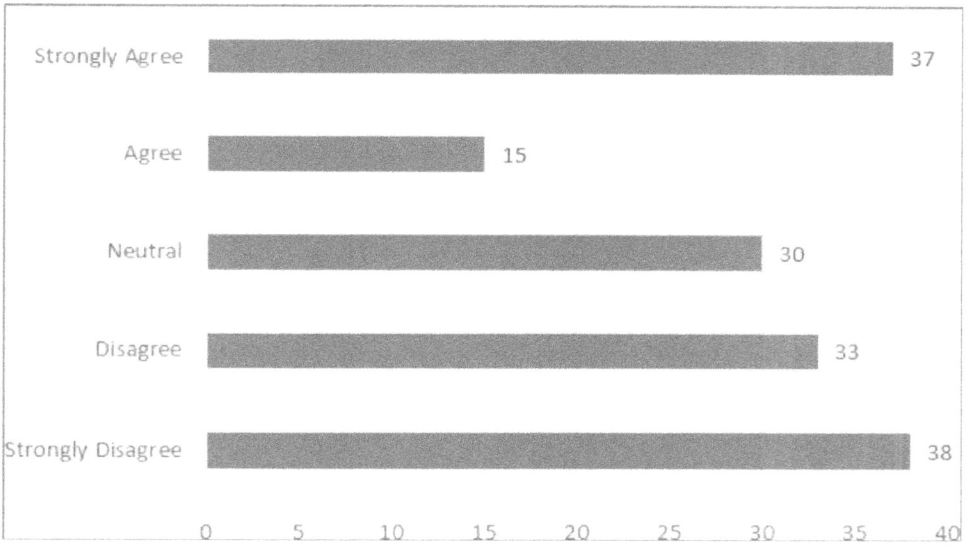

There are 54 (35%) participants who were not sure, 29 (19%) participants who strongly agree, and 23 (15%) participants who agree that they can increase their expertise by moonlighting, while 39 (26%) participants disagree and 8 (5%) participants strongly disagree (Table 9) (Figure 9).

Table 9. You are able to increase your expertise by moonlighting

	Frequency	Percent	Valid Percent	Cumulative Percent
Valid Strongly Disagree	8	5.2	5.2	5.2
Disagree	39	25.5	25.5	30.7
Neutral	54	35.3	35.3	66.0
Agree	23	15.0	15.0	81.0
Strongly Agree	29	19.0	19.0	100.0
Total	153	100.0	100.0	

Figure 9. You are able to increase your expertise by moonlighting

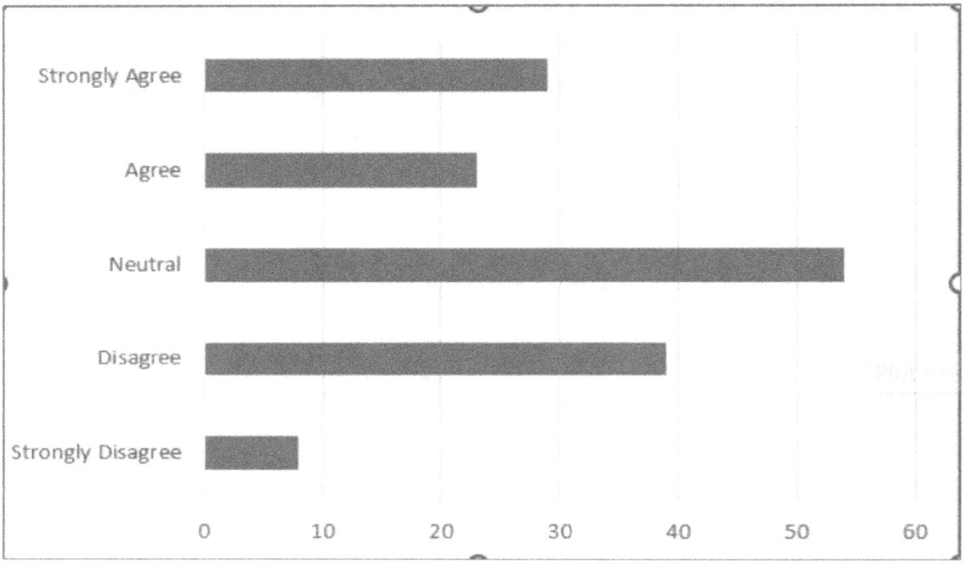

There are 44 (29%) participants who agree, 30 (20%) participants strongly agree, 32 (21%) participants who were neutral, 40 (26%) participants who disagree, and 7 (5%) participants who strongly disagree that moonlighting often make them exhausted and perform poorly in their organization (Table 10) (Figure 10).

Table 10. Moonlighting often makes you exhausted and perform poorly in your organization

	Frequency	Percent	Valid Percent	Cumulative Percent
Valid Strongly Disagree	7	4.6	4.6	4.6
Disagree	40	26.1	26.1	30.7
Neutral	32	20.9	20.9	51.6
Agree	44	28.8	28.8	80.4
Strongly Agree	30	19.6	19.6	100.0
Total	153	100.0	100.0	

Figure 10. Poor performance and exhaustion due to moonlighting

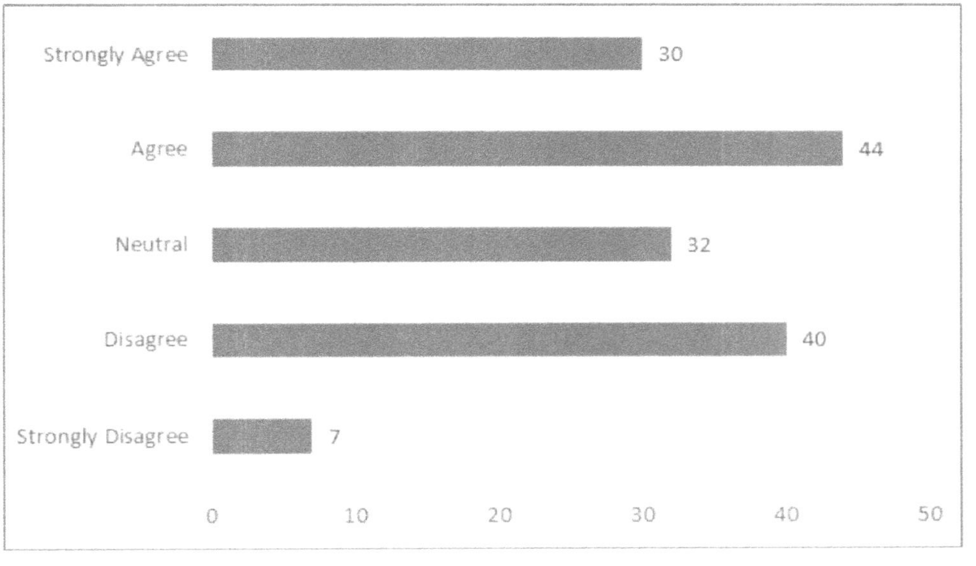

There are 36 (23.5%) participants who agree, 23(15%) participants who strongly agree, 39(26%) participants were neutral, 32 (21%) participants disagree, and 23(15%) participants who strongly disagree that they opted moonlighting due to job insecurity in their organization (Table 11) (Figure 11).

Table 11. You opted moonlighting because of job insecurity in your organization

	Frequency	Percent	Valid Percent	Cumulative Percent
Valid Strongly Disagree	23	15.0	15.0	15.0
Disagree	32	20.9	20.9	35.9
Neutral	39	25.5	25.5	61.4
Agree	36	23.5	23.5	85.0
Strongly Agree	23	15.0	15.0	100.0
Total	153	100.0	100.0	

Figure 11. Moonlighting because of job insecurity

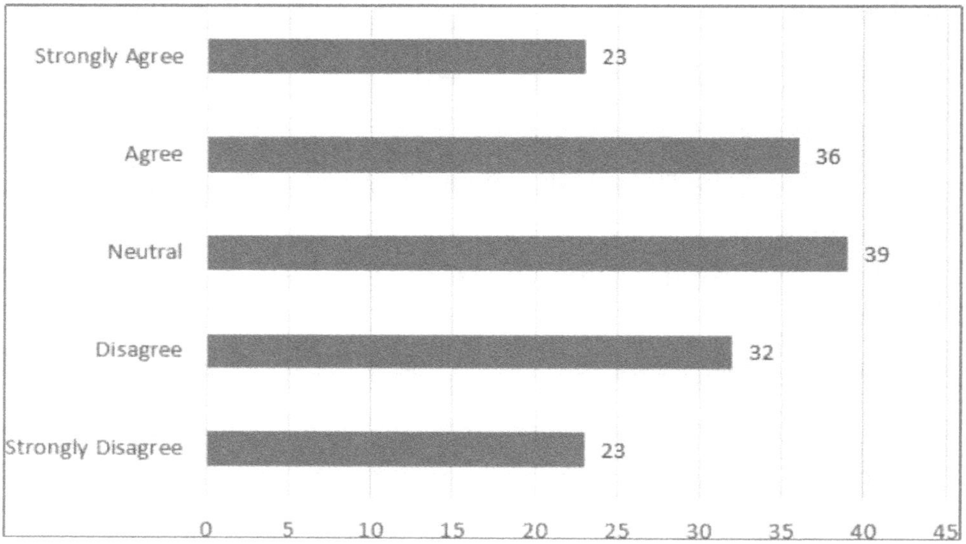

There are 60(39%) participants who strongly agree, 32(21%) participants who agree, 30(20%) participants who were not sure, 16(11%) participants who disagree, and 15(10%) participants who strongly disagree that they find it hard to take time for family, friends, and other aspects of their life because of moonlighting and working from home (Table 12) (Figure 12).

Table 12. You find it hard to take time for your family, friends, and other aspects of your life due to moonlighting and working from home

	Frequency	Percent	Valid Percent	Cumulative Percent
Valid Strongly Disagree	15	9.8	9.8	9.8
Disagree	16	10.5	10.5	20.3
Neutral	30	19.6	19.6	39.9
Agree	32	20.9	20.9	60.8
Strongly Agree	60	39.2	39.2	100.0
Total	153	100.0	100.0	

Figure 12. Problem in taking time for personal life due to working from home and moonlighting

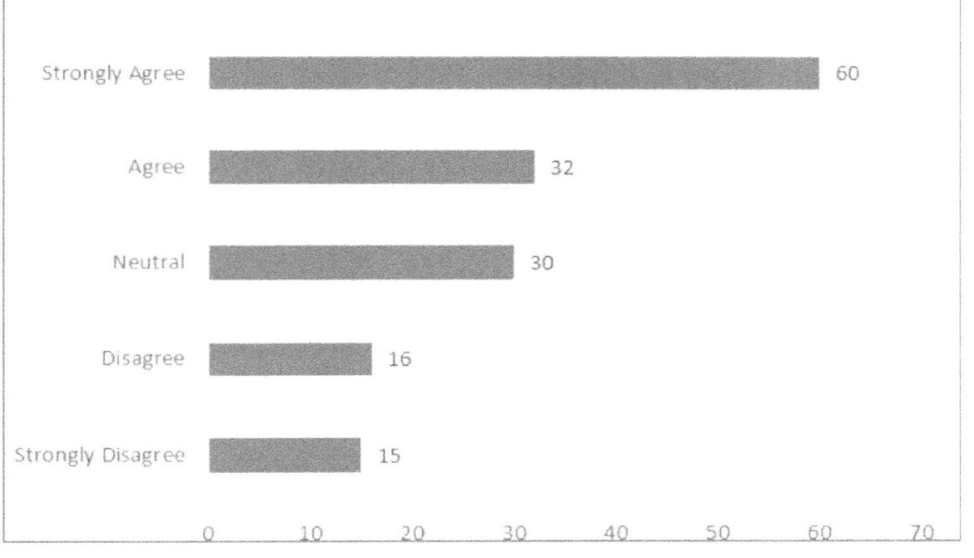

Hypothesis Testing

In order to solve the following hypothesis, one sample T-test was conducted using IBM SPSS software. It is observed that the value of significance (2-tailed) was 0.000 for all the variables, which means H1 is approved, i.e., there is "a significant impact of work from home and moonlighting on wellbeing of employees" ($p<0.005$) (Table 13).

H1 – There is a significant impact of work from home and moonlighting on wellbeing of employees

H0 – There is no significant impact of work from home and moonlighting on wellbeing of employees

Table 13. One-Sample Test

	Test Value = 0					
	t	df	Sig. (2-tailed)	Mean Difference	95% Confidence Interval of the Difference	
					Lower	Upper
Your productivity has been	32.973					
improved by working from		152	.000	3.706	3.48	3.93
home	26.492					
Your organization allows	23.544					
employees to moonlight		152	.000	2.366	2.19	2.54
after working hours	33.715					
You are not able to fulfill						
your financial needs by working for home only for	34.547	152	.000	2.869	2.63	3.11
your company	29.077					
You are able to increase						
your expertise by		152	.000	3.170	2.98	3.36
moonlighting	33.986					
Moonlighting often makes						
you exhausted and perform		152	.000	3.327	3.14	3.52
poorly in your organization						
You opted moonlighting						
because of job insecurity in		152	.000	3.026	2.82	3.23
your organization						
You find it hard to take						
time for your family,						
friends, and other aspects of your life due to		152	.000	3.693	3.48	3.91
moonlighting and working						
from home						

5. RESULTS

Moonlighting has become the matter of debate with rising trend of "work from home" in corporate world and in the job market. Though it is unethical for some, which breaches contractual terms, while it is supported by some employees as it is helpful for underpaid and young professionals who can diversify income and earn extra. It is believed that moonlighting trend was originated with "work from home" culture, considering the existing economic uncertainty in the job market because of arrival of automation, AI, robotics, machine learning, and other latest technologies. A lot of companies took strict measures to prevent moonlighting they considered it dangerous for their employees' productivity and company's profitability.

In this study, majority (44%) participants strongly agree that their productivity has been increased by working from home, and most of the employees disagree that their company allow moonlighting after working hours. In addition, we have observed significant impact of moonlighting and work from home on overall well-being of employees, as most employees find it hard to take time for their friends, family, and other personal aspects of daily life. There are also other negative effects of moonlighting. For example, employees' focus is divided. They face pressure to perform best for both jobs and their personal time is affected to manage their mental health. There is also a risk of data leakage because vital information of the company can be leaked unknowingly by employees working for other companies. They also cannot do justice to their employer because of loss of productivity.

All in all, there are both positive and negative aspects of moonlighting for employees. For example, they can meet their financial needs, diversify their income, and ensure their financial freedom in this age of inflation (Singh, 2022). In addition, employees can have a backup just in case they are laid off by their current employer due to any reason. Hence, organizations need to ensure that their employees don't have to worry about job security and financial obligations, so that they can serve their organization with dedication without losing their productivity and efficiency.

6. CONCLUSION

An employee does overtime and moonlighting (i.e., secretly doing two jobs) like layoffs, job insecurity, less pay, etc. Sometimes, this behavior affects employer's interest. A lot of companies face problems because of high turnover rates caused by moonlighting. Companies promote work culture on the basis of trust and distance. However, employers have to understand that work ethics are subject to change in tough times and work environment is not in everyone's control. It is the right time for companies to improve their work culture by introducing friendly policies to create jobs that employees would like to work for. Satisfied employees don't waste their time. A lot of companies monitor employees and detect data losses. Some companies add a clause in their employment contracts that keeps employees from doing other work, occupation, job, or profession under any circumstances, either through someone else or directly, without seeking approval. Companies can conduct some assessments to detect red flags associated with productivity and engagement of employees.

REFERENCES

Ashwini, A., Mirthula, G., & Preetha, S. (2017). Moonlighting intentions of middle level employees of selected IT companies. *International Journal of Pure and Applied Mathematics*, 114(12), 213–223.

Asravor, R. K. (2021). Moonlighting to survive in a pandemic: Multiple motives and gender differences in Ghana. *International Journal of Development Issues*, 20(2), 243–257. 10.1108/IJDI-08-2020-0180

Böheim, R., & Taylor, M. P. (2005). And in the evening, she's a singer with the band – second job holding in the UK. In ISER Working Papers (No. 2004–3).

Choudhary, V., & Saini, G. (2021). Effect of job satisfaction on moonlighting intentions: Mediating effect of organizational commitment. *European Research on Management and Business Economics*, 27(1), 100137. 10.1016/j.iedeen.2020.100137

Dickey, H. S., Watson, V., & Zangelidis, A. (2015). What triggers multiple job-holding? A stated preference investigation. In Economics (Discussion paper No 15-4).

Gaitonde, S., Lele, U., & Sayyed, S. (2022). Impact of Moonlighting on Industries During And Post Covid-19 Pandemic: Evidence From Pune, India. *Scopus*, 18(104).

George, A. S., & George, A. H. (2022). A Review of Moonlighting in the IT Sector And its Impact. *Partners Universal International Research Journal*, 1(3), 64–73.

Hausken, K., & Ncube, M. (2018). Service delivery weaknesses within education and healthcare: Applying empirics from Kenya, Uganda, Tanzania, and Senegal. *Review of Development Economics*, 22(1), 133–147. 10.1111/rode.12333

Heineck, G., & Schwarze, J. (2004). Fly me to the moon: The determinants of secondary jobholding in Germany and the UK. *Available atSSRN* 609905. 10.2139/ssrn.609905

Jaggarwal, S., (2022). Modeling Of Moonlighting Intention And Its Motives With Moderation. *ANVESAK; UGC Care Group 1 Journal, 51*(2).

Kawakami, A. (2019). Who Holds Multiple Jobs? Empirical Analysis of Multiple Job Holding Using a Japanese Online Survey. *Japan Labor Issues*, 3(12), 34–52.

Kim, J., & Garman, E. T. (2004). Financial stress, pay satisfaction and workplace performance. Compens. Benefit. *Compensation and Benefits Review*, 36(1), 69–76. 10.1177/0886368703261215

Sakyi, E. K., & Agomor, K. S. (2021). Moonlighting in Ghana's higher education institutions: Exploring lecturers' experiences at the Ghana Institute of Management and Public Administration (GIMPA). *Journal of Applied Research in Higher Education*, 13(1), 180–194. 10.1108/JARHE-01-2020-0006

Singh, A. (2022). Moonlighting in India: Another work trend that is sparking a lot of interest. Retrieved 16 June 2023, from https://smefutures.com/moonlighting-in-india-another-work-trend-that-is-sparking-a-lot-of-intere st/

Urwick, J., & Kisa, S. (2014). Science teacher shortage and the moonlighting culture: The pathology of the teacher labour market in Uganda. *International Journal of Educational Development*, 36, 72–80. 10.1016/j.ijedudev.2013.11.004

Williams, C. C., & Kayaoglu, A. (2020). COVID-19 and undeclared work: Impacts and policy responses in Europe. *Service Industries Journal*, 40(13-14), 914–931. 10.1080/02642069.2020.1757073

Yamini, G., & Pushpa, M. (2016). The Factors Affecting Employee Double Jobbing. *Imperial Journal of Interdisciplinary Research, 3*(1).

Chapter 9
Personal Development Plans (PDPs) Supporting Employee Learning and Job Satisfaction:
Lightening Role of Emotional Intelligence

Trupti Suresh Patole
https://orcid.org/0009-0003-7770-4874
Anjuman-I-Islam College of Hotel and Tourism Management Studies and Research, Mumbai, India

Swati Sharma
https://orcid.org/0000-0003-2949-2363
Amity University, Noida, India

Kshitiz Sharma
https://orcid.org/0000-0002-1145-8979
ABBS School of Management, India

ABSTRACT

This research examines the connection between emotional intelligence, workplace PDP, employee learning, job happiness. The study employed review research papers and finds the role of EI in PDP initiative and job satisfaction. It was found that emotional intelligence is a key factor that influences how learning activities affect job satisfaction. These findings have important implications for organizational strategies including better job performance, increased employee retention, and a

DOI: 10.4018/979-8-3693-7668-3.ch009

happier workplace environment, and have the potential to impact organizational practices significantly. To successfully fulfill the demands of a diverse workforce in changing business contexts, organizations should cultivate a culture that supports and fosters employee development and job satisfaction.

INTRODUCTION AND ORIGIN

Organizations must support their employees' continuous development through effective learning in the contemporary business environment characterized by increased globalization and competition. The ability to comprehend and regulate one's own emotions as well as those of others is known as emotional intelligence, and it is a key component in the interaction between learning activities, workers, and job satisfaction.

Origin of PDP

Learn where HRD came from if you want to know where the Personal Development program came from. Performance improvements may be achieved via the process of creating and releasing human knowledge through training and development programs for both organizations and their employees. Human Resource Development includes the following components: 1. Education and growth 2. Organisation Development. (Swanson, 1995). Swanson R.A., the author, has laid out the steps necessary to comprehend the connection between learning expertise and performance.

Figure 1. Relationship between learning, expertise, and performance

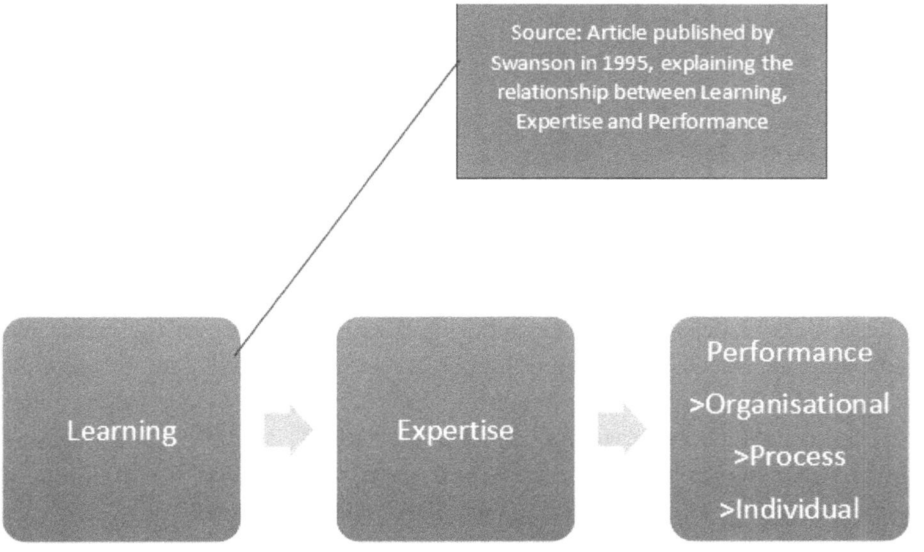

To uncover the performance factors at different levels of performance, author Swanson has devised a performance diagnostic matrix of enabling questions.

Table 1. Performance Diagnosis Matrix of Enabling Questions

	Organizational Level	**Process Level**	**Individual Level**
Mission/ Goal	Does the organizational mission/goal fit the reality of the economic political and cultural forces+	Is it possible to achieve organizational and personal objectives via the process goals?	Is there congruence between people's professional and personal aspirations and the organization's
Systems Design	Does the organizational system provide structure and policies supporting the desired performances?	Are processes designed in such a way as to work as a systems	Do they encounter any challenges that hinder their ability to do a good job?
Capacity	Is the organization well-equipped to carry out its purpose and achieve its objectives?	In terms of quantity, quality, and timeliness, can the process handle the job?	Can they perform to the best of their cognitive, physical, and emotional abilities?
Motivation	Do the policies, culture, and reward systems support the desired performances	Are the data and human elements needed to keep it running available during the process?	Does the individual want to perform no matter what

continued on following page

Table 1. Continued

	Organizational Level	Process Level	Individual Level
Expertise	Does the organization establish and maintain selection and training policies and resources?	Does the method of skill development adapt to the needs of evolving procedures?	Can they do the job because they have the necessary knowledge, training, and experience?

Source: Swanson 1994 p.52

The above figure explains employee development at various organizational, process, and individual levels. while understanding the organization's development we can slowly start to see the history of Personal Development Programs (PDPs). Abraham Maslow developed a theory of motivation with an eye on employee growth in the context of the business. According to Maslow's theory of the hierarchy of needs, there are five distinct levels of human desire: physiological, safety, belonging, esteem, and self-actualization. According to this belief, there are interconnected steps that people must do to attain the pinnacle of life. When the peak achieves self-actualization, he is able to go on to the next stage, unless a level is still unfulfilled. During this phase, people are able to meet their most fundamental needs while also experiencing a profound sense of self-actualization. (Lawson, 1943).

In 1993, Carl Rogers created the PSY (Personal Growth Process) concept. Shows how one's development is an embodied sociocognitive process. Many changes in perspective characterise this process, which a person undergoes throughout time. Starting from a point of psychological safety, these transformations are not a one-and-done deal, but rather an ongoing, ever-changing process. These qualities include knowing oneself, being receptive to new experiences and changes, having the guts to face the unknown, having faith in one's own abilities and those of others, being compassionate towards oneself and others, and having a strong sense of internal locus of control. These changes represent advancements in the pursuit of well-being, which is thus characterised as genuineness, harmony, and contentment with life. Notably, the concept emphasises the connectivity of personal development and the wider context, rather than focussing on individual psychological constructs. Instead, it relates the process to one's social surroundings and shared societal responsibility. People may develop into fully realised persons via the process of self-discovery. Everything from our actions to our feelings to our thoughts reinforces one another. Rather than being separate entities, we are all interdependent parts of one greater whole. This kind of thought had an effect on a large number of psychologists who worked under him. A term used to describe this was "self-actualization." What was once called "personality development" is now more often referred to as "self-actualization" by psychologists.. (Daukantaitė, Hoff, Maurer, & Maurer, 2023).

Figure 2. Carl Rogers's Theory of Personal Development

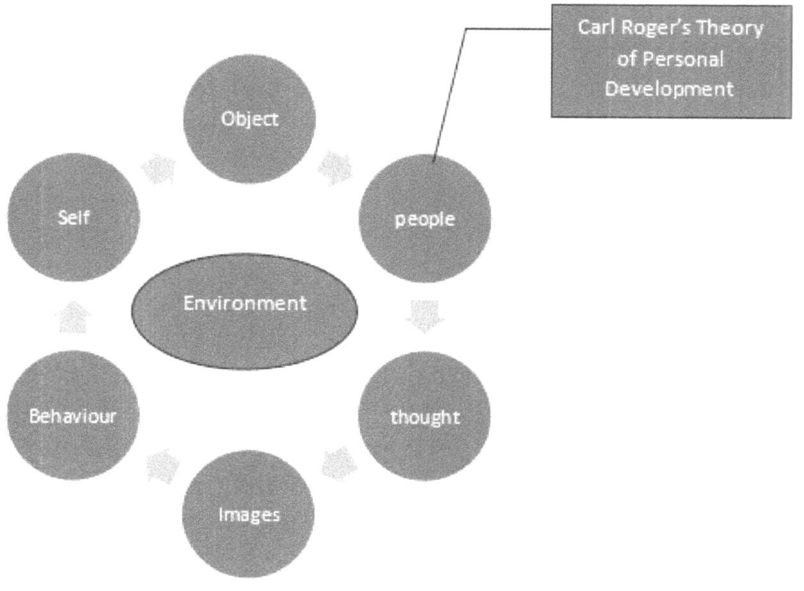

Personality Development Program (PDPs) helps in understanding the drawbacks in the work environment and improves it. The employees who are working in the organization experience events in life that alter social roles and surroundings and have an impact on personality development. Here, we summarize the distinctions between individual and societal life experiences that are important for the formation of personality. We examine the differences in thematic, social, geographical, and temporal elements of personal and collective life experiences, predicated on certain basic assumptions on the changeability of character characteristics as a result of life circumstances. In the lives of individual people, personal life events frequently pertain to the areas of health, employment, family, and love. By comparison, communal life events relate to conflicts over power and tragedies that impact a large number of individuals. Events in collective life are unique because they have the potential to (a) produce varying personal life events for various individuals and (b) have an indirect impact on a large number of people who identify with a group that is directly impacted. (Wundrack & Asselmann, 2021).

The definition of Personal Development is described as "the methodical upkeep, enhancement, and expansion of knowledge and abilities as well as the development of character traits required for the performance of professional and technical tasks throughout the practitioner's working life." (Evans, 2002).

Understanding Personal Development Plans (PDP) at the Workplace

Personal Development Plans (PDPs) are a fundamental tool for personal growth, often involving training but also encompassing much more. They align well with the current focus on employee involvement and empowerment, as well as initiatives like Total Quality Management that stress employee accountability, shared values, and continuous improvement (Beausaert, 2011). PDPs enhance and support this general strategy. PDP gives the individual complete control over their development, from creating the plan to implementing and updating it. While HR professionals and line managers have a role in this process, their primary function is to work with the individual to create and implement the growth plan or assist them in achieving it. (Tamkin, 1995). In view of recent shifts away from yearly evaluations, this research looks at the potential effects of instituting a personal development plan (PDP) that emphasises continuous feedback and improvement on both individual and organisational performance. Results The diagnostic stage of the research is often successful, according to the results; more has to be done to promote the PDP and development activities after diagnosis. This reduces the amount of time spent on diagnosis and may lower employee motivation and self-efficacy.

A PDP can be defined as a tool to present information about competencies the employee has been working on and is planning to further develop. In 1995, Stephen J. Brown (McMullan et al, 2003). As part of these evaluation cycles, several companies use a Personal Development Plan (PDP). The PDP, with its roots in portfolios used by brokers, painters, photographers, and architects, has a rich historical evolution. (Mathers, 2002). It was later adopted by students in secondary schools and higher institutions to aid in certification and enhance their studies (Lyons & Evans, 1997). A section of the document, the PDP, was dedicated to the student's professional development (Driessen, 2008). PDPs, available in various formats and known by several names such as logbooks, portfolios, (continued professional) development plans, and personal professional profiles, are most commonly associated with portfolio appraisal and professional or personal growth plans.

Furthermore, a systematic training program would be required to create a culture shift inside the firm for it to adopt a constant focus on development via PDPs. There was an unconscious predisposition to think that better individual and organizational performance would result from a PDP with written goals. While the appeal of Personal Development Plans (PDPs) is undeniable, there is a need for more extensive research on them. Like several similar endeavors, there is considerable ambiguity over their precise nature, introduction strategy, and necessary support. It can be challenging to discern how they differ from development action plans, which are sometimes essential to appraisal schemes. (Monishaa, 2023). However, PDPs are gaining traction and becoming increasingly popular. Some organizations, despite their appeal, are facing challenges in implementing them. (Kneale, 2015).

In this research, the term' development plan' is used instead of 'portfolio assessment' to avoid confusion with other meanings in the organizational literature (Fenwick, 2006). For instance, "portfolio (evaluation)" may be either a reporting system used to hold corporations accountable or portfolio management, which is concerned with developing and implementing plans for buying. Portfolio work is another name for this kind of flexible self-employment in which individuals construct a portfolio of their professional activities while contracting their expertise and talents to other persons and organisations (Schmitz, 2005).

The Use of PDPs in Organisation and Its Role

Organizations must prioritize strategic human resource development in today's rapidly evolving knowledge economy to acquire and preserve a competitive advantage. The outcome is an increase in the adoption of tools such as Personal Development Plans (PDPs). These plans play a crucial role in fostering employee growth, from developmental to evaluative. Even while the program's strength is in promoting workers' learning and professional development, it is debatable whether using a PDP for assessment reasons does not diminish its developmental potential. We focus on how the PDP's perceived mission impacts its functioning and personnel performance in this research. Furthermore, the employee's performance was influenced by their perception of their own developmental objectives, which were moderated by the PDP practice components of feedback quality and effort quality. In the ongoing quest for instruments that promote and evaluate learning and development, companies use personal development plans (PDPs) more and more frequently. A personal development plan (PDP) is a useful tool for employees to reflect on their job, keep track of the skills they've been honing, and map out their plans for professional advancement. Focussing on the PDP practice, this research included 287 workers from a Dutch government organisation. Inspiring supervision, information supply, feedback's critical function, and learning and reflection were the three pillars upon which it rested. The study looked at which traits improve performance, expertise growth, adaptability to changing conditions, and learning activities undertaken. According to the hierarchical regression analysis results, an employee's ability to contribute to the four output variables that are measured depends on several factors, including a motivated supervisor, information, feedback, and self-reflection based on their PDP. After that, mediation studies provide evidence that learning activities mediate the link between the practice and two outcome variables. (Beausaert, 2011).

Consequences for practice in the event of a supportive organizational culture, implementing PDPs as a substitute for annual appraisals may benefit a company. Uniqueness and Worth According to a recent article (Houghton, 2017) two-thirds of large organizations intend to reconsider their yearly appraisal approach. The need for more frequent feedback was one of the leading forces behind this. Considering the recent change in perspective, further investigation is necessary to determine a suitable replacement for the annual appraisal. (Greenan,2016).

Throughout evaluation cycles, PDPs detail the skills and knowledge that employees have been concentrating on and want to further hone. In a recent meta-analysis evaluating PDP effectiveness, 54 separate empirical research were included. The results of the analysis demonstrated that several ancillary elements influence the reaction of workers to a PDP. The employee's perception of the PDP as a tool for learning and reflection is an important personal factor to consider. The importance of self-evaluation and ongoing learning in the PDP process is highlighted by this viewpoint. Second, the PDP serves as a springboard for conversations between the employee and their coach or supervisor, who then provides feedback, promotes self-reflection, and ignites a need for knowledge. For a PDP to work, it's important for the supervisor to be both a motivator and a source of feedback. Van van Heijden developed and validated the five-factor model of professional competence based on previous work in the field: information, metacognitive knowledge, skill demand, social recognition, and development and adaptation. This second part of growth and flexibility is where most personal development plans focus their efforts. According to Van van Heijden, experts are those who can become very good at a number of various things, whether those things are closely linked or completely separate. They can also learn a lot about how to become experts in other areas. (Heijden & Beatrice,2018). These people are experts in their fields and can also adapt to new situations with ease. Able to show that they are adaptable to technological developments and that businesses can act promptly and strategically when possibilities arise in adjacent fields. It is believed that the employee consciously engages in learning activities, supported and guided by the PDP, and as a result, demonstrates development and adaptability, ultimately leading to improved performance. Theoretically, there is a significant relationship between distinct outcome variables; nevertheless, the theoretical cascade effect is compromised by working conditions. Research on 360-degree feedback emphasises the importance of individuals' reactions to comments on their emotional and cognitive performance. These responses have a substantial impact on managers' final responses. (Atwater, 2005).

In the PDP, four major questions are put forward: ‘What have I done so far?’, ‘What do I want to achieve?’, ‘How am I tackling that?’ and ‘What is the next step to take?’. When an employee's competencies diverge from those they are expected to possess, it may indicate that learning needs to be encouraged. This underscores the pivotal role of the PDP in empowering professionals to grow and perform better. Many people see a sharp contrast between certification, responsibility, and professional development. The use of a PDP to foster growth places a strong emphasis on learning and receiving feedback. However, it's important to be aware that self-presentation becomes more significant when a PDP is employed for certification, making the audience feel prepared for this aspect. (Bunker, 2004). “Using PDPs [portfolios] effectively means putting the goals for learning and reflection first,” according to Tillema and Kari Smith. A learning environment where the employee takes charge of their education is one way the manager can inspire motivation in their staff member. The manager's role is not just to outline the tool's purpose and explain its use, but also to trust and respect the worker's autonomy. This involves frequent meetings, continuous support, and providing excellent feedback. These conclusions are supported by the research on portfolios. According to Tillema's findings, reflective portfolios that centre on professional growth and learning objectives greatly enhance performance and foster learning. In addition, a favourable link was shown between teachers' utilisation of PDP and reflection (Tigelaar et al., 2006). However, the study also uncovered a gap in teachers' deep reflection on their motivation, emotions, thoughts, and personalities (Smith, 2001).

Implication for Practice

Through Personal Development Programs (PDP), people can step up into new responsibilities that can increase their level of satisfaction while also significantly increasing the organization's ability to meet its strategic objectives. In addition to producing high marginal productivity, PDP activities also enhance organizational efficacy and efficiency, which raises organizational performance. Employee knowledge, skills, and capacities are enhanced through these programs, which can result in a significant rise in labor productivity and higher earnings. This essay's primary goal is to address the difficulty of developing a methodological framework for an empirical investigation of PDP's influence, which is a critical first step in comprehending the effects of PDP on people and organizations (Cristina, 2020). Elena Cristina Cernătescu employed tools from both inductive and deductive reasoning, synthesis, and idea generalisation in one of her research projects.. (Elena Cristina, 2020) The theoretical framework is derived on the Learning Transfer System Inventory (LTSI) measuring model proposed by Reid (2012) and developed by Holton (1996). Personal development plans (PDPs) are positively correlated with competences, skills, professional performance, and job satisfaction, and negatively correlated with employee attrition.

A. Performance at Work. Research from empirical studies often shows a beneficial association between PDP and performance at work. Employee development not only raises the knowledge and skill level of the workforce but also instills a sense of accomplishment, producing a synergistic effect that enables them to carry out particular activities more effectively.
B. Job Satisfaction. PDP leads to the acquisition of information and skills, which not only boost productivity and reduce working hours but also significantly enhance job satisfaction, morale, and self-esteem.
C. Staff Retention. When it comes to keeping employees from leaving, PDP programs are vital. These initiatives lead to a more dedicated and consistent workforce by encouraging employees to want to make a difference in the company's bigger picture.

Figure 3. The conceptual framework of the research

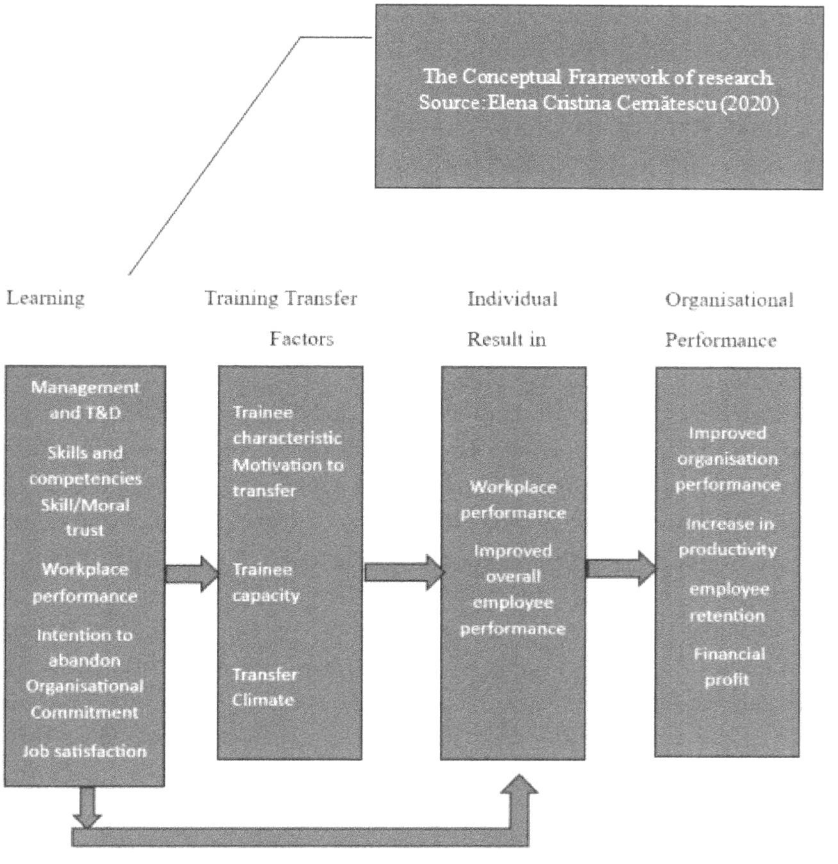

Source: adaption from research done by Cernătescu (2020), and Holton and Bates (2008).

The tool's potential to aid in the pharmacy assistant's professional development may be greatly enhanced if it were also used as a feed-forward device. In other words, more people should use the program to plot out their career paths, get a bird's-eye perspective of their ideal goals, and then do the kinds of learning that would get them there. With these PDPs in mind, pharmacy assistants may make progress towards their professional objectives. Value and uniqueness: The use of personal development plans (PDPs) to facilitate employee learning and development is on the rise across various companies. One of the reasons why companies succeed is because their workers learn to adapt to a constantly changing internal and external environment. This is because competitive advantage is inseparable from personnel. Companies are using PDPs as a means to motivate workers to take part in training and education programs. One of the most important factors in a company's success is its employees' ability to learn and adapt to the ever-changing business world. Employees are a company's most valuable asset, so it seems sense that PDPs would be a good way to keep them engaged in learning and development programs that would help them become more flexible and productive. This study is poised to significantly influence both research and practice. While existing research supports the widely accepted notion that managers need training in performance management, this study underscores the need for specific training on how to use a PDP for formative staff development and training transfer. It has also revealed a gap between the diagnostic and support phases of PDP use, with the former seemingly complete and the latter not. This need more research to completely comprehend its consequences for people and businesses, along with the possible effects of thorough PDP adoption (Greenan, 2016). Regular, rather than biannual, recognition of continuous professional development (CPD), which is "a method by which individual establish ownership of their learning and development by participating in an ongoing process of thought and action," should be the norm in the company. Instead of being used mandated today to evaluate performance, the PDP, in this case, would subsequently be utilized formatively voluntarily for self-directed learning and reflection (Megginson, 2003).

Using a PDP as a tangible document and anticipated that it would be modified in response to changes in progress. Its findings may have been limited by its emphasis on documented "proof" that goals have been met. Future research could examine the advantages and disadvantages of verbal and written goals concerning the subconscious priming effect. Research and practice are both impacted by this study. Research backs up the widely accepted theory that managers require instruction in the art of performance management in general. However, it emphasizes training specifically on how to use a PDP to facilitate formative staff development and training transfer. It has brought to light a division that exists between the diagnostic and support phases of using a PDP; while the former seems to be completed, the latter is not. More investigation is needed to fully grasp the implications of this for the individual and the organization, as well as the impacts of an organization implementing a PDP in its totality.

Figure 4. Conceptual framework for assessing the relation between TPD and organizational performance

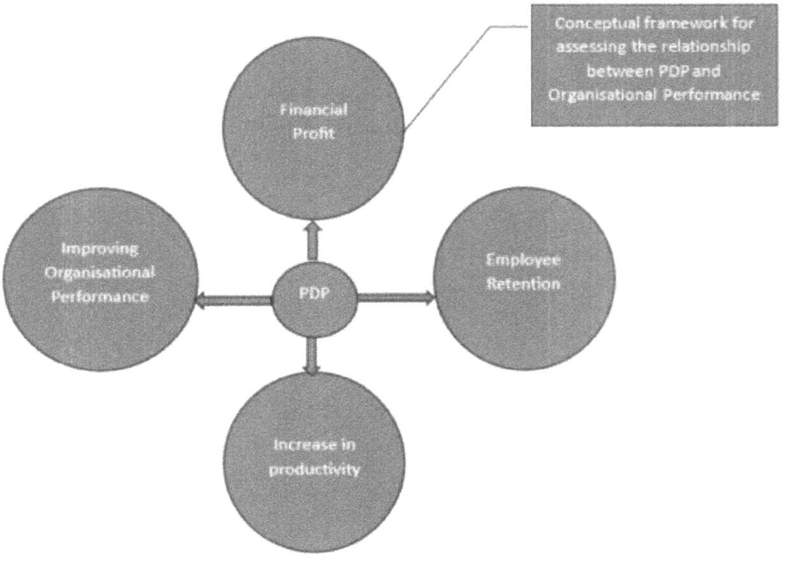

The above figure 4 explains how PDP is beneficial for organizations regarding performance, job satisfaction, and staff retention.

A. **Performance at Work:** Research from empirical studies often shows a beneficial association between PDP and performance at work. Employee development not only raises the knowledge and skill level of the workforce but also in stills a sense of accomplishment, producing a synergistic effect that enables them to carry out particular activities more effectively.
B. **Job Satisfaction**: PDP leads to the acquisition of information and skills, which not only boost productivity and reduce working hours but also significantly enhance job satisfaction, morale, and self-esteem.
C. **Staff Retention**: PDP programs play a crucial role in reducing staff turnover rates. These initiatives aid in the development of a loyal and consistent workforce by encouraging participants to feel that their job matters in the grand scheme of things.

Role of Emotional Intelligence in PDPs

The ability to recognise, label, and appropriately respond to one's own and other people's emotional states as well as to interpret and act on nonverbal clues is known as emotional intelligence. John D Mayer and Peter Salovey have proposed three EI models, with the 'ability model' highlighting the importance of emotional information in positive connection and empathy in social interactions.

Figure 5. A model of Emotional Intelligence and organizational effectiveness. Figure adapted from Cherniss (Cherniss, 2010).

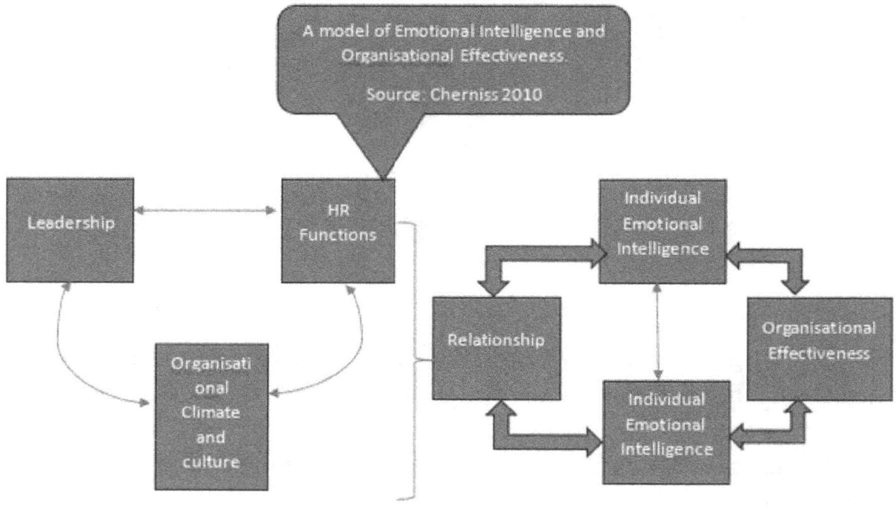

The accompanying picture illustrates how Emotional Intelligence (EQ) contributes to both personal growth and professional fulfilment in the workplace.

A. Perception of others' emotions: others who have good social perceptions may find out a lot about the beliefs, goals, and interests of the others in their group. They should be able to use this information to persuade others by listening to and responding to members' unstated needs and by establishing reasonable, attainable objectives.

B. Emotional intelligence (EI) significantly contributes to the success of a team by empathising with its members. One of the most important factors influencing group success is the ability to control one's emotions in accordance with what the group needs.
C. Control your emotions: Those who are adept at controlling their emotions wield a significant influence on the way other group members feel about certain behaviors; they do this by using lessons learned from the past.

Since the 1995 publication of the best-selling book on emotional intelligence (EI) by Goleman, there has been much discussion on its influence on human performance. Unfortunately, a lot of it is based more on conjecture than on solid scientific evidence. This chapter aims to provide empirical evidence that EI does affect performance. We aim to present the facts, and this chapter presents the empirical data we have on how emotional intelligence affects performance (Ria Pita, 2007). These results may help explain why some people excel, take up leadership roles, and even volunteer for highly demanding and potentially dangerous duties, while others struggle in these areas. The contrast between these two groups underscores the practical implications of our research. The ability of a person to recognize and express their feelings is known as self-emotional assessment. Because service occupations involve a lot of customer engagement, self-evaluation is always necessary in this sector. Before comprehending the emotions of others, it is preferable to understand one's feelings. (Bar-On, 2006) Other's emotional appraisal has to do with a person's capacity to recognize and comprehend the emotions of those around them. As it enables people to decipher both internal and external clues, contend that the capacity to recognize and comprehend emotions directly affects social interactions. The capacity of an individual to control their emotions is known as the Regulation of Emotion. (Hennig-Thurau, 2006) Emotional control affects the quality of social relationships, which can have an impact on task performance in customer-facing roles in service industries. (Lopes, 2005) Ultimately, the use of emotions describes people's capacity to make use of their feelings by coordinating them with constructive endeavors. A key element of providing excellent customer service is setting expectations and evoking good feelings. To foster a great customer experience that increases customer satisfaction and future loyalty, service personnel must elicit emotions from their clients properly. Researchers have discovered that service personnel must control their emotions to control the emotions of their clients. This finding applies to both the hospitality and education sectors. (Côté, 2006).

It can be intimidating, for example, to decide to quit a safe career for an uncertain entrepreneurial endeavor. The processing of the exhilaration of possible achievement and the dread of failure can be aided by emotional intelligence, leading to the making of a decision that is both practically sound and emotionally rewarding, bringing a sense of accomplishment.

It is possible to identify blind spots in emotional awareness and regulation by asking for constructive criticism on decision-making processes. This input may come from peers, mentors, or simply introspection. The power to modify decision-making tactics by one's understanding of personal inclinations, such as a tendency toward hasty decisions or an excessive degree of caution stemming from fear, is empowering. Team members who score high on the emotional intelligence scale are good at controlling and monitoring their own and others' emotions, according to the idea that links EQ to team effectiveness. This heightened emotional sensitivity and regulation significantly influences team performance, fostering the development of healthy and productive working relationships. It is recommended that individuals with high emotional intelligence engage in open communication, as this can lead to significant benefits within the team, enhancing group success outcomes. Effective communication among emotionally intelligent individuals can amplify the benefits of EI, making it a crucial component for successful outcomes in the increasingly prevalent virtual work environment. (Murmu, 2022).

Relationship Between EI and Leadership Through PDP

There were five main ideas that linked PDPs with organisational leadership:

1. The development and application of skills
2. Self-awareness
3. Self-efficacy
4. Self-exploration
5. Continuous Improvement (Coopersmith, 2022)

Modern culture places a premium on the ability to continuously improve oneself, which is more than simply a personal journey. Being able to adjust to unexpected events and comprehend one's own and other people's needs are both improved by this. Being mindful entails being cognisant of one's own mental processes, feelings, principles, and deeds. Executives who excel in self-awareness and emotional intelligence are highly sought after by businesses and society alike. One of the most important results of maturing as a person is vertical growth, which is the capacity to take stock of one's viewpoint and incorporate it with other people's. *Self-efficacy*, the capacity to effectively utilize one's resources and skills to accomplish tasks, plays a pivotal role in personal growth. Individuals with high self-efficacy levels can maximize their abilities and achieve their objectives. As one of the five pillars of this study's theoretical framework, self-efficacy is a key component that facilitates personal progress by helping individuals bridge the gap between their knowledge competence and their actual skill application.. (Lyons, 2019). The principles of self-efficacy also guide the development of vertical leadership, as an individual's ability to recognize and capitalize on their growth is a key indicator of vertical capacity. The process of formulating an answer to the question, "What do you want to become?" is the journey of *self-exploration*. This journey, often associated with personal development programs, leads to growth on an intellectual, moral, and artistic level. It's a quest for acceptance, awareness, and understanding. Discovering fresh viewpoints, relational settings, and objectives is a significant step in the vertical leadership development process. (Jones, 2020). It's a journey that encourages individuals to take the time for introspection, to reflect and make sense of their lives, and to understand who they are and what they want out of life. The definition of *personal development* is "a continuous lifelong process of nurturing, shaping, and improving skills and knowledge to ensure maximum effectiveness and ongoing employability." It is regarded as a continuous activity as opposed to a one-time occurrence. Plans for personal development work best when they emphasize the idea that people can keep improving and growing throughout their lives. Vertical leadership, where development is viewed as a lifetime activity, reflects this process of continuous progress. (Akram, Haider, & Akram, 2016).

While Implementing the PDPs in the Organization Following Activities Are Conducted

1. Workplace training. As a formal on-the-job training program, participants are fortunate to be paired with peers who are both influential and experienced. Your time and energy will be well-spent thanks to these "mini-training programs" that are arranged around certain learning goals. This might be in the form of an impact seminar or workshop for PhD candidates, further enhancing the exclusivity of this opportunity.
2. Evaluating research for 'prescriptive readiness.' Your role in this process is integral. Not all research will be immediately useful or affect societal problems and their commercial implications. For instance, exploratory research is less likely to produce unambiguous conclusions and recommendations. To assess prescriptive readiness, researchers need to learn about the many stages of prescriptive evidence, such as exploratory, preliminary, option, guideline, and standard. Your engagement in this process is crucial.
3. Associating with the other employees. This gives you the chance to meet others, share skills, and take advantage of great learning opportunities. It's an exciting prospect that will broaden your horizons and enhance your learning experience. (Aguinis, 2021).

Impact of PDPs on Employees and Organisation

Personal development plans (PDPs) are commonplace in many companies, but little is known about their effectiveness or how they influence workers' capacity for professional advancement. More research and data-driven insights are needed in Human Resource Development, since this is evident. While personal development plans (PDPs) are intended to foster workers' knowledge and skill acquisition, practical experience reveals that this is not always the case. Workers may view the tool as ineffectual, as a component of a yearly ritual dance, or as a cover-up to hide the fact that organizations do not care about personal growth. Even though personal development plans (PDPs) are widely used in the workplace, there needs to be more understanding about how the tool is used or how it affects employees' learning and growth. This underscores the need for more research and data-driven insights in Human Resource Development. There is some evidence that PDPs do not always help people learn and advance in their careers, despite the fact that this is the principle behind them. Workers may view the tool as ineffectual, as a component of a yearly ritual dance, or as a cover-up to hide the fact that organizations do not care about personal growth. This potential misuse of PDPs highlights the need for better implementation and oversight (Ketels, 2012).

Our research highlights the impact of PDP on the perceived efficacy and efficiency of coaching. The PDP, as we discovered, not only broadens the coaching repertoire but also significantly enhances the perceived efficacy and efficiency of coaching. It helps in organizing the coaching activities and provides an additional avenue for feedback. The coach gains more access to the PDP-maker's background, education, and points of view, which, in turn, enriches their coaching exercises. (Firssova, 2006). Our research highlights the impact of PDP on the perceived efficacy and efficiency of coaching. The PDP, as we discovered, not only broadens the coaching repertoire but also significantly enhances the perceived efficacy and efficiency of coaching. It helps in organizing the coaching activities and provides an additional avenue for feedback. The coach gains more access to the PDP-maker's background, education, and points of view, which, in turn, enriches their coaching exercises (Johnston, 2005).

According to the findings of this study conducted at educational institutions, principals only viewed portfolio work (PDP) as professional development when it was placed inside a more comprehensive, supportive professional practice social network. Due to its use in facilitating professional growth, the portfolio has the potential to serve as both a learning tool and a tool for policy compliance. Value and individuality This research adds to what is known about principal professional development by showing that school leadership may be fostered via the use of portfolios for incoming principals. For this to be realised, it specifies the types of social and professional support that are necessary. In order to guide organisations in improving the effectiveness of PDP assessment, greater research into the individual and contextual factors that enable PDP assessment in real-world organisational settings is required. Despite the PDP's marketing claims to the contrary, it often has the opposite effect on learning. Inspiring self-reflection and continuous professional development, questions about the pros and cons of using a PDP inevitably arise. Future research on the contextual supporting process conditions discussed in this literature review might address these specific issues that this brings to light. How can we best support the PDP evaluation process? How should a supervisor conduct themselves? In order to discuss the PDP, how often should formal meetings like monthly team meetings or quarterly performance evaluations be held? What is the required format for the supervisor's feedback on the PDP? How can the supervisor persuade the employee to use a personal development plan? Relevant research issues for the individual supporting process conditions are as follows: Is the PDP assessment appropriate for all categories of employees? Does this apply to employees who work for profit or nonprofit organizations, whether they are knowledge workers or not? What abilities does an employee require to use a PDP effectively? Can we train staff members to use a PDP? (Lejeune, 2006).

The findings underscore the crucial need for organizations to cultivate a culture that supports and actively encourages employee learning and job satisfaction as part of their PDP initiatives. This is not just a recommendation but a necessity in today's business environment.

LIMITATIONS AND SUGGESTIONS FOR FUTURE RESEARCH

The degree of the study's external validity must be taken into consideration while interpreting the results. However, the factor patterns may change depending on who fills out the questionnaires—experts versus novices, for example—what organizational setting—profit or nonprofit—different disciplines or branches—different organization sizes—small versus large significant—different competency focuses, different learning structures, and different goals. The items' interpretations can vary depending on the context due to their various applications (Maurer, 2002).

For example, whether the parts are still applicable to PDP practice depends on whether the tool is used for learning and continuous professional development or not. Subsequent studies are crucial to further examine the tool's factor structure in many contexts and concentrate on evaluating the efficacy of those approaches in carrying out educational tasks, developing knowledge, and enhancing performance. The two ideas that follow could account for the variation in response rates. Organizations encourage their staff members to put in much effort in their careers to grow, which in turn increases their drive to use a PDP to study and accomplish the same goal (Boud, 1994).

It is equally crucial to reflect on one's personal qualities as it is on one's professional knowledge. Utilizing opportunities and resources more effectively may result from teachers and students working together and reflecting on shared experiences. By combining information and skills, such initiatives should improve the effectiveness of the PDP process. Reflection can be aided by evidence from audits, patient and colleague feedback, portfolios, and other quality improvement data, as this is what motivates performance changes and is essential to a successful PDP. To fully employ a PDP, it's crucial for supervisors to continuously learn and grow, gaining more experience in applying these skills. (Patel, 2013). Poor PDP preparation and progress could be caused by inexperience with the PDP process, ignorance (e.g., curricular gaps), a lack of resources and opportunities at work and in the larger working or learning environment, infrequent progress monitoring, and a lack of reflection. There is significant variation in PDP policies and practices both within and between institutions. This variation may be attributable to different interpretations of the PDP idea itself.

CONCLUSION

An organization's overall performance benefits from a work environment that encourages personal development planning (PDP) since it enhances employees' skills, knowledge, and abilities. The foundation of increasing workplace productivity is personal development programs (PDPs), which have the ability to greatly improve workers' abilities, expertise, and knowledge. By presenting the crucial training and personal development methods that were created in response to the literature's research, we hope to in still hope about the potential of PDP programs. The relationship between PDP programs and organizational success is a relatively contentious subject of debate. However, instead of seeing PDP initiatives as an investment that would pay off, most firms instead see them as a cost. An organization's performance may not increase proportionally despite a vital investment in PDP initiatives if the management is insufficient.

Lots of studies all around the world have stressed the significance of emotional intelligence (EI) in getting organisations to function at their best. However, there is still a pressing need to resolve the assessment and predictability of EI leading to success. Aspects of emotional intelligence (EI) are congruent with the model for achieving organisational goals and, in the end, resulting in work happiness, according to the study. The ability to be aware of, make sense of, and manage one's own emotional responses in different situations is called emotional intelligence (EI). Handling one's emotions requires being able to identify the specific situations or factors that cause us to become distracted or ineffective. Being able to control your emotions and moods and having faith in your abilities is crucial for making the kind of decisions that will propel your career forward. Critical psychosocial skills for a successful adjustment to work life, including career management, include emotional intelligence and career development. Compared to others who lack established emotional skills, employees with these abilities can act more appropriately and make wiser decisions. The current study aims to investigate the connection between public employees' career progression and emotional intelligence. The study's findings, which are novel and insightful, demonstrate how crucial it is for workers to increase their emotional intelligence to manage their professional growth effectively. The comprehension of the relationship between career advancement and emotional intelligence may be helpful to upper management in crafting public organizations' HR policies that guarantee the performance of public sector workers meets high standards (Poliak, 2020). These days, employers are placing a greater emphasis on employability and talent management as they understand that employees' skills and abilities are what ultimately fuel their success. Personal development plans (PDPs) are becoming more popular among businesses as a means of evaluating workers' progress towards skill and competency goals. These records detail the employee's progress on and future goals for strengthening several abilities. The purpose of this research was to examine previous work that addressed PDP objectives (research question 1), PDP effectiveness (research question 2), and potential enabling factors (research question 3). We accomplished this by reviewing the relevant literature in light of the three research topics. We are unaware of any other study that has systematically reviewed the empirical literature on PDP assessment in the workplace before this one.

Moreover, the majority of papers needed a thorough explanation of the data analysis or research methodology. Regarding the initial research question: Which objectives are served by PDP assessment?

We said that professional growth and certification/selection/accountability are generally distinguished from one another. Nine goal clusters were established based on the literature reviewed in this study. The following are some of these clusters: (1) external mobility; (2) coaching; (3) evidence-based learning; (4) documenting; (5) certification and selection progression; (6) coaching; (8) confidence-boosting; and (9) organising (Lawson, 2008)

Furthermore, several authors classify the previously listed PDP objectives based on the assessment's objective. The PDP assessment has distinct formative and summative purposes. Some authors argue that data gathered during the learning or development process can be useful for summative evaluation, while others maintain that the audience for formative and summative assessment differs, with the former being colleagues (learning) and the latter being the organisation (assessment). The results of the empirical investigations do not provide a conclusive response to the second study question, which concerns the efficacy of PDP evaluation. According to specific research, PDP assessments are helpful for professional practice improvement, reflection, and personal or ongoing professional development, emphasizing the ongoing nature of professional development and engaging the audience in a commitment to their growth (Leggett, 2004).

We do a thorough literature review on the benefits of training and development for individuals, groups, companies, and society as a whole. Our multidisciplinary, multilevel, and global perspective demonstrates how workplace development and training initiatives may be very beneficial to all of these parties. We provide several research articles as examples to support these claims. To identify the conditions that maximise these benefits, we thoroughly examine the literature on requirements assessment and pretraining states, training design and delivery, training evaluation, and training transfer (Aguinis, 2009).

In conclusion, the results of our research provide further insight into the use of PDP evaluation in the workplace. However, further investigation is needed to fully analyse the effects of PDPs and the supportive process conditions.

REFERENCES

Aguinis, H., & Kraiger, K. (2009). Benefits of training and development for individuals and teams, organizations, and society. *Annual Review of Psychology*, 60(1), 451–474. 10.1146/annurev.psych.60.110707.16350518976113

Akram, T., Haider, M. J., & Akram, M. W. (2016). Emotional intelligence and personal development in employees: A case study of Zigron corporation, Pakistan. *Humaniora*, 7(3), 293. Advance online publication. 10.21512/humaniora.v7i3.3582

Alison Bunker, M. L. (2004, Jan). Being wise about teaching portfolios: Exploring the barriers to their development and maintenance. *Research Gate*. doi:https://api.semanticscholar.org/CorpusID:167878329

Bar-On, R. R. (2006). The Impact of Emotional Intelligence on Performance. *Linking emotional intelligence and performance at work: Current research evidence with individuals and groups*, 3-19.

Beatrice, I. J. M., & Van der Heijden, G. N. (2018). 6). Development and validation of the short-form employability five-factor instrument. *Journal of Vocational Behavior*, 106, 236–248. 10.1016/j.jvb.2018.02.003

Brown, S. J., & Goetzmann, W. N. (1995, June). Performance Persistence. *The Journal of Finance*, 50(2), 679–698. 10.1111/j.1540-6261.1995.tb04800.x

Cazacu, E. C.-N. (2020). Research on the Impact of Training and Personal Development on Individual and Organizational Performance. *"Ovidius" University Annals. Economic Sciences Series*, 20(2), 619–623.

Cazacu Elena Cristina, C. A.-N. (2020, February). Research on the Impact of Training and Personal Development on Individual and Organizational Performance. *Ovidius University Annals. Economic Sciences Series*, 20(2), 619–623.

Cherniss, C. (2010, June). Emotional Intelligence: Toward Clarification of a Concept. *Industrial and Organizational Psychology: Perspectives on Science and Practice*, 3(2), 110–126. 10.1111/j.1754-9434.2010.01231.x

Connie, C., & Schmitz, B. A. (2005, December). Report Carding: A Model for Foundation Portfolio Assessment. *The American Journal of Evaluation*, 26(4), 518–531. 10.1177/1098214005281326

Coopersmith, K. (2022). Prsonal Development:planning and vertical leadership developemnt in a VUCA world. *The Journal of Values Based Leadership*, 15(1), 1–10. 10.22543/0733.151.1380

Daukantaitė, D., Hoff, E., Maurer, J., & Maurer, M. M. (2023, August). What is the process of personal growth? Introducing the Personal Growth Process Model. *New Ideas in Psychology*, 70, 101024. 10.1016/j.newideapsych.2023.101024

David Megginson, V. W. (2003). Continuing Professional Development. Chartered Institute of Personnel and Development.

Davis Boud, R. K. (1994). *Reflection:Turning Experience into Learning*. New York: Nicolas Publishing Company.

Elena Cristina, C. A.-N. (2020). Research on the Impact of Training and Personal Development on Individual and Organizational Performance. *"Ovidius" University Annals. Economic Sciences Series*, 20(2), 619–623.

Erik Driessen, J. V. (2008, January). Portfolios in medical education: why do they meet with mixed success? A systematic review: portfolios. *Medical Education*, 41(12), 1224–1233. 10.1111/j.1365-2923.2007.02944.x18045373

Fenwick, T. (2006, January). Contradictions in portfolio careers: Work design and client relations. *Career Development International*, 11(1), 65–79. 10.1108/13620430610642381

Firssova, O. (2006). ePortfolio as a coaching support tool for workplace learning of teachers. *Estrategies for Empowering*, 1-12.

Greenan, P. (2016, July). Personal development plans: Insights from a case based approach. *Journal of Workplace Learning*, 28(5), 322–334. 10.1108/JWL-09-2015-0068

Hannah Jones, J. C. (2020, Jan). Helping Leaders Grow Up: Vertical Leadership Development in Practice. *Research gate*. 10.22543/0733.131.1275

Herman Aguinis, K. P. (2021, May). If You are Serious About Impact, Create a Personal Impact Development Plan. *Business & Society*, 61(4), 818–826. 10.1177/00076503211014482

Houghton, E. (2017). *Creating and capturing value at work: who benefits?* CIPD.

Kari Smith, H. T. (2001, June). Long-term Influences of Portfolios on Professional Development. *Scandinavian Journal of Educational Research*, 45(2), 183–203. 10.1080/00313830120052750

Kneale, P. (2015, December). Personal Development Planning (PDP) really happens at work. *Planet-Taylor and Francis*, 18(1), 33–35. 10.11120/plan.2007.00180033

Lawson, M. G. (2008, January). Making talent a strategic priority. *The McKinsey Quarterly*, 1(1), 48–59.

Leanne, E., & Atwater, J. F. (2005, June). Antecedents and consequences of reactions to developmental 360 degrees feedback. *Journal of Vocational Behavior*, 66(3), 532–548. 10.1016/j.jvb.2004.05.003

Leggett, A. B. (2004). Being wise about teaching portfolios: Exploring the barriers to their development and maintenance.

Lejeune, D. M. (2006, July). Learning and perceived performance: the moderating role of self-directedness. *Business, Psychology Human Resource Development International*, 307-328. 10.1080/13678868.2016.1203639

Lopes, P. N., Salovey, P., Côté, S., & Beers, M. (2005). Emotion Regulation Abilities and the Quality of Social Interaction. *Emotion (Washington, D.C.)*, 5(1), 113–118. 10.1037/1528-3542.5.1.11315755224

Lyons, N., & Evans, L. (1997). *Portfolio: A Tool for Self Directed Learning at Work.* U.S. Department of Education Office of Educational Research and Improvement (OERI) National Library of Education (NLE) Educational Resources Information Center (ERIC).

Marilyn Johnston, M. T. (2005, August). Riding the wave of administrator accountability: A portfolio approach. Journal of Educational Administration. *Journal of Educational Administration*, 43(4), 368–386. 10.1108/09578230510605423

Maslow, A. (1943). A Theory of Human Motivation. *Psychological Review, 50*, 370-396.

Mathers, M. C. (2002, January 4). Portfolios in continuing medical education – effective and efficient? *Medical Education (Association for the Study of Medical Education-ASME), 33*(7), 521-530. 10.1046/j.1365-2923.1999.00407.x

Maurer, T. J., Mitchell, D. R. D., & Barbeite, F. G. (2002). Predictors of attitudes toward a 360-degree feedback system and involvement in post-feedback management development activity. *Journal of Occupational and Organizational Psychology*, 75(1), 87–107. 10.1348/096317902167667

McMullan, M., Endacott, R., Gray, M. A., Jasper, M., Miller, C. M. L., Scholes, J., & Webb, C. (2003, March). Portfolios and assessment of competence: A review of the literature. *Journal of Advanced Nursing*, 41(3), 283–294. 10.1046/j.1365-2648.2003.02528.x12581116

Milos, P. (2020). The Role of Emotional Intelligence in the Career Development of Employees in Public Organizations. *Proceedings of the 3rd International Conference on Economics and Social Sciences* (pp. 709-715). Bucharest University of Economic Studies. DOI: 10.2478/9788366675162-070

Monishaa, N. (2023, October). Importance of improving personal growth in workplace. *Journal of Emerging Technologies and Innovative Research*, 10(10), 59–73. Retrieved 2024, from

Nina Ketels, S. B. (2012). The Use of Personal Development Plans (PDPs) in Organizations and the Role of Its Perceived Purpose. *Learning at the Crossroads of Theory and Practice*, 71-92. 10.1007/978-94-007-2846-2_6

Paul Evans, V. P.-L. (2002). The Global Challenge: Frameworks for International Human Resource Management. McGraw-Hill.

Paul Lyons, R. P. (2019, May). Self-efficacy: Core of employee success. *Development and Learning in Organizations*, 33(3), 9–12. Advance online publication. 10.1108/DLO-04-2018-0045

Reid, B. E.-P. (2012). A revised learning transfer system inventory: Factorial replication and validation. *Human Resource Development International*, 15(5), 37–41. 10.1080/13678868.2012.726872

Ria Pita, F. K. (2007, May). The location of trait emotional intelligence in personality factor space. *British Journal of Psychology*, 98(2), 273–289. 10.1348/000712606X12061817456273

Santosh Patel, G. K. (2013, August). Personal development plans – Practical pitfalls. *Trends in Anaesthesia & Critical Care*, 3(4), 220–223. 10.1016/j.tacc.2013.04.003

Simon, A. J., & Beausaert, M. S. (2011). Using a Personal Development Plan for Different Purposes: Its Influence on Undertaking Learning Activities and Job Performance. *Vocations and Learning*, 4(3), 231–252. 10.1007/s12186-011-9060-y

Simon, A. J., & Beausaert, M. S. (2011, June 8). Using a Personal Development Plan for Different Purposes: Its Influence on Undertaking Learning Activities and Job Performance. *Vacationa abd Learning, 4*, 231-252.

Simon Beausaert, M. R. (2011). Nivember). The use of a personal development plan and the undertaking of learning activities, expertise-growth, flexibility and performance: The role of supporting assessment conditions. *Human Resource Development International*, 14(15), 527–543. 10.1080/13678868.2011.620782

Stéphane Côté, C. T. (2006, March). Emotional Intelligence, Cognitive Intelligence, and Job Performance. *Administrative Science Quarterly*, 51(1), 1–28. 10.2189/asqu.51.1.1

Susan Murmu, N. N. (2022). Impact of emotional intelligence and personality traits on managing team performance in virtual interface. *Asian J Bus Ethics.*, 11(1), 33–53. 10.1007/s13520-022-00154-1

Swanson, R. (1995). Human Resource Devepoment: Performance is the Key. *Human Resource Development Quarterly*, 6(2), 207–213. 10.1002/hrdq.3920060208

Tamkin, L. B. (1995). *Personal development plans: Case studies of practice*. The Institute for Employment Studies.

Thorsten Hennig-Thurau, M. G. (2006, July). Are All Smiles Created Equal? How Emotional Contagion and Emotional Labor Affect Service Relationships. *Journal of Marketing*, 70(3), 58–73. 10.1509/jmkg.70.3.058

Tigelaar, , Dolmans, D. H. J. M., De Grave, W. S., Wolfhagen, I. H. A. P., & Van derVleuten, C. P. M. (2006, June). Portfolio as a tool to stimulate teachers' reflections. *Medical Teacher*, 28(3), 277–282. 10.1080/01421590600607013 16753728

Wundrack, R., & Eva Asselmann, J. S. (2021, August). Personality development in disruptive times: The impact of personal versus collective life events. *Social and Personality Psychology Compass*, 15(9), e12635. 10.1111/spc3.12635

Chapter 10
Ensuring Privacy in Genomics and Transhumanism: The HRM's Imperative

Gurdish Pal
Independent Researcher, Australia

ABSTRACT

In the time of the genomics and transhumanism, protection of genetic and enhancement information represents an obvious necessity for human resources management (HRM). As identified in this chapter, there are ethical, legal, and practical issues in incorporating such information into the policies of an organisation. This increases the focus on the creation of sound protocols for the informed consent procedures, the standards of data access and control, and the policies against discrimination. This chapter shows how organisations are able to tackle the associated dilemmas of handling genetic and enhancement data responsibly by respecting employee privacy and encouraging truthful conversation. The goal of this chapter is to describe the key role of HR function in enhancing the level of trust in today's world of big data.

INTRODUCTION

Genomics is the analysis of the genomes which include the processes of sequencing, analysis and even manipulation of the genetic materials. It has the trans-formative capacity in medicine, agriculture, and other branches of sciences among others. The fact that manipulating the code of life means controlling the destiny of organisms

DOI: 10.4018/979-8-3693-7668-3.ch010

and their diseases may result in tremendous advancements in treating ailments or altering the fundamental features of life (figure 1).

Figure 1. Overview of Genomics

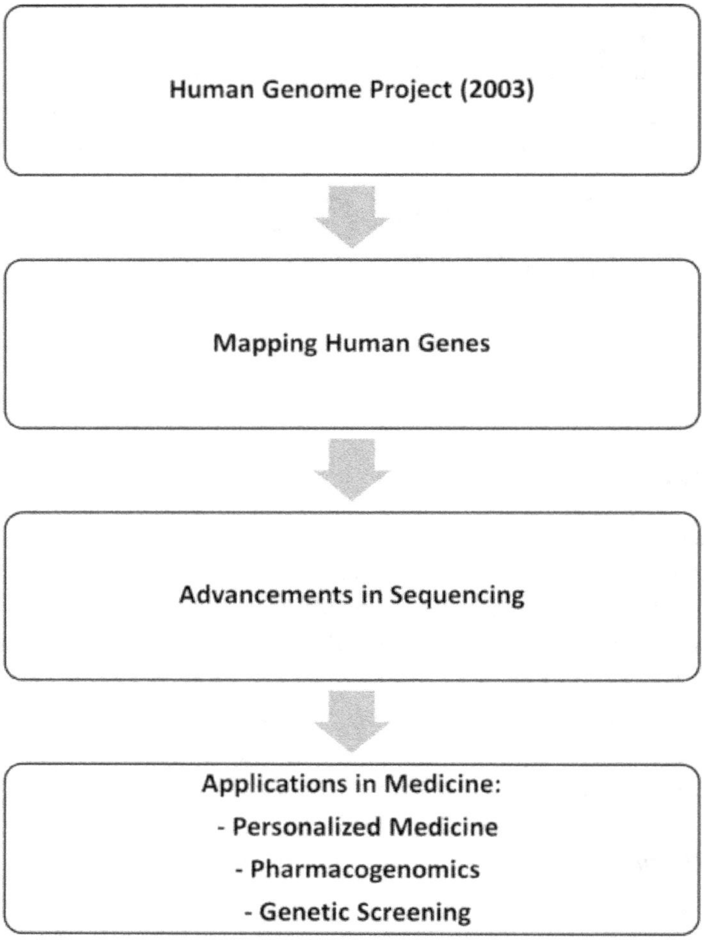

Similar to this, another movement dubbed transhumanism endorse the utilization of advanced technologies to augment scientifically every aspect of the human body and mind is quite in harmony with Genomics. The merging of these fields anticipates significant transformations in people's perception of possibilities and disabilities (Rueda, 2024).

Transhumanism in general is based on the desire to go beyond biology and cybernetic restrictions of human organisms (Klyczek, 2024). As a result of the application of advanced technologies, trans-humanists expect that people could control the pace of aging or even reverse it, get rid of practically any disease at the genetic level, and enhance human capacities multiple times beyond physiological limits. Some of these innovations are genomics technologies include the services like the CRISPR-Cas9 for gene editing. What these tools offer is the ability to delicately alter the DNA code and raise prospects for the boost in one's intelligence, muscle mass, and immunity to sickness. Nevertheless, a relatively swift growth of Genomics and the inclusion of the concept into trans-humanist objectives raises many ethical and privacy concerns. The challenges observed with the collection, storage, and use of Genomics data are different from other data types. While other categories of personal information are not always related to family members, genetic data are interlinked and may contain stressful facts about a person's health, his/her likelihood of getting sick, and propensity to behave in a certain way. Hence, the confidentiality of Genomics data remains an essential aspect because it affects not only the involved people but their relatives and future generations as well (Madden,2024).

WHAT IS GENOMICS AND TRANSHUMANISM?

Genomics is a comparatively recent area with an exceptionally rapid development pace observed after the accomplishment of the Human Genome Project in 2003. This was a historic project that allowed for determination of the human genome identification that is the genes that comprise the DNA and the sequence. Since then, improvements have been made in the technologies used in sequencing with the idea of sequencing individual genomes taking less time and being cheaper (Hocquette, 2005).

This has led to realization of genetic medicine where treatments as well as preventative measures can be prescribed based on an individual's genes. Genomics can in the case of Personalized medicine be used to identify how a particular patient would react to certain drugs, his/her tendency to certain diseases and whether or not such diseases are hereditary. For instance, pharmacoGenomics which focuses on relations between genes and drug reactions gears to developing safer and efficient drug intervention plans. Also, it contributes to perceiving the implication of screening and counseling on possible odds of inherited diseases in prospective parents (Madden,2024).

Transhumanism is not limited to disease control and curing whereas Posthumanism seeks to control and even eradicate diseases. It includes the process of increasing human capacities and characteristics by attainments of technology (figure 2).

Figure 2. Overview of Transhumanism

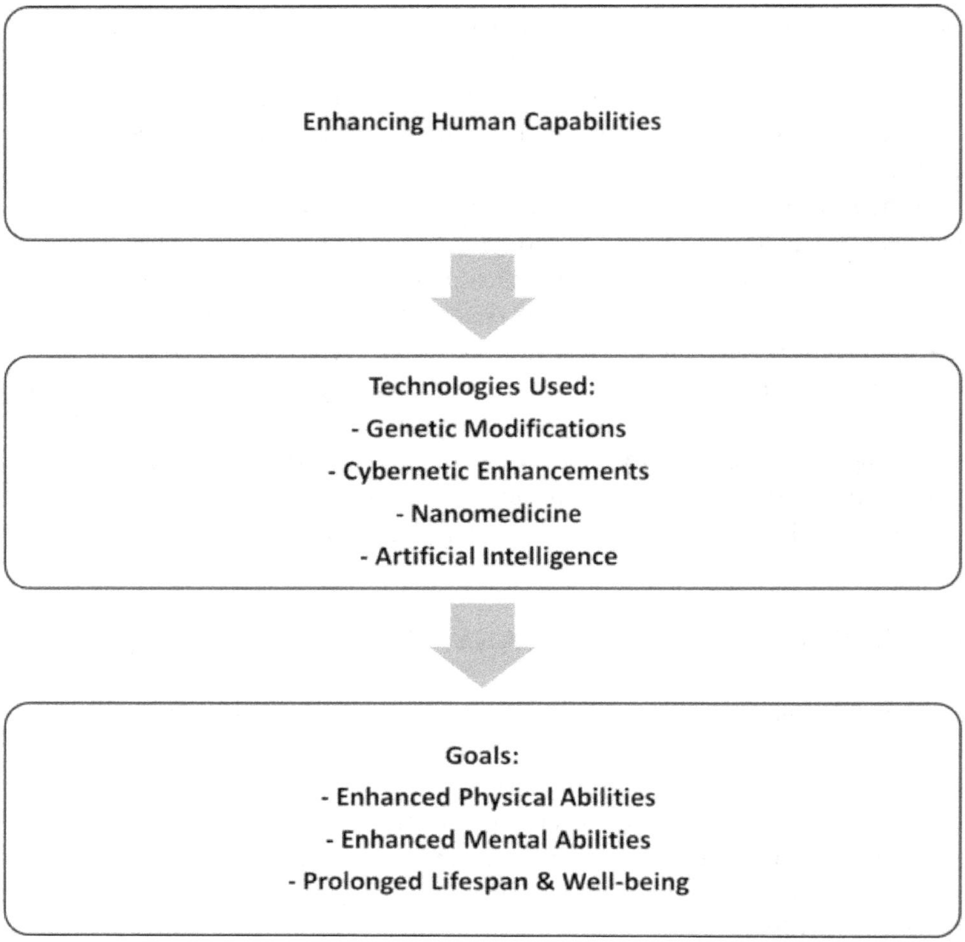

This also involves the modification of a human's DNA or genetic code and also cybernetic enhancements, nanomedicine, and Artificial Intelligence systems (Baelo-Allué & Calvo-Pascual, 2021). The prospective of use is to increase the effect of muscle and bones, brain power, and lifetime well-being. Thus, such enhancements as genetic one may include the improvement of memory and intelligence, while such enhancements as cybernetic may include enhancement of the sensations and movements. Transhumanism when integrated with genomics results in the creation of 'designer humans,' where the persons can select the characteristics of the off-spring like intelligence, physical strength, or immunity against diseases. This, in turn, poses social philosophical questions regarding the meaning of human beings, fairness, and the emergence of new patterns of unfairness.

It also raises the obvious issues of consent especially for changes that would throw into reckoning ensuing generations (Kriman, 2019).

DIFFICULTIES FACED AND SIGNIFICANCE OF PRIVACY IN THE FRAMEWORK OF HRM

Regarding the activities of Human Resource Management (HRM) as well as the benefits that may be provided by Genomics or threats that these applications may pose, the situation is still rather ambiguous. The integration of Genomics data in to the designing and implementation of HRM activity can offer the opportunity of altering how organizations handle issues relating to health, well-being and productivity of their people. For instance, through genetic data, it is possible for employers in the formulation of work place wellness programmes, prediction of the approximate work-related health risks or demarcation of the suitable job classification in accordance with one's genetic characteristics (Kaur & Arora, 2023).

But such advancement should be made in a way that does not infringe the rights of the employees, and at the same time, does not abused the genetic information. Thus, it is possible to state that the aspect of privacy is of great importance in this case. Genetic identity on a micro level is an extremely serious factor through which one can understand the future health prognosis of the patient involved, although it does not stop there. It can be a means by which some of the family backgrounds and perhaps behavioral too and possibly predisposition to some form of mental disorders could be appreciated. Someone might suffer discrimination, stigmatization, and loss of trust in case the use or disclosure of the such information releases it. And that is why, it proves to be essential to protect the information regarding the genetic data of individuals and this is only possible with the existence of efficient protection tools (Twine, 2010).

However, the information concerning genetics integrated in HRM raises questions about consent as well as autonomy. They should also be able to regulate disclosure of genetic information that they have, furthermore they should have the capacity to understand uses of the information disclosed. A main idea of informed consent should be introduced and effectively implemented in order workers thought about all the implications of providing their genetic information. Besides, there should be well developed policies on the use of genetic information in employment to prevent cases of misuse and fairness. The application of genetic information in organizations also has some of ethical issues which organizations should address (Mali, 2022).

Despite the fact that genetic information can be useful, it should not be used as the basis for decisions involving employment measures, promotions, or remunerations. Criticisms have arisen in relation to the fact that the use of genetics information will breed new forms of prejudice, whereby people are believed to be hired or fired based on probabilities of failing or succeeding genetically, not probabilities of behaving or performing as a result of their genes. In addition, storage and Genetic data are unique in a way that requires their data to be well secured. Hence, genetic databases are vulnerable to cyber criminals because they contain highly sensitive data. That is why organizations have to spend more money and resources in implementing better security to prevent leakage of this genetic information. This includes as for instance data encryption, restricting access to the data, and periodical system security scans (Kaur & Arora, 2023).

In summary, the idea of including Genomics and transhumanism into the best practices of HRM might produce many positive outcomes, yet it should be done with proper precautions. Finally, it is imperative that privacy be respected especially with regards to the employees and use of genetic information must be done ethically. However, strong regulations, well-articulated consent procedures as well as security standards are laudable essentials to make sure that these advancements' efficacy is fully realized without infringing on the rights of the individuals. Therefore, in the subsequent weeks, it will be vital to aim at achieving a middle ground, which will sufficiently harness the value of Genomics technologies while protecting individuals' right to privacy to guide the evolution of the field of HRM.

THE IMPACT OF GENOMICS AND TRANSHUMANISM ON HRM

Due to the fast growing technologies such as Genomics and transhumanism, there is a high possibility for great disruption in the field of HRM. These technologies provide one-of-a-kind possibilities of improving the condition of employees and performance of organizational tasks. But they also come with new ethical, privacy, and regulatory issues which are unique and inevitable for the HR to deal with (Urgan, 2019).

Genomics is widely known as the study of genomes and their structures, traits, evolution, and location. A genome is therefore the entire quantity of DNA in the organism as well as all the genes in the organism. This field has expanded tremendously since the Human Genome Project which coded for the human body map and modern genetic research. The importance of Genomics can be given on the fact that this field holds the ability to revolutionalize multiple fields ranging from Health care to agriculture, Environmental science, and now to Human resource Management. In the field of healthcare, Genomics can be useful in the practice of precision medicine, a approach that delivers medical care based on an individual's genotype. Thus, the prospects for greater effectiveness of treatments and fewer side effects are possible when employing this approach (Kaur & Arora, 2023).

For instance, pharmacoGenomics which is a branch of Genomics discusses the genes that determine a patient's reaction to certain drugs so as to influence better production of efficacious and safer medicine. Within the farming practices, Genomics has potential to increase crop productivity and its ability to resist diseases, which impacts on food security positively. In the same manner, in environmental science Genomics assist in the discovery of the species' distribution and effects of climate change on diverse species (Urgan, 2019).

From the perspective of HRM, therefore, genomics may be viewed as potentially effective tool for altering the process of managing the problem of employees' health in the organization. The employers also come to know the potential risks pertaining to their workers' health through such mechanisms to prevent such cases by other measures like the differential or special prosperity-indicator promotion and test. Of this, such measures assist in lowering the costs of health, the number of truancy from work and in improving the overall health of the work force. Also, Genomics can be re commendatory in establishing hereditary influences of employment that may help in placing the right employees in various jobs. However, there are strict ethical and privacy considerations when Genomics tools are to be used in the context of HRM. Any such information is considered private and contains not only alternative disease and health condition of the individual but also the root, inclination towards some kinds of behavior, and response to some sort of mental illnesses (Sanghavi et al., 2021).

It is thus at risk of being abused, being used possibly in discriminating someone or in the infringement of his or her right to privacy; hence, such data should be protected and those handling them should exercise professional integrity. In this sense, certain conditions should be met, and the most important ones are: employees should be able to have control on their genetic info; and secondly, the employees should have complete information about how the genetic info contained in genes would be utilized.

OVERVIEW OF TRANSHUMANIST TECHNOLOGIES

Transhumanism may be regarded as vision that primarily rests on the merger between technology and human beings with an eye on the improvement of the physical and cognitive qualities among individuals. It depicts what transhumanism brings to the next level for example genetic engineering, cybernetics particularly information input and processing circuits, artificial intelligence, and nanotechnology.

Transhumanism is that branch of posthumanism which concerns with the process of overcoming of the subject's biological given conditions and transformation of the latter to the physically, intellectually, and existentially better one. It mainly relates to the change of the specific part of the DNA code and is associated with the recent developments like CRISPR-Cas9 gene editing system. It can be applied to correction of congenital abnormalities and at the same time for expression of the beneficial wanted traits or addition of new abilities. For example, genetic engineering can therefore be utilised to cause change like; making of human beings to be smarter, to be able to store information much better and even to be made stronger. These applications are still quite humble and dangerous for now; but if it is indeed possible to elevate human beings' powers by several dozen or even hundred fold, these are true marvels (Fletcher, 2014).

Transhumanist technology also includes more like Cybernetic Implantation and prosthetic tools. It can be used to function as backup of biological organs; used to mean the improvement of resolution and acuteness of the perception or a change in the motor and the sensory attributes. For example, for the deaf there is most likely an opportunity to return hearing through cochlear implants and at least the prosthesis can be more useful than the removed limbs. In the future, a person could have electrodes implanted on the brain to boost the presumed congenital attributes, for instance, memory and the ability to decide, which can also be referred to as cognition (Hurlbut & Tirosh-Samuelson, 2016).

Implications for the Workplace

Genomics and transhumanist technology's infusion in the workplace provides meaning and goal ideas to the HRM. These changes present possibilities to improve the well-being, efficiency, and satisfaction of the employees but can also present new problems in the realms of privacy, ethical issues, and equal opportunity.

Health and Wellness Programs: Genomics support the development of an effective lifestyle modification plan of health and wellness as defined based on a person's genotype. For example, employees who are hereditarily predisposed to certain ailments are allowed to undergo special tests and exercising to prevent the childhood diseases from befalling them. An effective nutrition and exercise plan can also thus be created based on one's genetic tendencies resulting in enhanced health and decreased expenditures on health care. Still, it is necessary to guarantee that people are willing to participate in such programs and genetic information of employees cannot be disclosed or misused (Kaur, 2024).

Hiring and Placement: There are facilities that Genomics could offer about an individual's likelihood of being fit for certain occupations depending on his/her specific genes. For instance, some people any possess specific genes that suggest future propensity to stress-related diseases and can be useful for stressful occupations. Likewise, the results of a genetic test can be applied to select those people that are more suitable for specific jobs and conditions. Although the above information is useful in making better placement and employment decisions, the information can only be used legally and fairly. Employers need to therefore not only ensure that they consider genetic data as only part of other factors while taking these decisions but also that genetic discrimination of an individual is not practiced (Kaur & Arora, 2023).

Performance Enhancement: There will be cognitive provokers or cybernetic enhancements tools that will improve the operations of the employees in the organization. For example, the neural implant may enhance the memory as well as decision making capacity and on the other hand the exoskeletons may enhance the physical strength and stamen power. These could potentially give a boost to work rates and efficiency and at the same time, the satisfaction levels of the employees. Thus, their usage remains dubious with regards to ethical questions as those of fairness and equality. Managers must ask the question whether enhanced employees would be more privileged than other fellow employees and if there will be a solution to it (Kaur & Arora, 2023).

Privacy and Consent: The Genetic/transhumanist has a major contribution in the area of work means security and the approval process of the privacy act. This data has to be owned by the employees and they must be updated on further use of their genetic data. Consent must be a proper informed one so that the employees have a real adequate notion of what they are conforming to as to regards their genetic data. In addition, the employers need to establish proper measures of security so as to curb early release of the genetic files (Kaur & Arora, 2023).

Regulatory and Ethical Considerations: Thus, the specifics of the interaction between Genomics and transhumanism that happen within the HRM context require regulation, and there should be some established ethics regarding it. A new area of concern for the government and regulatory authorities will be to establish standards to control bodies and gene alterations to be used in organizations. This regulation is very important and should be of interest to the HR specialists as they should ensure its implementation within the concerned companies. Furthermore, it is imperative for the organizations to enhance their internal ethical standards and compliance regarding these technologies' use (Hurlbut & Tirosh-Samuelson, 2016).

DATA PRIVACY AND SECURITY

Genomics along with enhancement technologies are creating a new set of questions concerning data privacy and security in general and more specifically in the area of HRM. Since genetic information and transhumanist upgrades are used in enhancing employees' health, efficiency, and productivity in organizations, this data is sensitive in nature and requires robust security measures to guard against theft and misuse. This section looks at the issues of genetic and enhancement data, the dangers involved in the leakage and misuse of the data, and ways through which data of employees can be protected efficiently (Carter, 2019).

Genetic and enhancement data is considered to be personal information as it follows an individual's biomarkers and in many cases may even unveil many personal details about an individual. For example, other forms of data include genetic, which can offer information on a person's likelihood to genetic ailments, his or her ability to develop diseases, ethnicity, and his or her manners. This data type is not only personal, but social as well, as it involves relatives, which implies that misuse of it will have repercussions not only for the subject but also their family (Sharma, & Makhija, 2024). In the same way, backed data from enhancement technologies like neural implants or cybernetic organs can disclose cognitive and physical functionalities and transformations that may influence views of the person's privacy and independence. To make matters worse, this type of data is very much likely to have long-term consequences, hence, the sensitivity of the information. As opposed to the other personal data which changes with time and does not anticipate the future, genetic data is unique and stays the same throughout one's life and sometimes can even anticipate future health complications. This is a major concern in matters of discrimination, stigmatization and possible psychological effects that the individual might incur in case that information is disclosed willingly or otherwise. These issues raise concerns that this information might be used by employers, insurers or other organizations to deny people jobs, promotions, insurance or treatment as the case may be thus the need to protect this data (Mohammed Yakubu et al., 2020).

POSSIBLE CONSEQUENCES OF DATA INCIDENTS AND MALPRACTICES

The general risks that can originate from data breaches and misuse regarding genetic and enhancement data are clearly diverse and evident. Discrimination of genetic information is one of the biggest risks that may follow people across different institutions. Collecting, for instance, genetic data by employers or insurers implies that people might be discriminated based on some of the detected signs, and thus, employers or insurers may refuse participants, for example, offering a job to someone because he or she has genetic tendency to some illness, or raising insurance rates for a family based on some genetic risks. They are as follows: This form of genetic discrimination results in a negative change in opportunities and an increase in stress among people whose genetic records are compromised (Greenbaum et al., 2011).

The second major threat is that of identity theft and related fraudulent activity. Genetic info is not transferable or alterable like a credit card number or a password and therefore should not be compromised. On this data, synthetic IDs can be produced or fraud is committed which as major long-term security risks regarding individuals are involved. Moreover, the improper usage of enhancement data will enable surveillance and control within the society, violating some of the freedoms of the individuals. For example if through chips implanted in the human brain, employers were able to follow the mental processes of workers in their organizations; this is a very gross violation of the rights of such individuals. Further, data breaches underlined in the given context can diminish the relationship between employers and employees. If employees think that their genetic and enhancement information is not properly safeguarded, they are likely to be less accommodating to the wellness programs or the enhancement technologies that can be advantageous to their overall health and productivity. Such mistrust results to non-use or minimal use of the technologies a result that goes against the effort being put in to implement the technologies in industries (Naveed et al., 2014).

PLANS FOR PROTECTING THE EMPLOYEE INFORMATION

- ✓ Unfavorable consequences of gathering, storage, and processing of the genetic and enhancement information can be minimized only with the help of the definite protection measures to save the personal data of the employee. Adopting complex protection measures, which are considered as measures of the highest protection level, is crucial to safeguard information from leakage. This includes applying encryption which may include data in motion and data at rest, which means only authorized personnel have access to genetic and enhancement data with multi-factor authentication. Annual and/or semester security check and risks assessment can be of great help to discover existing loopholes in the system. Furthermore, organizations should incorporate the use of intrusion detection and prevention systems (IDPS) which scans the traffic to the network for any suspicious activity and take action on a possible threat. Other data optimization strategies may also be applied to mask specific individuals involvement in the analysis or use of particular databases by eradicating any PPI details (Vavekanand, 2024).

- ✓ Director and managerial production of clear and strong personal privacy rules are necessary to avoid mistreatment of individuals' genes and enhancements. Some of the aspects that these policies should contain include; The nature, use, and security of the collected data. Organizations should be able, and it is most probable, to differentiate basic rules with regard to when and how the employees' genetic information can be collected or used, and when an informed consent can be legally requested from the employees. Employers must also come up with the policies concerning the transfer of information to third parties, and such transfers should only be made with the consent of an employee (Oestreich et al., 2021).
- ✓ Employers should adhere to specific ethical principles in the employment of genetic and enhancement data to make certain that such technologies will not harm the basic human rights of an individual and that fairness will be the characteristic guiding the utilization of such technologies. This is related to the formation of ethical review boards on use of sensitive data and Workplace civility and respect. More on the accountability, organizations should design and uphold a culture of openness and the manner in which employee's information is processed and utilized should be communicated in detail and employees should also have control to be able to see their data as well as correct it. In addition to this, offering training sessions on data policies in relation to the particular company and the ethical approaches to the use of the data can be useful for the employees to grasp the significance of the protection of the valuable information and their contribution towards the provision of the data security (Vavekanand, 2024).
- ✓ Laws protecting genetic and enhancement data before and after they are stored must be followed to the latter to promote their security. It is recommended that organizations have continuous knowledge and updates of the liberate laws regarding data and information privacy and protection like GDPR in Europe and the HIPAA in the United States. These regulations contain strict measures on the protection of personal data and entail severe consequences for failure to adhere to the rules. Besides compliance with current legislation, organizations should call for enactment of new rules and regulations in the management of genetic and enhancement data. To ensure that regulatory frameworks are up to date to contain novel risks, additional recommendations include healing relations with policymakers and other industry groups (Oestreich et al., 2021).

ROADMAP ON WHO SHOULD USE THE DATA AND WHEN, WHAT, HOW AND WHY; ALONG WITH THE QUESTION OF WHO HAS THE RIGHT TO USE THE DATA.

There exist very significant issues that relate to designation, evaluation of measures for the conservation of genetic and enhancement data from different accesses, uses and disclosures of the data. Countless should include those covering the care of the patient information concerning the handling policy of the law and the ethical practice. It should be noted that several fold guidelines must be incorporated for data access and control for acquiring optimum security to data. First, there are access controls that are also obtained by using the means of encryptions (Balagurunathan & Sethuraman, 2024).

Thus, to reduce the genetics and enhancement data within an organization, certain measures should be taken so that such data will be less available. Its presence should be limited towards the particular employer who desires to look at the same for some particular requirements such as health or superior researches. This minimize the chances of other people who are not allowed to access the data to be given a chance and also assure the rightful management of people to the data. Second, encryption and security measures come in place when answering the 'who' and 'how' of the actual putting into operation of the plan. This means there should be regulation of the information itself and the genetic and enhancement data requirements for them to contain provisions on how to secure the same in transit and at rest from Interception and theft. Managers have also to install firewalls, the intrusion detection system, and security auditing in the hunt for the threats. Such measures help to address issues of change, deletion, or modification of the data and reports by unauthorized persons. Thirdly, the organisations should adopt rights in data minimization. This is done only collecting the genetic data and the data related to enhancements that allow some tasks, and storing this data in the shortest possible time. Sometimes, material that is no longer useful or important should be removed, or else its details should be masked so that it cannot be viewed by undesirable people. When you minimize data, you essentially end up working with less data and also you dodge virtually all Data breaches (Sharma, & Makhija, 2024).

Fourth, data sharing should be controlled by guidelines that should state on sharing data with third parties. This includes giving releases to third parties with respect to one's genetic and enhancement information to healthcare clinics, insurers or research facilities. Companies must also ensure they have the employees' consent to use their data both internally and externally and that whoever processes the data also agreed to abide by the same standards of privacy and protection. This way the flow of data is done responsively legally and in a dignified manner. Last but not the least, rights of the employees are an aspect that must be clearly defined. Staff members should be aware of the employee's rights for retrieving and regulating the genetic and enhancement information, such as the right to obtain the data, right to rectify information, and right to withdraw the consent of the data processing or sharing. The behaviours should include having clear protocols that addresses the employees' requests including the time taken to process them, and how the responses are to be channeled back to the employees. Esteeming the rights of employees makes them follow rules that have been set down by virtue of the law hence improving on ethic (Balagurunathan & Sethuraman, 2024).

Guidelines on How to Communicate with Employees in the Most Transparent Way

The information about data collection and use of genetic and enhancement data should be clear, which will promote trust and optimum relations with the employees. The organizations should therefore ensure that the best practices of openness and transparency in data management are followed when communicating to the employees about the practices in place, the privacy measures taken in handling the data, and all the issues that pertain to the ethical use of data. First of all, it is essential to have fixes and standards being clear; organizations need to establish easily understandable policies on the acquisition of genetic and enhancement data as well as their application and protection. These policies should be developed and distributed in an easily understandable language and also through different means like Company handbooks, Company websites, and training sessions (Kaur & Arora, 2024).

Secondly, detailed information should be shared with the employees, regarding the use of genetic and enhancement data, their collection, processing, storing and sharing within the organization. These notices are often required to be revisited in small or frequent basis to adopt to any changes on data practices, to satisfy legal requirements such as GDPR's principles concerning transparency and accountability. However, engaging in educational work, in the form of workshops or training, associated with data privacy and matters of ethics regarding this area is helpful (Oliva, 2024).

These sessions also help the workers to find ways of valuing genetic and enhancement data through discussions of the breaches and employee right to access and control the data and ways of protecting the data. The other practice is that there should be no hindrances to the discussions and the feedback mechanisms that are also required should also be actively exercised. It is recommended that organisations' management should create organizational communication platforms through which workers should be encouraged to ask questions, report issues, and provide feedback on data handling. This feedback can then be used in the formulation of the policies and procedures on how to address the employees' expectations and values within the given organization (Sharma, & Makhija, 2024).

Last of all, in the aspect of communication transparency, there has to be the proof that an organization cares for ethical values together with the aspects like the right to self-management, justice, and no harm. It is equally necessary for the employees to have an assurance that GEN and enhancement data will not be utilized for forbidden usage as per the law and guidelines. The application of the above best practices makes sure that the management is able to gain the confidence of its workers as regards issues concerning the handling of their sensitive information. The integration of genetic and enhancement data into the Human Resource Management and practices will be an enhancement and has lots of prospects for the organizations. When it comes to addressing these problems using such approaches, one has to accept the fact that there can be some resistance by the employees and the complexities associated with the advancement of new technologies; on the positive side, it is possible to view the following opportunities for the further enhancement of the employee's health and productivity (Oliva, 2024).

Also, the staff may report ethical issues and apprehensions about being discriminated in the company based on genetic vulnerabilities or enhanced abilities. To address such concerns the following measures need to be taken: Adoption of policies against discrimination and ensuring compliance with such policies, raising the awareness of managers and subordinates on the ethical and legal considerations of non-discrimination, formation of ethical review committees that would oversee the use of genetic/ enhancement data and lastly offering the employees a chance for making decisions concerning their data. The issues arising from the technological advancement of genetic and enhancement technologies include; integration of data, compatibility of data, compliance with ever changing laws and policies, and data security. However, these challenges also hold promise in forming new strategies such as using the genetic information in designing customized health management programs, using all the data collected both for recruiting and retaining talents, employing enhancement technologies for boosting further productivity, and finally placing the organization as the industry pioneers in all aspects ranging from employee health management to emergence of the new technologies (Kaur, 2024).

Combining genetic and enhancement data in the HRM practices opens up significant possibilities to improve the welfare and productivity results of the employees. Genetic knowledge of diseases allows diagnosing a fatal condition in its infancy which can prevent complications and in rare cases, death, Personalized prevention and wellness plans based on the clients' genotypes contribute to a healthier life. Enabling technologies may be used to enhance mood, cognition, and executive functions, improve performance and proficiency in occupational tasks, and decrease rates of injuries in the workplace, as well as increase motivation at work and productivity through personalization of positions based on workers' aptitudes. The consideration of genetics and enhancement information within the strategy of HRM practices holds the potential for a number of risks and opportunities for organisations that are endeavoring to improve the quality of human resource, the efficiency of employees, as well as organisational performance. The main challenge is ensuring that employees do not resist such initiatives and this can be achieved by ensuring that privacy measures are put in place, where there is proper code of conduct and where there is honest and open communication about the practices of collection and usage of data. Immersing in the dynamics of transformational technologies implies accepting change while paying attention to the legal and ethical standards (Sharma, & Makhija, 2024).

Employees have unprecedented potential for improvement in word wellness and productivity through wellness programs customized to employees' profile; protective health-screening; stress management; skill development through enhancement technologies; job satisfaction and organizational commitment. The solid middle line that underscores ethical principles and empowers employees' rights accompanied by the welcomed innovations in the field of HRM enables organizations to make the best of genetic and enhancement data to produce a beneficial effect on employee health, performance, and organizational success. In the future, it is expected that Genomics, transhumanism and HRM will be the key steps towards revolutionizing the human resource department, employee health and performance and productivity of the organization. Thus, changes in the industry and IG as well as the advent of new technologies in Genomics and transhumanism, pave the way for new opportunity for HRM along with increased challenges (Charnysh et al., 2024).

Genomomics, which entail the study of an organism's genome, is continuously seeing developments through the application of technological developments in areas like next-generation sequencing and CRISPR-Cas9 genome editing tools. Such developments allow for targeted care of an individual's well-being, protection against illnesses, and enhancement of abilities. There is an increasing focus on precision medicine that is, exercises for an individual based on a patient's genetic makeup. In the case of the HRM, this may result in developed excellent corporate wellness programs that boast of high standards in the aspects of health of employees as well as cutting on expenditure on health care costs. This is possible due to the rising availability of genetic testing services that could identify the risks of developing certain diseases that are hereditary; thereby enabling the organization to screen the employees and take adequate preventive measures as soon as they are identified hence increasing their productivity due to improved health. But as genomics data is collected, shared and made more readily available organizations face the challenge of ethical issues relating of privacy, consent and discrimination with frameworks and policies holding the responsibility of preventing injustices against employees (Kurec, 2024).

Transhumanism, a field which examines the human capacities' enhancement with the help of technologies like neural implants, wearable gadgets, and drugs as well, is also vital for the analysis of the HRM position. Neurotechnology and cognitive computing could mean that organisations can upgrade the cognitive state of the employee; memory retention, concentration span, and skills of problem solving necessary for the decision-making processes and procreative creativity in the workplace. Occupational Robotic Assistance such as wearable devices and exoskeletons that provide additional strength to the body may help the employees who are involved in manual tasks, decrease cases of workplace injuries and maybe boost production. Nevertheless, the advancement of transhumanist technologies brings into existence numerous moral issues with regards to the civil liberties of a human being, and the probability of enhancing a human being to cause extreme inequality between 'up-graded' and 'non-up graciously' people. These are emergent challenges that HRM policies need to address in order to attain organisational justice/reform, diversity, and moderate usage of enhancement technologies at the workplace (Martins & Martins, 2024).

Their implementation in HRM also brings possible future prospects The integration of these technologies into HRM also presents certain future developments. Privacy issues, voluntary consent, no discrimination and benefits pertaining to risks associated with genetic and enhancement data have to be resolved. Therefore, getting society to accept the use of genetic testing and enhancement technologies, especially in the workplace, may be somewhat difficult because of issues of invasion of privacy and rights, as well as the general issue of equity as we saw in society division under the rule of the elites. Changing regulations like GDPR in Europe and GINA in the United States may be unable to meet the dynamic technological aspects; therefore, HRM policies have to solve these twists in order to create compliance, but at the same time encourage innovation. Employment legislation concerns may include data privacy, discrimination, and employee rights that might cause some legal issues, it is important for the Human resource management standards to align itself with legal provisions and protect employees' rights by availing legal comprehensible policies in organizations (Sharma, & Makhija, 2024).

Applying Genomics and transhumanist technologies to today's frameworks of or within HRM requires a devotion to infrastructures, training's and specialization to address integration issues and data compatibility in Human Resource Management systems. Another closely related problem which might pose challenges in the strategic application of genetic testing and enhancing objective among the employees relates to cost and access together with the realistic question of how to proactively establish and demonstrate justice and fairness in the distribution of the technology's gains across the employees within the predefined organizational goals and objectives of the corporation in its operations through HRM policies. Thus, to be ready for such changes in advance in regard to the unification of these trends, organizations have to set proper ethical and governance requirements (Guerreiro et al., 2022).

Devolving the general ethical standards as to the correct application of genetics and improving data in the sphere of HRM and constituting the committees for ethical impact of these technologies as the ways to reduce adverse effects of such technologies, as well as ethically regulating. By having proper data policies, safe storage technologies and methods, permission details for data, signing reviews and audits, and enhanced cyber security, and employees' awareness and knowledge. Every type of educational and relevant training programs that are still to be provided to the employees with regard to genetic and enhancement technologies, utilization, result, plus points and minuses and all the related implications should also be effectively imparted to the employees. Communicating with employees more freely and asking them their opinion is a good way to introduce the reliance in organization. The systematic assessment of the future evolution of Genomics and transhumanism and experimenting with this combined with research and pilot testing would presuppose the methodical regulation of future advancement and the implementation of legal and ethical solutions in the management of human resources (Bloom & Bloom, 2020).

Thus, the future of HRM cannot be discussed without mentioning the relation of Genomics and transhumanism, which represents threats and opportunities for organizations at the same time. Some new tendencies in Genomics concern possibilities of individual health control and prevention of diseases in the future; transhumanist technologies may improve people's cognitive and physical performances in the process of work. Ethical issues, regulatory issues, technological issues, and social issues are points that needs to be tackled by HRM in relation to technologies. Thus, due to the given challenges, by adapting HRM policies and practices, organizations can address them and, at the same time, reap the advantages of genetic and enhancement data in a proper manner. Creating standards of ethical behavior, increasing data protection and information security, training and soaring employees, and identifying technological changes are the key approaches to making Genomics and transhumanism ethical, legal and efficient in engagement with HRM. In this way, organizations can create a sense of innovation to increase employee satisfaction, work on well-being, and gain sustainable competitive advantage due to the growing role of technology in today's world (Gevorgyan & Baghdasaryan, 2021).

CONCLUSION

Genetic and enhancement data regarding the case of HRM show that innovation and ethics are rather in harmony within change and converge at this strategic point. In relation to the current state of Genomics and transhumanism in the HRM, it has been necessary for the present study to examine the pros as well as the potential 'situation' concerning large organizations' misplaced belief in the religion that is science. Genomics makes it possible for the management to understand and identify the possibilities of the overall health of the employee; this in turn assists the management in designing efficiency improvement training, health risks mitigation plans as well as delivery of specific health enhancement interventions at the place of work. It is as a result clear that transhumanism means possibilities of enhancement of the cognitive, physical aspects, and productivity through incorporation of technology. Therefore, one can identify several ethical considerations in the integration of both genetic and enhancement information within an organisation's HRM. The Herman company's consent for privacy interest of its employees, employment non-discrimination, and proportional appointment of gains and burden regarding the GDPR and GINA compliance must be addressed appropriately. Some of the requirements that have to be implemented are code of conduct, ethical code, and measures on dealing with special data.

Future issues in HRM include: relating to the dynamics of regulations, managing questions of privacy and fairness in the society, and incorporating new technologies into the aforementioned frameworks of HRM. Issues that may be derived out of possible use of embryonic and enhancement data like fairness which can be reduced to discriminating between enhanced and non-enhanced persons are clear signs that there are hard questions which in themselves call for policy making and engagement of all stakeholders. Here are some possible steps to overcome these difficulties: The organization should have extensive policies in the sphere of personal data protection, further enhancement of the protection of personal information, along with raising awareness of the employees in the sphere of genetics and enhancements, promoting the idea of an open dialogue.

REFERENCES

Baelo-Allué, S., & Calvo-Pascual, M. (Eds.). (2021). *Transhumanism and posthumanism in twenty-first century narrative*. Routledge. 10.4324/9781003129813

Balagurunathan, Y., & Sethuraman, R. R. (2024). *An Analysis of Ethics-Based Foundation and Regulatory Issues for Genomic Data Privacy. Journal of The Institution of Engineers*. Series B.

Bloom, P., & Bloom, P. (2020). Introduction: Preparing for a "Transhuman" Future. Identity, Institutions and Governance in an AI World: Transhuman Relations, 1-29.

Carter, A. B. (2019). Considerations for genomic data privacy and security when working in the cloud. *The Journal of Molecular Diagnostics*, 21(4), 542–552. 10.1016/j.jmoldx.2018.07.00930703562

Charnysh, E., Pal, S., Reader, J., Uhlmann, W. R., McCain, S., Sanghavi, K., & Roberts, J. S. (2024). Health Care Utilization and Behavior Changes After Workplace Genetic Testing at a Large US. *The Health Care System*.38733246

Fletcher, D. J. (2014). Transhuman perfection: The eradication of disability through transhuman technologies. Humana. *Mente Journal of Philosophical Studies*, 7(26), 79–94.

Gevorgyan, S., & Baghdasaryan, K. (2021). Toward a transhumanist transformation of human labor potential in the context of global challenges. *Futurity Economics&Law*, 1(4), 36–45.

Greenbaum, D., Sboner, A., Mu, X. J., & Gerstein, M. (2011). Genomics and privacy: Implications of the new reality of closed data for the field. *PLoS Computational Biology*, 7(12), e1002278. 10.1371/journal.pcbi.100227822144881

Guerreiro, J., Loureiro, S. M. C., Romero, J., Itani, O., & Eloy, S. (2022). Transhumanism and engagement-facilitating technologies in society. *Journal of Promotion Management*, 28(5), 537–558. 10.1080/10496491.2021.2009615

Hocquette, J. F. (2005). Where are we in genomics? *Journal of Physiology and Pharmacology*, 56, 37.16077195

Hurlbut, J. B., & Tirosh-Samuelson, H. (Eds.). (2016). *Perfecting human futures: Transhuman visions and technological imaginations*. Springer. 10.1007/978-3-658-11044-4

Kaur, J. (2024). Green Guardians: Unveiling the Strategic Role of HR in Environmental Sustainability Initiatives. In Building Sustainable Human Resources Management Practices for Businesses (pp. 125-143). IGI Global.

Kaur, J., & Arora, R. (2023a, June). Exploring the dimensionality of employee silence in healthcare sector. In AIP conference proceedings (Vol. 2782, No. 1). AIP Publishing. 10.1063/5.0154178

Kaur, J., & Arora, R. (2023b, June). Exploring the impact of employee silence in private hospitals-a structural equation modeling approach. In AIP Conference Proceedings (Vol. 2782, No. 1). AIP Publishing. 10.1063/5.0154173

Klyczek, J. (2024). The Precision Eugenics Initiative (Part 2): The Precision Transhumanism Initiative. *Hand (New York, N.Y.)*, 6, 19.

Kriman, A. I. (2019). The idea of the posthuman: A comparative analysis of transhumanism and posthumanism. Russian journal of philosophical sciences, 62(4).

Kurec, A. S. (2024). Employee selection. Clinical laboratory management, 159-181.

Madden, E. B., Hindorff, L. A., Bonham, V. L., Akintobi, T. H., Burchard, E. G., Baker, K. E., & Green, E. D. (2024). Advancing genomics to improve health equity. *Nature Genetics*, 1–6.38684898

Mali, F. (2022). Human genome editing technologies in the center of current transhumanistic discourse on human enhancement. Analele Universită ii din Bucure ti–Seria Filosofie, 71(1), 69-86.

Martins, I., & Martins, A. (2024). Back to the future: multidimensional transhuman influence on creativity and innovation. In *Artificial Intelligence in Production Engineering and Management* (pp. 193–206). Woodhead Publishing. 10.1016/B978-0-12-819471-3.00007-0

Mohammed Yakubu, A., & Chen, Y. P. P. (2020). Ensuring privacy and security of genomic data and functionalities. *Briefings in Bioinformatics*, 21(2), 511–526. 10.1093/bib/bbz01330759195

Naveed, M., Ayday, E., Clayton, E. W., Fellay, J., Gunter, C. A., Hubaux, J. P., ... & Wang, X. (2014). Privacy and security in the genomic era.

Oestreich, M., Chen, D., Schultze, J. L., Fritz, M., & Becker, M. (2021). Privacy considerations for sharing genomics data. *EXCLI Journal*, 20, 1243.34345236

Oliva, A., Kaphle, A., Reguant, R., Sng, L. M., Twine, N. A., Malakar, Y., Wickramarachchi, A., Keller, M., Ranbaduge, T., Chan, E. K. F., Breen, J., Buckberry, S., Guennewig, B., Haas, M., Brown, A., Cowley, M. J., Thorne, N., Jain, Y., & Bauer, D. C. (2024). Future-proofing genomic data and consent management: A comprehensive review of technology innovations. *GigaScience*, 13, giae021. 10.1093/gigascience/giae02138837943

Rueda, J. (2024). Genetic enhancement, human extinction, and the best interests of posthumanity. *Bioethics*, 38(6), 529–538. 10.1111/bioe.1308536134439

Sanghavi, K., Feero, W. G., Mathews, D. J., Prince, A. E., Price, L. L., Liu, E. T., Brothers, K. B., Roberts, J. S., & Lee, C. (2021). Employees' views and ethical, legal, and social implications assessment of voluntary workplace genomic testing. *Frontiers in Genetics*, 12, 643304. 10.3389/fgene.2021.64330433815477

Sharma, K., & Makhija, T. K. (2024). Transcending Boundaries: Evolving Transhumanist Technologies in International Business and Associated Consideration.

Twine, R. (2010). Genomic natures read through posthumanisms. *The Sociological Review*, 58(1, suppl), 175–195. 10.1111/j.1467-954X.2010.01917.x

Urgan, S. (2019). Transhumanism and Positive Psychological Capital in Organizational Behavior. In Handbook of Research on Learning in the Age of Transhumanism (pp. 328-345). IGI Global. 10.4018/978-1-5225-8431-5.ch020

Vavekanand, R. (2024). Data Security and Privacy in Genomics Research: A Comparative Analysis to Protect Confidentiality. *Studies in Medical and Health Sciences*, 1(1), 23–31. 10.48185/smhs.v1i1.1158

Chapter 11

The Impact of Technological Advancements on Corporate Culture and Employee Well-Being: A Comprehensive Analysis

Navpreet Kaur Sidhu
https://orcid.org/0000-0003-3963-6623
Chitkara University, India

Swati Gupta
https://orcid.org/0009-0005-5795-8353
Chitkara University, India

Dixit Kalra
https://orcid.org/0000-0002-1744-6717
Parul University, India

ABSTRACT

The study focuses its analysis on how advanced technology impacts organizational culture and employees' well-being. In today's world, technology with its rapid growth has been brought into the workplace wherein it plays a significant role in influencing organizational development and the employees' health. It explores the dualistic nature of technology's impact and its possibilities to enrich an organizational culture and match its norms and standards with the employee's needs. It outlines and discusses the nature and extent of advancement and its impact in aspects that cuts across the

DOI: 10.4018/979-8-3693-7668-3.ch011

workplace based on the review of literature and case studies. New technologies like collaborative software, artificial intelligence, and remote working platforms have critically enhanced new ways of working that are more open, inclusive, and efficient. This has brought about a culture of lifelong learning which is very appropriate in the current complex business environment thereby leading to sustainability.

INTRODUCTION

Studying various key concepts according to their applied perspective through the lens of the relatively short history of the digital age, it is now possible to determine how the corporate world has shifted and reformed businesses as well as people's perceptions of the workplace. The changes that have taken place due to Technological innovations such as automation, Artificial intelligence, remote work technology and digital communication platforms have modified corporate gospel and effected employee's health. Implementation of these technologies has caused organizational change in terms of efficiency and unforeseen productivity making organisations transform into innovation and global competition (Brynjolfsson & McAfee, 2014). However, this technological revolution also brings difficulties that are excessive volumes of digital information, the fast erosion of the boundary between the work and the private spheres and changes in the frameworks of organisations. Such problems should be discussed by referring to the possibilities of the change of the organizational culture and the workers' individual results.

Corporate culture encompasses the beliefs, assumption and outward behaviour that people have regarding activities within an organization and determines the working behaviour of the employees. Neo-culture has been discussed as resulting from technological developments such as collaboration, flexible working environment, culture of creation and constant feedback among others by Tarafdar et al. (2007). Thus, the COVID-19 pandemic has led to the emergence of radical changes in the traditional structure of office work because of the development of remote work technology. This has meant creation of both. prospects and issues towards the maintenance of a coherent McIntosh corporate culture (Contreras et al., 2020).

Employee welfare is a core part of any organization covering mental, emotional and physical health and advancements in technology have a direct bearing on these assets. High technology has ensured that work-life balance is achieved, health care resources are easily accessed and the feeling of independence and satisfaction in a given work is improved (Bailey et. al., 2002). On the other hand, issues such as over reliance on ICTs; constant connectivity to the internet; and integration of the workplace and personal life have become crucial concerns that cause stress and burn out among the workers (Tarafdar, Pullins, & Ragu-Nathan, 2015).

This far-reaching research aims to examine how technological advances affect the company culture and the welfare of employees, what positive and negative outcomes are observable. Knowing these effects, organisations can develop ways of using the technology for effectiveness and avoid the negative impacts of technology hence creating a formidable and productive structure that is sustainable for the environment.

Conceptual Framework

The variables of interest consist of technical change as an antecedent that is expected to impact the culture of organisations and the health of staff members, with sustainability being the moderating factor in the conceptual framework. Recent innovations are a game changer for organizations, altering the organizational culture and employees' well-being. Currently, the imperatives of sustainability feature as the key considerations as far as the organisational architectures and the integration of the technologies are concerned.

Independent Variable: Technological Innovations

Technological advancements include the adoption and use of new technologies, processes or strategies that dramatically enhance efficiency, performance or organisational efficiency (Masriadi et.al., 2023). These developments mean advancement in the area of artificial intelligence, machine learning, automation and technological solutions related to digital environment.

Dependent Variables

1. **Corporate Culture**

In today's world organisations have developed the notion of corporate culture which mean system of values, beliefs and processes. This has a very large influence on how employees conduct themselves, whether they participate and even the productivity of the business or company (Kumar, 2023). That is why present-day technologies can foster a reformist corporate culture primarily based on the principles of continuous improvement, innovation and adaptability.

2. Employee Well-Being

Workplace health and wellness means the state of health of employees in all domains: psychological, physical, and emotional. This paper also showed that every innovation in technology can either have positive and negative impact to the employees and organisations through modification of work procedures, alteration of working duties and also change in the working environment (Lee et.al., 2016).

Key Element: Sustainability

In this context sustainability is the ability of an organisation to operate in a responsible manner to the environment, society and in a sound economical way for a long-term. It has the regulating function because it keeps checking whether new technology is good for the organizational culture and employees' standards and its impact is positively aligned with a sustainable vision (Safarli, 2024).

Figure 1. Model-Impact of technological Innovations on Corporate Culture and Employee Well-being

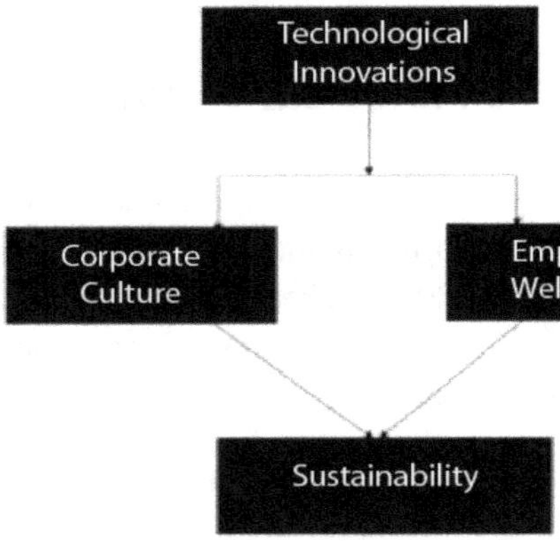

Source- Author's Own Work

The model's explanation is as follows:

- Technological Innovations act as the primary independent variable influencing both corporate culture and employee well-being.
- Corporate Culture and Employee Well-Being are the dependent variables directly affected by technological innovations.
- Sustainability therefore serves the purpose of a balancing variable, controlling the extent up to which the technologies interact with the dependent factors so as to ensure that any changes in the course of the interaction are sustainable.

This conceptual framework emphasizes the interconnectedness between technological innovations, corporate culture, and employee well-being, highlighting the role of sustainability as a crucial control factor. Realising these relationships, organisations can deliberately incorporate emerging technological advancements to foster a desired corporate culture and enhance the standards of their employees' lives while being environmentally conscious and sustainable.

Influence of Technological Innovations on Organizational Culture

This paper identifies that due to the evolution of technology over the recent past, the culture in organisations has been greatly impacted on the way organisations operate as well as their interaction with employees and the global society. In this case, technology affects various aspects in the corporate culture inclining to communication, cooperation, employee engagement and overall organizational behaviour.

1. **Communication and Collaboration:** New and advanced forms of communication tools such as Slack, Microsoft Teams, and Zoom have transformed the approach to communication across organizational employees. These technologies provoke the instant information exchange, making distance work and world cooperation more possible. About the possibility of the negative impact of digital communication technologies on the work of organisations, Poltoratska et.al (2024) investigated how such technologies led to improved co-ordination and faster decision-making.
2. **Employee Engagement:** Technology has also axed over strategies of handling employees, where before workers were given general incentives, new technologies have enabled companies to make selective rewards which targets individual workers. Employees' motivation and satisfaction in the workplace are being enhanced by the application of gaming mechanics and artificial intelligence in feedback. Research established that organisations that took advantage of the engagement tools in gamification experienced a 30% rise in productivity among the employees and a 25% reduction in staff turnover rates (Ergle et. al., 2018).
3. **Organizational Behaviour:** Integration of artificial intelligence (AI) and machine learning (ML) in HR activities has transformed the organisational dynamics by improving practices like recruitment and performance management, as well as, personal development. The use of robotic automation HR practices increases capability and equals opportunities regardless of gender (Albassam, Wael. 2023).
4. **Adaptability and Innovation Culture:** The dynamic environment continuously presenting innovations and new ways of technology calls for change adaptability and embracing knowledge as a lifelong process at an organisational level. A

study focusses on the fact that organisations belong to the culture that fosters innovations and adaptability hence are more likely to thrive when faced with technological disruptions (Chen et. al., 2019).

Various technologies are imposing a significant impact in modifying the culture of organizations. Organizations that attain effective implementation of emerging technology in their organizational structures not only improve on their communication and cooperation but also create engagement among employees and foster learning and discovery. Hence, the organisations must be ready and willing to embrace the change chasm that is precipitated by the enhancement of the technology.

Influence of Technological Innovations on Employee Well-Being

The technological changes have been realized to have influenced the lives of employees of organizations. COVID-19 epidemic has channelled the work direction towards digital transformation and changes in work environment and health status among people.

1. **Ergonomics at work** and health observing equipment and devices have helped improve the physical health aspect of people. Wearable gadgets and health applications allow employees to track their health indicators; hence, they adopt positive attitudes toward physical activity. An application of the health technologies in workplaces has led to an improved physical health and qualitative lives among the workers (Núñez-Sánchez et al., 2023).
2. **Psychological Well-being**: Technology use and interaction also increases continuously so that possibility to have mental health support is also important. Telephone therapy and mental health applications are in existence to ensure that mental health services are easily obtainable. Effective implementation of digital solutions for employees' mental health significantly enhances the mental resilience of workers (Muñoz-Ortega et al., 2024).
3. **Achieving a balance between work and personal life:** Due to the advancement in technology, management of working conditions such as telecommuting has been made easier hence helping to balancing between work and personal life. However, it should be marked that the integration of several types of borders in the chosen context can lead to the Some problems at times. Hence, in accordance with the Shukla's (2024) study, it is seen that due to continuity of connection by means of technology there is a need to set clear boundaries and guidelines regarding the staff's working hours and not allowing them to get burned out as a result.

4. **Adoption of New Technologies:** New technologies, especially automation and artificial intelligence, have made machines to work and people to get less stressed since many monotonous tasks can be performed by machines and not individuals. It has been evidenced that this transition experience boosts job satisfaction and probably productivity levels. The research carried out by Nurohmat et al. (2024) depicts how the advancement of technology within Industry 4. This has led to enhancements of the general state of employees' health and the conditions of work since 0 provided more encouraging workplace environments.

The consequences of technological developments hinder and affect the quality of life of employees in multifaceted ways. Even though they offer significant benefits as one's physical health, mental toughness, productivity and more, they come with challenges touching on balance and fatigue (Kaur, 2021). These technologies must be integrated purposely in organisations to ensure that the workplace is helpful and accommodating.

Technological Innovations and Sustainability

Over the last few years, new technologies have continued to shape managerial strategies towards sustainability, which has considerably affected corporate culture along with the health of workers. The incorporation of technological features that include AI, IoT, and blockchain has led to the promotion of sustainability within organizations through minimizing the company's ecological impact and encouraging sustainable habits across the firm. For instance, AI analytics in conjunction with IoT sensors have enhanced timed-based energy pro councils in corporate building hence minimising emissions and cutting costs substantially (Bibri 2018). In addition, through the use of blockchain, there has been improvement of accountability of each supply chain partner in ensuring sustainable sources and fair labour practices, thus being instrumental in developing ethical organizational culture and boosting the morale of the employees according to Saberi et al. (2019).

These technological advancements also enabled the practice of telework and flexible working schedules that promote employees' health through better work-life balance and less time consumption and stress on traveling (Wang et al., 2021). With current technological advancements such as cloud computing and the use of collaborative platforms, workers have been able to effectively work remotely, organizations are able to maintain productivity during periods of environmental constraints while at the same time reducing the impact of congestion by providing work from home options (Gartner, 2021). Also, other green-progressive technologies like Virtual Reality (VR) and Augmented Reality (AR) are being used for training and development; which are effective experiences that allow for drastic cuts on traveling to other locations, resulting in lesser emissions (Pantelidis, 2023).

In addition, the firms that adopt technological solutions mostly targeting sustainability end up seeing increased levels of their employee engagement and staff satisfaction. However, employees are appreciating their organizations' sustainability efforts and are more driven by the positive effect of their undertaking on the environment (Deloitte, 2020). Thus, the alignment of corporate sustainability initiatives to the employees' supreme values improves organisational culture as well as increases staff loyalty with focusing on talented personnel, securing the company's competitive advantage.

Future Implications of Technological Advancements on Corporate Culture and Employee Well-Being

1. **Escalated utilisation of remote work and hybrid models:**

The COVID-19 epidemic has performed a great service to distantly carried out work through significant technological advancement. Further consequences include long-term changes in company policies in order to provide for the employees' alternating home and office work schedules (Bartsch et.al.,2021). While this idea may help in improving work-life interfaces, it also has problems in maintaining a single corporate culture across the organization.

2. **Employee surveillance and the issue of privacy:**

To it, since modern organisations subject their employees to high-level monitoring devices in an effort to determine their rate of output, the in evitable worry that comes with the concept of privacy loses ground. In the future, the human resource policies and legal guidelines protecting the employee's data are expected to be progressively enhanced while searching for the ways to optimize the business processes drives (Ball, 2010).

3. **Implementation of employee well-being programmes:**

Implementation of employee well-being programmes will help in the development of vast systems on employee health. Such programmes could include stress management tools, health check-ups and organizational activities aimed at improving the quality of employees' health, which greatly affects the levels of satisfaction and performance (Pfeffer,2018).

4. **Flexibility in Skills Called for and Continuing Education:**

As technology increases in the market, so do the skills required in the employment market. The constant learning and the continuation of training and development, to attain new knowledge as well as new competencies, will be required for the employees to remain relevant. Employers will have to set some amount of money and time for training and development of its human resource.

5. **AI & automation's influence on corporate culture:**

AI and automation are expected to takeover routine and boring tasks thereby allowing one to focus on creative and problem-solving tasks (Autor,2018). It will need a shift in business culture particularly on how people deal with each other to ensure that they come up with enhanced cooperation and better innovative solutions.

6. **Ethical and Social ramifications:**

New technologies introduced in organizations create ethical and social issues. Therefore, in order to offer benefits that are at the same level for every worker, it is necessary to focus on the problems which have relation to the fairness, problem of differentiation of the AI algorithms, and the problems involving digital divide (Floridi et.al., 2016).

7. **Improving Freedom and Authority of Employees:**

The expectations of future technology advancements are expected to provide employees more freedom and power in the workplace (Spreitzer et.al.,2005). One of the implications of this is that there are positive impacts that power and authority have on the employees as they are made to feel satisfied and motivated in their duties hence promoting a healthy work environment within the firm.

CONCLUSION

Technological advancements play a critical role of spearheading the sustainability processes within corporate contexts, as they play a crucial role in transforming organizational culture as well as improving the health of the workforce (Dawra et.al.,2024). Thus, the application of IT solutions and innovative tools helps organizations to achieve environmental objectives, promote the culture of sustainability among the workforce and increase the efficacy of its functioning. With these technologies evolving over time, they stand out as becoming even more important in driving sustainable corporate practices and improving employee health.

REFERENCES

Albassam, W. (2023). The Power of Artificial Intelligence in Recruitment: An Analytical Review of Current AI-Based Recruitment Strategies. International Journal of Professional Business Review. 8.

Autor, D., & Salomons, A. (2018). Is automation labor-displacing? *Productivity growth, employment, and the labor share* (No. w24871). National Bureau of Economic Research.

Bailey, D. E., & Kurland, N. B. (2002). A review of telework research: Findings, new directions, and lessons for the study of modern work. *Journal of Organizational Behavior*, 23(4), 383–400. 10.1002/job.144

Ball, K. (2010). Workplace surveillance: An overview. *Labor History*, 51(1), 87–106. 10.1080/00236561003654776

Bartsch, S., Weber, E., Büttgen, M., & Huber, A. (2021). Leadership matters in crisis-induced digital transformation: How to lead service employees effectively during the COVID-19 pandemic. *Journal of Service Management*, 32(1), 71–85. 10.1108/JOSM-05-2020-0160

Bibri, S. E. (2018). The IoT for smart sustainable cities of the future: An analytical framework for sensor-based big data applications for environmental sustainability. *Sustainable Cities and Society*, 38, 230–253. 10.1016/j.scs.2017.12.034

Brynjolfsson, E., & McAfee, A. (2014). *The second machine age: Work, progress, and prosperity in a time of brilliant technologies*. WW Norton & Company.

Chen, Viardot, & Brem. (2019). Innovation and innovation management. .10.4324/9781315276670-1

Contreras, F., Baykal, E., & Abid, G. (2020). E-leadership and teleworking in times of COVID-19 and beyond: What we know and where do we go. *Frontiers in Psychology*, 11, 590271. 10.3389/fpsyg.2020.59027133362656

Dawra, S., Sharma, S., & Tennin, K. L. (2024). Creating a Neuro-Inclusive Workplace: Challenges and Opportunities for Leaders. *Building Organizational Resilience with Neuroleadership*, 1-13.

Deloitte. (2020). 2020 Deloitte Global Human Capital Trends. Retrieved from https://www2.deloitte.com/global/en/pages/human-capital/articles/introduction-human-capital-trends.html

Ērgle & Ludviga. (2018). Use of gamification in human resource management: impact on engagement and satisfaction. .10.3846/bm.2018.45

Floridi, L., & Taddeo, M. (2016). What is data ethics? *Philosophical Transactions. Series A, Mathematical, Physical, and Engineering Sciences*, 374(2083), 20160360. 10.1098/rsta.2016.036028336805

Gartner. (n.d.). How HR Leaders Can Successfully Manage the Future of Work. https://www.gartner.com/smarterwithgartner/how-hrleaders-can-successfully-manage-the-future-of-work/

Gupta, M., Arora, R., & Taneja, S. (2023). Bibliometric analysis on employee engagement and human resource management. In *Enhancing Customer Engagement Through Location-Based Marketing* (pp. 185–205). IGI Global. 10.4018/978-1-6684-8177-6.ch013

Kaur, N. (2021). Digitalization and Work Behaviour: A Paradigm Shift. In *Big Data Analytics for Improved Accuracy, Efficiency, and Decision Making in Digital Marketing* (pp. 102-123). IGI Global. 10.4018/978-1-7998-7231-3.ch008

Kumar, M. (2023). Digital Transformation and Organizational Change.

Lee, H. W., Pak, J., Kim, S., & Li, L.-Z. (2016). Effects of Human Resource Management Systems on Employee Proactivity and Group Innovation. *Journal of Management*, 45(2), 819–846. Advance online publication. 10.1177/0149206316680029

Masriadi, Dasmadi, Ekaningrum, Hidayat, & Yuliati. (2023). Exploring the Future of Work: Impact of Automation and Artificial Intelligence on Employment. Endless: International Journal of Future Studies. 6. 125-136. .10.54783/endlessjournal.v6i1.131

Muñoz-Ortega, S., Santamaría-Guayaquil, D., Pluas-Borja, J., Alvarado-Villa, G., Sandoval, V., Alvarado, R., Cherrez-Ojeda, I., & Faytong-Haro, M. (2024). Mental Health in Healthcare Workers Post-COVID-19: A Latin American Review and Insights into Personalized Management Strategies. *Journal of Personalized Medicine*, 14(7), 680. 10.3390/jpm1407068039063934

Núñez-Sánchez, J. M., Angosto, S., Díaz-Benito, V. J., & Chacón, R. G. (2023) Use of technologies for the creation of new hybrid services based on physical.

Nurohmat, Khuzaini, & Shaddiq. (2024). The Influence of Industry 4.0 on Work Behavior and Employee Productivity in Technology Companies.

Pantelidis, C. (2023). *Exploring Place Attachment Theory in VR of a Rural Destination: The effect of VR Experience on domestic tourists' attachment to places* (Doctoral dissertation, Manchester Metropolitan University).

Pfeffer, J. (2018). Dying for a paycheck: How modern management harms employee health and company performance—and what we can do about it.

Poltoratska, A., Garmash, O., Marchenko, I., Garmatiuk, O., & Stefinin, V. (2024). Analysis of the Impact of Digital Communication Tools on Organizational Culture and Communication Effectiveness within Enterprises. *International Journal of Religion.*, 5(10), 1684–1692. 10.61707/zaa7yf77

Saberi, S., Kouhizadeh, M., Sarkis, J., & Shen, L. (2019). Blockchain technology and its relationships to sustainable supply chain management. *International Journal of Production Research*, 57(7), 2117–2135. 10.1080/00207543.2018.1533261

Safarli. (2024). the Role of green technologies in shaping sustainable cities: Socio-economic benefits and challenges.

Shukla, M. (2024). Talent management strategies in Indian organisation: Trends and challenges. *Ajasra, 13*(6), 325-335.

Spreitzer, G. M., & Doneson, D. (2005). Musings on the past and future of employee empowerment. *Handbook of organizational development, 4*, 5-10.

Tarafdar, M., Pullins, E. B., & Ragu-Nathan, T. S. (2015). Technostress: Negative effect on performance and possible mitigations. *Information Systems Journal*, 25(2), 103–132. 10.1111/isj.12042

Tarafdar, M., Tu, Q., Ragu-Nathan, B. S., & Ragu-Nathan, T. S. (2007). The impact of technostress on role stress and productivity. *Journal of Management Information Systems*, 24(1), 301–328. 10.2753/MIS0742-1222240109

Wang, B., Liu, Y., Qian, J., & Parker, S. K. (2021). Achieving effective remote working during the COVID-19 pandemic: A work design perspective. *Applied Psychology*, 70(1), 16–59. 10.1111/apps.1229033230359

Chapter 12
Navigating the Future: A Comprehensive Study on Robo-Advisors in Wealth Management

Ravneet Kaur
https://orcid.org/0009-0008-3101-5738
Chandigarh University, India

Kapil Kumar Aggarwal
https://orcid.org/0000-0002-2752-9495
Chandigarh University, India

ABSTRACT

Artificial intelligence is revolutionizing various industries, including travel, education, health, and investment. Robo advisory models (RAs) have increased assets under management (AUM) due to price, transparency, and competitiveness. Although not widely used, robo-advisors are becoming more cost-effective than traditional human advice services. This study explores the current state of robo-advisors and their potential to transform wealth management by facilitating effective decision-making processes and introducing a new era of financial advising services.

1. INTRODUCTION

AI and robotic has transformed all the sectors and fields in economy, be it health-care, education or Investment. Wealth Management is the professional management of an individual's financial assets and investments, has undergone exponential transformation with inclusion of AI and Robotics in investments. The integration of finance and technology, commonly known as "fintech," has fundamentally transformed the operations of financial institutions and advisers, presenting novel

DOI: 10.4018/979-8-3693-7668-3.ch012

opportunities for increased productivity, tailored services, and enhanced decision-making. The emergence of robo-advisors (RAs) in recent times has posed a threat to the conventional financial management sector. The AUM of RAs have increased significantly due to their competitive pricing, transparency, and services, as well as the use of financial technology, which leads to higher predicted returns and reduces subjective human interference.

This chapter focuses on the rise and evolution of robo advisory, its need and future potential in wealth management. This chapter also tries to draw difference between conventional and automated system of wealth management.

2. LITERATURE ANALYSIS OF RELATED TOPICS

2.1 Wealth Management

Wealth management is composed of two words 'Wealth' meaning money and money assets and 'Management' meaning to organize and distribute, therefore it can be said that wealth management is the term used for organizing or distributing money of the organization or individual in such a manner that maximum profits can be generated by reducing risk. The notion of Wealth Management developed in the United States throughout the 1990s. It is an investment advice service that offers individuals private banking, asset management, taxes advising, and portfolio management for their financial planning needs(Kapil Kumar Aggarwal & Taneja, 2024).

Wealth Management refers to the practice of developing customized strategies to manage a client's assets, including both liquid and non-liquid assets, in accordance with a pre-established financial plan. This involves executing the strategies with varying levels of customer involvement. Essential assistance provided under wealth management encompass economic and legitimate advice, management of assets, tax strategizing, offshore solutions, and future planning. (Ślązak, 2018).

Put simply, it is essentially guidance or support in managing an individual's financial requirements, sometimes referred to as investment advice. Investors are provided with bundles of services that aim to achieve two primary objectives: the growth and safety of their current assets(Poonam & Aggarwal, 2023). Wealth management providers, often offer services provided by private banks. In addition to monetary services, these firms also offers services in investment and management of luxury goods, products, diamonds, numismatics, and condo hotels(K.K. Aggarwal et al., 2023). These non-financial offerings are customized, and their terms are determined through talks with individual clients. They typically involve higher costs and a greater level of associated risk (Dziawgo, 2013; Kozińska, 2019). When assessing uncertainty, financial management and banking services provided by private sector, one must examine five primary risk categories: The risks that a bank faces can be categorized into several types: macroeconomic risk, which is influenced by macroeconomic factors; business risk, which arises from internal issues specific to the bank; reputation risk, which is associated with the bank's image; model risk, which occurs when incorrect financial models are used; and non-compliance risk (Niedziółka, 2018).

Therefore, it can be stated that financial and wealth management is associated with creation of financial portfolio and management of various assets of the organization, such that risk can be minimized and profits are maximized through investments. Conventional services offered by financial institutions and other established financial service providers are currently encountering substantial obstacles from emerging technology-driven solutions that offer enhanced accessibility, reduced prices, or improved operational procedures(Kapil Kumar Aggarwal & Singhal, 2024).

2.2 AI and ROBO in Wealth Management

Robo-advisors, as defined by the Sovereign Wealth Fund Institute (SWFI), are financial advisors that offer web-based portfolio management with minimal human involvement. These advisors primarily rely on algorithms and formulae to make investment decisions (SWFI [2015].

According to (Fisch, Laboure, &Turner,2019), RA framework can be divided into four stages: (i) gathering relevant customer's investment and financial income and savings data (ii) estimating risk and uncertainty using uniformed surveys to divide customers into different categories of risk averseness, (iii) creating financial portfolios for managing wealth of customers based on collected data, and (iv) monitoring and maintaining continuous financial investment management. The typical RA workflow consists of four main stages: (i) Designing AI enabled system, (ii) Customizing through adaption, (iii) personalization and (iv)Adjusting and Restoring (iv) (Jungetal.,2019).

Evolution of robotic-advisory can be attributed to the practice of trading equities through telephone communication between brokers and their clients(K.K. Aggarwal & Agrawal, 2024). Since then, it has exponentially grown into financial advisory services. Robotic Advisors can fulfill the desired benefits that customers expect. In conventional service encounters, the social contact and relationship between human advisors and clients have a substantial impact on customers' sense of advisor trustworthiness and their inclination to accept their advice (Soderberg, 2013). Most research studies show that, the efficacy of recommended financial products is equally dependent on the design of digital advisory services. When assessing digital systems that include human participation, it is expected that an automated system will offer more clarity regarding expenses and the advice procedure (Tertilt

& Scholz, 2018). Hence it can be stated that ROBO-Advisors has positively transformed the facet of financial sector and has empowered and eased more efficient system for financial advisory services.

3. METHODOLOGY OF RESEARCH

This paper is based on qualitative data collected through secondary sources. The study is exponential survey of already existing research. This chapter focuses on establishing conceptual understanding of AI and RA enabled wealth management, its need, use and benefits. This paper also tries to foresee the future growth of automated technological advisors in financial sector.

4. CONCEPTUALIZING WEALTH MANAGEMENT

Financial Management revolves around three fundamental activities:

i. **Wealth Development and Goal Setting:** The financial advisor instructs the customer and employs a strategy centered on diversification and asset allocation, taking into consideration the clients' objectives and preferences.
ii. **Portfolio Management:** Wealth preservation is achieved through meticulous portfolio management that mitigates risks (risk management), Asset management strategies customed to meet the needs of customers, safeguarding against unforeseen circumstances (such as illness or disability), and regularly rebalancing the portfolio to ensure stability and instill confidence in the client.
iii. **Tax optimization**: techniques that aid in effective tax planning and reducing tax liabilities.

5. FACTORS AFFECTING WEALTH MANAGEMENT

Figure 1. Showing Factors Affecting Wealth Management Decisions

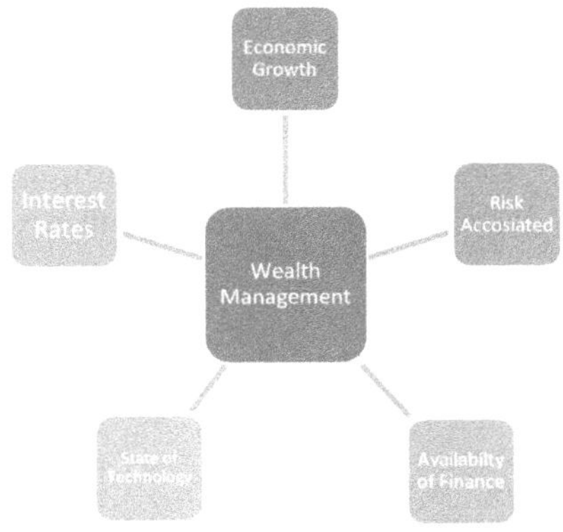

Source: www.goole.com

6. ROBOTIC ENABLED WEALTH MANAGEMENT

The 2023 Global Asset and Wealth Management Survey by PwC provides insights into the sector's current state and future trends until 2027. The survey highlights that the industry is undergoing significant transformation due to factors such as extensive consolidation, market instability, and disruptive technological advancements. One of the survey's most notable discoveries is the expected surge in robo-advice. According to the report, the assets overseen by these digital platforms, which are powered by algorithms and are becoming more advanced with artificial intelligence, are expected to reach about US$6 trillion by 2027. This is a nearly twofold increase from the figure recorded in 2022 (Pwc research forecast, 2023).

6.1 Automated Wealth Management in Asia

Our research indicates that the Asian robotic-advisors in Asian nations are still in the early stages of development. Due to their delayed growth, the assets under management used by the Asian nations are comparatively of smaller size than used by developed countries like USA. Furthermore, the limited availability of ETF's traded by Asian exchanges has led to a slower advancement of the latent management strategy in comparison to U.S. A's robotic-advisors. Nevertheless, as per development prospects for Asian countries robotic-advisors are exceedingly promising due to their delayed entry into the market.

However, robotic-advisors in Asia, by and large, extend their advises to global investors. For instance, Clipper enables investors in China to access global portfolio management with certain restrictions. The robo-advisor services are transforming significantly in Chinese economy. Consequently, the economy is poised to witness an increase in online sales of financial advisory and management services for the affluent population of China by major financial institutions and key players such as Yu'e Bao, in addition to substantial robo-advisors equipped with robust distribution capabilities.

6.2 Steps in Investment Management

Figure 2. Showing Steps in Investment Management using Automated Technologies

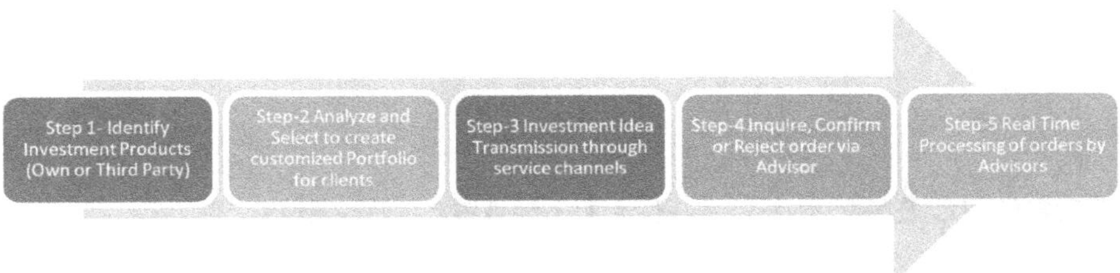

Source: Phoon, 2018

6.3 Famous Robotic- Advisors used in financial and Wealth Management

The following information shows the use of RAs in financial wealth management:

A. WEALTHFRONT (USA):
- It is a private firm, established in 2008.
- Its main clientele consists of individuals in their twenties and thirties, a significant portion of whom are novice investors. The company also serves affluent individuals, philanthropic institutions, and businesses.
- The firm uses advanced algorithm techniques to provide financial guidance to its clients.
- It utilizes passive investing combined with mean-variance analyzing to generate portfolios. Investment options encompass a range of securities such as traded stocks, municipal bonds, real estate etc.

- It has best performance of portfolio against S and P 500 Index.

B. NUTMEG (UK):
   - Established in 2011.
   - It incorporates a simplified process of getting started by assessing a client's investing objectives and risk tolerance. A diversified portfolio, consisting of various assets across multiple nations and industry sectors, is created based on the provided inputs to fulfill the client's requirements.

C. CLIPPER ADVISOR (CHINA):
   - It was the fir Robo Advisor in China, launched in 2015.
   - It streamlines the process of opening accounts by using straightforward approach to assess clients' goals for returns and risk preference. Based on the above assessment, Clipper offers customized investment portfolio for its clients.
   - It uses automated Risk management system to analyze market situation.

D. YOUYU (HONG KONG):
   - Established in 2017.
   - Customers undergo a straightforward risk-assessment process to determine their willingness to take on risks. Following the evaluation, clients are advised to consider a model portfolio.
   - Youyu caters to retail investors seeking easy mobile-based fund investing services, as opposed to in-person consultations provided by banks and other fund management companies.

E. BAMBU/CROSSBRIDGE CAPITAL (SINGAPORE):
   - It was launched in 2016.
   - It is a B2B Robo Advisor.
   - It provides tailored solutions to cater to High-Net-Worth Individuals (HNWIs), affluent investors, and retail investors. The system utilizes exclusive algorithms for managing portfolios and analyzes third-party data and investor information to offer goal-oriented solutions for various types of investors.

7. MODES OF ROBO ADVISORS

A. **Auto Pilot Mode: -** This mode is designed to provide a starting point for novice investors. Several fundamental packages are provided, and investment is commenced based on individual requirements. Scripbox exemplifies this mode.

B. **Direct plan-based Mode:** This mode aims to eliminate biased recommendations, undisclosed charges, and commissions. It entails minimal guidance and a low cost associated with one's transactions. Organizations such as Invezta.com utilize it.
C. **Goal-based advisory:** Within this mode, the investor is granted permission to input their personal details, following which it offers fundamental advising assistance to aid them in attaining their long-term investment objectives. In addition to providing advice services, it facilitates transactions involving shares, company deposits, and corporate bonds.
D. **Full-service:** This method aims to offer comprehensive advisory services by gathering extensive information from the investor through a detailed questionnaire completed by the investor. The questionnaire is used to evaluate the investor's risk tolerance, financial objectives, saving and spending habits, and current asset and obligation status.

8. MODELS OF ROBO ADVISORS IN WEALTH MANAGEMENT

A. **D2C Model: -** These platforms are web-based services that offer automated portfolio management using algorithmic methods, without any involvement from humans.
B. **B2B Model: -** refers to digital platform providers that assist conventional advisors in delivering a digital wealth management solution.
C. **Hybrid Model: -** Hybrid firms encompass businesses that offer customized services and actively oversee portfolios, while also incorporating computer-generated recommendations for portfolio management.

9. BENEFITS OF USING ROBO ADVISORS IN WEALTH MANAGEMENT

- **Enhancing Efficiency and Productivity:** AI and Robo Advisors makes human advisors more effective by automating the time-consuming tasks such as creating notes, filling out paper work etc. It reduces manual work so that human advisors can create deeper customer relationships, as they can spend more time enhanced client interaction.

- **Storage of Institutional Memory:** AI and Robbo Advisors can effectively store large information as per interaction with clients. Hence retaining better record of institutional memory.
- **Augmented Advisory and Next Best Action (NBA):** are two concepts that include using advanced technology to provide enhanced guidance and provide optimal recommendations. With the emergence of technologies that enable accessing a larger amount of customer data, artificial intelligence (AI) can assist financial institutions (FIs) in utilizing this continuously expanding resource to generate personalized recommendations for individual clients. Morgan Stanley has created an NBA system that utilizes machine learning to analyze clients' life events and produce highly tailored investment recommendations within a very short period of time.
- **Cost Effective:** These advisers provide cost-effective services in comparison to conventional financial advisors. Due to the absence or limited presence of human supervision, the online platform can provide services at a much-reduced cost compared to the expenses associated with human resources.
- **Easy and Fast Processing:** AI in wealth management is mostly utilized for customer request handling. In order to optimize the handling of inquiries, AWM enterprises might employ chatbots powered by artificial intelligence. These chatbots not only offer immediate responses to user queries, but also guarantee continuous help at all times. AI chatbots, unlike human professionals, possess the ability to operate continuously without experiencing fatigue and handle several requests concurrently.
- **Customized Processing:** By implementing artificial intelligence in wealth management, firms have the ability to greatly customize the customer experience to an exceptional degree. Through the utilization of AI and machine learning, businesses examine extensive sets of data, including in-app user behavior, previous actions, feedback, and transaction records. An artificial intelligence system has the capability to divide the audience into several segments and generate a comprehensive understanding of each client from all angles. As a result of this, AWM firms are able to provide personalized advice about investing, financial health, and asset management. AI enables institutions to gain a comprehensive understanding of client needs, so allowing them to optimize marketing efforts, enhance lifetime value, and mitigate customer attrition.
- **Tax planning:** AI is extensively utilized in wealth management for tax planning purposes. Given that intelligent tax planning allows consumers to reduce expenses, it should be a significant component of an individual investor's strategy. Nevertheless, tax planning is commonly perceived as an intricate procedure that necessitates considering various factors, such as the

timing of income and purchases. Furthermore, accountants are required to acquire knowledge in many tax planning options, compile tax statements, and seek advice from financial specialists. Consequently, they invest significant amounts of time and resources, with the possibility of not ultimately identifying the optimal choice.

10. ROAD BLOCKS IN USING RAS IN FINANCIAL WEALTH MANAGEMENT

- AI and Robotic advisors cannot provide personalized advice beyond the data fed into them. Life altering situations, special circumstances or events or intricate investment goals are better addressed by human advisors.
- Complying with dynamic financial laws becomes challenging for AI and Robo advisors as updating or upgrading them with continuously changing legal environment can reduce its efficiency.
- Security of sensitive financial data stored on AI poses a great deal of cyber threats and unauthorized access to client's confidential information.
- Resistance from employees as automated technologies create threat to employee's job, by replacing human resources.
- AI systems may have inherent biases in their algorithms due to the training data they are exposed to. These biases can result in less-than-ideal suggestions and decision-making. Maintaining equity and rectifying algorithmic prejudices is a continuous and complex task.
- Building trust among clients about the effectiveness of automated technologies in financial advisory services is a challenging task, specially where clients lack digital education.
- Scalability becomes essential as the clientele utilizing robo-advisory services grows. The problem lies in ensuring that the system can accommodate an expanding user population while maintaining optimal performance and response times.

11. CONCLUSION

Based on the above exponential study of inculcation of AI and Robo-Advisors in wealth management, it can be concluded that use of automated technologies in financial advisory services has been globally used and adopted, and its use has fostered the evaluation of risk and opportunities in market and has resulted in creating more effective, productive and customized portfolio of investments for clients. The impact of wealth tech is far reached and positive. Surely the Wealth Tech technologies are navigating future of financial services towards cost and time effective system of providing algorithm-based asset allocation, portfolio creation etc, by optimally utilizing the available technologies. Many companies are also adopting these automated technologies to step up their game, retain clients and remain updated with the latest trends of digitalization and automation. It is not wrong to say that Robotics is the new face of service providers in the market.

REFERENCES

Aggarwal & Taneja. (2024). Impact of organizational culture on employee satisfaction, motivation, performance, commitment, and engagement in educational institutions. In *Building Sustainable Human Resources Management Practices for Businesses*. IGI Global. 10.4018/979-8-3693-1994-9.ch012

Aggarwal, K. K., & Agrawal, S. (2024). Artificial intelligence and its role in financial market. In *Global Financial Analytics and Business Forecasting*.

Aggarwal, K. K., Kaur, R., & Lakhera, G. (2023). Green finance: Addressing environmental challenges through sustainable investments. In *Sustainable Investments in Green Finance*. 10.4018/979-8-3693-1388-6.ch010

Aggarwal, K. K., & Singhal, S. (2024). The effect of perceived philanthropic activities of corporate organisations on consumer's purchase intention. *International Journal of Business and Globalisation*, 36(2–3), 212–224. 10.1504/IJBG.2024.137401

Fisch, J. E., Laboure, M., & Turner, J. A. (2019). The emergence of the robo-advisor. In Fisch, J. E., Laboure, M., & Turner, J. A. (Eds.), *The disruptive impact of Fin Tech on retirement systems* (pp. 13–37). Oxford University Press. 10.1093/oso/9780198845553.003.0002

Jung, D., Dorner, V., Weinhardt, C., & Pusmaz, H. (2018). Designing a robo-advisor for risk-averse, low-budgetconsumers. *Electronic Markets*, 28(3), 367–380. 10.1007/s12525-017-0279-9

Jung, D., Glaser, F., & Kopplin, W. (2019). Robo-advisory: Opportunities and risks for the future of financial advisory. In V. Nissen (Ed.), Advances in consulting research (pp. 405–427). Springer International Publishing. https://doi.org/.10.1007/978-3-31995999-3_20

Phoon & Koh. (2018). Robo-advisors and wealth management. *Journal of Alternative Investments*. 20, (3), 79-94.

Poonam, & Aggarwal, K. K. (2023). Analysis of the impact of COVID-19 on the stock market and capability of investing strategies. *AIP Conference Proceedings, 2782*. 10.1063/5.0154174

Pwc Research Report, 2023 https://www.pwc.com/gx/en/issues/c-suite-insights/the-leadership-agenda/ai-and-wealth-management-a-new-era.html

Ślązak, E. (2018). FinTech i jego znaczenie dla usług finansowych (FinTech and its' importance for financial services). Bank Spółdzielczy no 4/590.

Tertilt, M., & Scholz, P. (2018). To advise,or not to advise—How robo-advisors evaluate the risk preferences of private investors. *The Journal of Wealth Management*, 21(2), 70–84. 10.3905/jwm.2018.21.2.070

Chapter 13
Enhancing Employee Engagement in Genetic Research through Effective Communication Strategies:
HRM Insights

Anupal Mongia
https://orcid.org/0009-0002-4224-7725
Mody University of Science and Technology, India

Aparna Saluja
Independent Researcher, Canada

Harwinder Singh
York University, Canada

ABSTRACT

Employee engagement in genetic research is critical for fostering innovation and productivity within organizations. This study explores the role of effective communication strategies in enhancing engagement among employees involved in genetic research. Human resource management (HRM) insights are leveraged to identify key factors influencing engagement, such as clear communication of research goals, transparent sharing of findings, and fostering a collaborative work environment. The chapter highlights the importance of aligning organizational objectives with individual aspirations through targeted communication efforts. By promoting open

DOI: 10.4018/979-8-3693-7668-3.ch013

dialogue and knowledge sharing, organizations can cultivate a culture of engagement that motivates employees to contribute actively to research endeavors.

INTRODUCTION

Employee engagement is the extent to which employees can identify with the organization for which they work and this limits their ability and willingness to go the extra mile in their work (Kwonn & Kim, 2020). It is characterized by passion, commitment and possessed ties with what one does and the people one works with. When it comes to genetics research, an application of the discussed concept is especially important in terms of the facts that the work process can be difficult, challenging, and even revolutionary (Yadav et al., 2022). The employees who are motivated are more likely to present a higher degree of creativity, productivity, and persistence in their work that could bring progressive changes in the existing knowledge for the development of this scientific field. Engagement in the workplace has its merit for the following explanations (figure1).

Figure 1. Importance of employee engagement in the context of genetic research

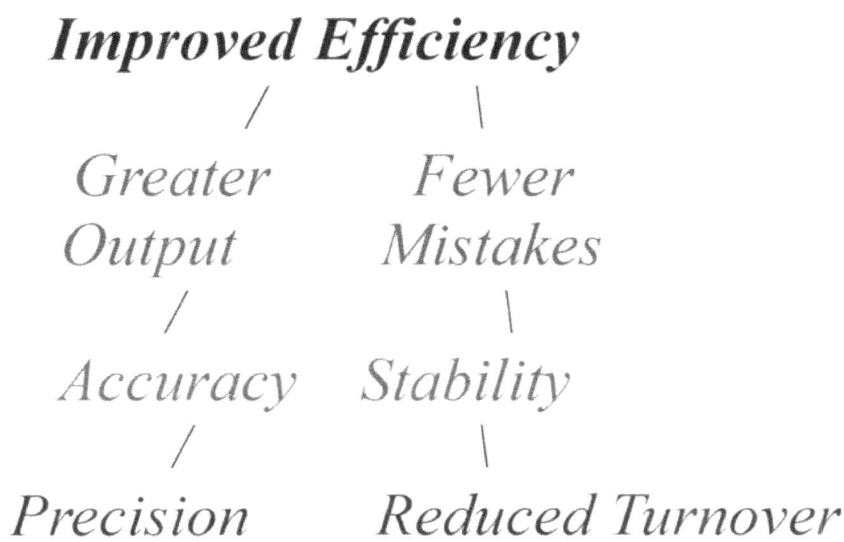

Firstly, it results in improved efficiency of individual and organization. Employees who are committed and enthusiastic about their job will offer a better output, carry out their assignment with greater accuracy and with fewer mistakes (Saks, 2022).

This is where the advantages of engagement can be summed up in the role of engagement in genetic research where the levels of accuracy and precision in the results are critical. Secondly, engaged employees will think about the stability and future of their organization and thus will not quit hence cutting down on turnover. This is especially the case in genetic research where the specialized knowledge and skills needed are different, and losing key personnel proves disastrous to ongoing projects and in the advancement of research. In addition, engagement construct a positive organizational climate (Verčič, 2021). Generally, the general working environment is enhanced by better relations that employees demonstrate with their fellow employees/associates as well as their seniors. This is particularly important in the genetic disease research because such work frequently requires the combined efforts of individuals and teams to solve the problems and come up with the discoveries that can help cure some diseases. Finally, there is a relationship between engagement and innovation among the employees. Employees with commitment are more likely to perform their duties beyond expectations and come up with new ideas and solutions to the problem, thus propelling genetic research (Nienaber & Martins, 2020).

UNIQUE CHALLENGES IN THE GENETIC RESEARCH FIELD

However, promoting EE in the genetic research field, while seemingly crucial, isn't without its difficulties. A main issue is that activities in this area are, as a rule, technical and require a qualitative and individual approach. Genetic research therefore requires many elaborate procedures, expensive tools and equipment, and rigorous knowledge in molecular biology and genetics (Pemmasani et al., 2020). This can be a challenging experience for an employee to maintain his or her focus on the job because he or she is likely to face repeated setbacks or because the work starts to feel repetitive at times. The other problem is that there is often great pressure of result performance. Research in genetic is mostly motivated by the quest for breakthroughs that will help promote the understanding of genetic diseases hence enhancing patenting and fund raising. It is not uncommon for such pressure to lead to higher stress levels where organizational concern may be more inclined to the result than to employee satisfaction. Also, the competitiveness of the situation can lead to the complete severing of team era where there is more emphasis put on achievement or a personnel job title (Hardcastle et al., 2024).

Funding is also another factor that affects the research on genetic disorders and is also a challenge in employee engagement. Since research projects are the primary income source for universities in many cases, their funding is unpredictable and may decrease. This can create feelings of insecurity and anxiety among the personnel, which in turn negatively impacts the staff's dedication towards organizational tasks. In addition, the ethical and/or legal issues involved in genetic research can present a total level of difficulty on their own. Deciding on the ethical issues of genetic engineering, privacy issues relating to genetic information, and the rules and regulations of the law can be rather complicated and can take its toll on the motivation of the working researchers due to what may be perceived as a mountain of paperwork (Leroy, 2020).

THE ROLE OF COMMUNICATION IN EMPLOYEE ENGAGEMENT (FIGURE 2)

Figure 2. Role of efficient communication in enhancing employees' participation in genetic research

It is imperative to have efficient communication as a key enabler of employees' participation in genetic research. Clarity and simplicity of communication, together with the continuous use of communication, in management–employee relations foster the creation of trust that is vital for engagement. When the employees are highly involved in the decision making processes of the organization, their feelings of the organization's inclusion are heightened. (Imam et al., 2023).

Especially in genetic research, tasks can be rather large and multifaceted, and it is crucial to develop a framework that is precise when it comes to conveyances between the working groups. It is important to set the proper frequency of the meetings, review the relevant project's status, and provide constant feedback pathways for the team. Another feature of communication is the identification of inputs or outcomes of the employees and responding to them. Praise, which may include acknowledging accomplishments, encouraging the client to continue with the work, and giving positive remarks regarding his or her performance or work, increases motivation and engagement besides that Criticism on the other hand, may entail offering negative feedback on the work done by the client, and solving any issues that may have arisen during the work done; thus an improvement on motivation and engagement (Jiang & Shen, 2023).

However, there is the need to enhance communication in the management of the issues connected with genetic research. For example, effective communication on aspects such as the projected outcomes, requirements, and time frames of a particular project may act as measures to reduce tension and pressure when it comes to delivering the same. Likewise, the proper discussing of funding options and employment stability issues would help ease employees' concerns (Almerri, 2023).

This is especially appreciated in the genetic research since excellent leadership communication will enhance the achievement of organizational goals. Recalling the above-mentioned key factors of employee engagement, it can be stressed that leaders who are effective in communicating the research's goals and objectives to the team and who are receptive to the opinions of their subordinates as well as who show concern for employee well-being can improve employees' engagement levels. Leadership communication encompasses not only the process of transmitting information to the subordinates but also perceiving the views of other employees, and promoting the culture of recognition in the organization (Kang & Sung, 2017).

External communication also affects the levels of engagement of its employees in an organization not forgetting the internal communications. Passing on the information about the consequences and relevance of the research to other people including funding agencies, the scientific communities, as well as the members of the general public can trigger the sense of pride and purpose among the staff. It is crucial to note that understanding that one's work is useful in making contribution to science and the society helps in boosting engagement. To sum up, it is necessary to state that this work has identified employee engagement as one of the most important contributors to the improvement of genetic research (Mishra et al., 2014).

UNDERSTANDING THE WORKFORCE IN GENETIC RESEARCH

Genetic workforce studies people who perform different tasks in genetic research, whereby civilians are quite different in demographic nature and psycho-graphic variables. In terms of demographics, genetic researchers are educated to a high standard, many have PhD or similar in genetics, molecular biology, bio-informatics or related subject areas. This educational requirement means high level of intelligence and willingness to obtain considerable academic background. The field seems to encompass people from all age brackets but particularly attracts young people in their early to mid-twenties and thirties because the area of genetics and genetic research has only recently started to boom and expand (Pemmasani et al., 2020).

Bias in the genetic research with regard to gender has slightly been rectified in recent past though the problem is still featured. Though the number of women choosing computer science has risen, there is still prejudice with reference to career progression and numbers of women in high ranking positions. Measures for gender sensitivity and diversity in organizations have continued and are focused on eradicating such differences while providing proper career development for all the concerned researchers. Moreover, gene research offers the world talent attraction as human capital is U.S.' most valuable asset. The participants of research studies can indeed hail from culturally diverse backgrounds; thus, there are numerous cultures as well as ethnicity involved. It promotes diversity, which is important so that individuals can devise different ways of doing things, integral to science development (Hardcastle et al., 2024).

The genetic researchers are intrinsically motivated and are motivated by passion for discovery. It is frequently characterized by interest and the need to comprehend the molecular nature of life in general. These intrinsic motivations are supported by the primary altruistic motive: the development of knowledge in the topic area and enhancement of people's well-being, which is typical for the given field. The common personal characteristics of genetic researchers include aspects of work style such as attention to detail, analysis, and systematization when solving problems. They have good critical thinking and are capable of analyzing large volumes of scientific information and experiment processes (Leroy, 2020).

Furthermore, most of the persons engaged in genetic research possess inherent dental core solidity and persistent commitment, which helps to counterbalance the constant problems and failures, typical for scientific projects. For this reason, they also often display a growth mindset and treat mean failure as a possibility to learn from it. This attitude is particularly critical in a domain, one can hardly guarantee the outcomes of experiments or implement changes quickly (Pemmasani et al., 2020).

MOTIVATIONS AND EXPECTATIONS OF GENETIC RESEARCHERS

By dissecting the various incentives for people involved in genetic research, this paper shows that the incentives are both intrinsic and extrinsic in nature. Arguably, the majority of researchers, intrinsically at least, have the love for science and the want to partake in the enhancement of knowledge. The ability to contribute to scientific advancements and progress in terms of research or ability to come up with new treatments for genetic diseases or better produce yield in crops in a powerful motivation (O'daniel & Haga, 2011). These are profound sources of motivation; the 'sense of calling,' the 'calling system' or the knowledge that the scholar's work may help influence the course of society. Professionally, extrinsic benefits influence genetic researchers because they undertake their research to gain social appreciation, job promotions among other incentives. An award, grant, a chance to publish an article in a reputable journal, present a paper at a conference, all of them serve as important checkpoints for their work and development. Finally, chances to advance, for example, to tenure-track positions, senior positions in the research group or department, or to manage research units and become key science officers in the institutions or in the industry, are considered major incentives (Davies et al., 2020).

However, financial rewards remain one of the reasons but they mostly stand as additional to intrinsic and professional rewards. But such factors as; salary that can be matching those of counterparts in other organizations, sponsorship of research projects, and other resources that can be used to support career development are other reward programs that can increase job satisfaction and organizational commitment. In other aspects of job expectations, the genetic researchers expect to have a favorable environment that will challenge them to the utmost. They demand to have the rights to use ultra-modern facilities and technologies as these are vital ingredients for sound research. Access to funds to finance their projects also matter a lot. Academic advisors require organizational support in providing guidance regarding the ethical and other legislative measures within genetic research and, therefore, guarantee socially beneficial outcomes of their work ((Fisher et al., 2020).

It was also evident that genetic researchers form networks of cooperation and prefer to act in the capacity of mentors. They also expect chance to work with other students/peers within the institution as well as though the scientific community in general. There are also some specifics of professional growth and career promotion which are also necessary in the process of the work (O'daniel & Haga, 2011). The colleagues' help and tips on how to improve the present working methods and develop new ones are also important. Also, in the genetic research field, there are assumptions placed on organizations to enhance learning as well as to innovate. They look for top operating conditions that allow them to explore new techniques and solutions and promote their innovation. This culture does not only increases participation but contributes to scientific advancement as well (Davies et al., 2020).

THE IMPACT OF ORGANIZATIONAL CULTURE ON ENGAGEMENT

Figure 3. Impact of organizational culture on engagement

Culture can be seen to have considerable influence on the level of engagement of the human resources in genetic studies. By adopting and promoting positive organizational culture that matches the values and the goals of genetic researchers, their spirits can enormously be boosted, and their levels of work commitment and satisfaction ought to improve substantially as well. On the other hand, a negative culture may result to reduced productivity and altitude change whereby many employees quit their duties (Al Shehri et al., 2017).

Leadership is one of the sub-elements of the organizational culture that affects engagement. Leadership that is open, and receptive creates confidence and rapport among the workers. Managers who care about the people they lead, strive for employees to enhance their knowledge and skills, share their ideas and vision, and listen to their subordinates' opinions create positive working conditions that engagement brings (Hasan et al., 2020). When people see that their managers and leaders care about them and appreciate their efforts, they feel motivated, and such leadership can create a massive difference in employees' productivity. The process of genetic research typically implies cooperation between individuals from different fields and in most cases, it is based on large-scale projects. It is acknowledged that the organization culture that supports collaboration, free flow of information and respect to each other fosters team work and sharing of new ideas. Besides, the given environment increases the effectiveness of studies and enhances the notions of togetherness and unity among researchers, which are crucial for engagement (Sarangi & Srivastava, 2012).

Another underline feature of organizational culture is the support to professional and personal development and consequent lifelong learning support. The genetic researchers are optimistic to be able to develop new sets of skill, to learn how the science progresses and to be able to improve their job status. Those companies that provide training sessions, seminars, conferences, and similar events are concerned with their staff's development, which promotes their dedicated work. Employees' satisfaction and health at work and outside work have become elements of organizational culture (Nurcholis & Budi, 2020). One of the weaknesses often identified with genetic research includes strains such as hard work and stress, long hours of work and social pressure. Managers who encourage reasonable working hours increase welfare services, and foster a good working environment should help in eliminating stress and burnout. This in turn improves the level of engagement of the employees and their productivity as well as retaining them (Jiony et al., 2015).

Diversity as well as inclusion are also attributes of organizational culture. Having a culture that appreciates talent and embraces people's individuality in terms of gender, ethnic origin, background or approach to problem solving fosters variety and growth in the organization. Recruitment affirms that all the staff is valued therefore employed bearing the fundamental principle of integration that leads to staff engagement. The organizations that demonstrate the policies, programme, and culture that support diversity and inclusion are most likely to experience the increased involvement of staff members in work processes (Abduraimi et al., 2023).

Intra-organizational partnerships indicate that the engagement of the staff depends with the cultural patterns of recognition and reward. A researcher's efforts, productivity, and commitment may be improved by appreciation for their work through awarding programs, promotions, and other incentives. Thus, rewarding validates and gives a boost of morale while at the same time promoting the behaviors values that the organization encourages (Al Shehri et al., 2017; Hasan et al., 2020).

Finally, the last factor that creates a positive attitude toward genetic research among the participants of the study is the perceiving of organizational culture in terms of ethical standards and social responsibility (Nurcholis & Budi, 2020). Ethical culture and other standards of research and development can contribute to employees' motivation, specifically when they allow defining the goals of investigations and their impact on society as the ultimate purpose of the project. Thus, the organizations which provide all the information on ethical principles, conduct themselves responsibly in research or other processes, and who are socially and ecologically responsible are likely to gain the motivated and committed employees (Abduraimi et al., 2023).

Therefore, as part of understanding workforce, there is a need to identify demographic and psycho-graphic characteristics of the genetic research workers; their expectations and motivation will also be necessary to consider as well as the impact of organizational culture or environment in influencing the workers. Hence, cultivating a favourable, welcoming, and progressive company environment, offering possibilities for development, and corresponding with the values and goals of genetic researchers, an organization can increase interest passionate results and make remarkable advances in the field of genetic research (Sarangi & Srivastava, 2012).

FOUNDATIONS OF EFFECTIVE COMMUNICATION IN HRM

The communication in Human Resource Management (HRM) is therefore crucial to the achievement of purposes that aim at creating productive and positive working environment of the organization. Free and open communication is critical to the health & safety management, and the guiding principles include; Clarity and Consistency in communicating full messages in HRM Transparency through being honest by showing legitimate understanding for workers' situation and participating in their circumstances Actively listening to the workers (Kaur & Arora, 2023). There is no doubt that simplicity is always the key; messages have to be simple and should not contain any jargon or confusing vocabulary. They must make it in simple language, so that all the employees in that organization or company can understand it. Consistency addresses the problem of keeping the same message for all the intended employees and all the channels where it is to be conveyed. This particular principle is of paramount importance when it comes to relating the various aspects of organizational life and especially in change management (Kellner et al., 2016).

Of them, the clarity or transparency is also one those must be followed. Transparency is an effective way of creating trust with the other party as it establishes credibility with the organization. It implies that when employee regards himself as informed and not left in the dark, he is likely to be more satisfied and committed. The company requires appreciation of employees' attitudes, their apprehensions, wants, and feelings, which is referred to as empathy (Bessy et al., 2024). That is all about how one can use words to demonstrate the acknowledgement of the employees' opinions and concern for their well-being since it may greatly improve the staff's performance and job satisfaction. Listening is deemed to be equally as important as speaking or even more important than speaking in alleviating communication problems. They defined it as a process of listening to the employees with passion and trying to grasp what they are saying. It establishes a two-way communication through which employees feel that they are recognized as well as their opinions on some issues are considered (Kaur & Arora, 2023).

COMMUNICATION CHANNELS AND TOOLS

HRM uses different means of communication to transmit information and for the purpose of carrying out two way communication. These channels can be broadly classified into verbal, non-verbal, written, and others which includes digital. Verbal communication is direct and normally more controlled as compared to written and normally occurs through meetings, calls or video conferencing. It enables feedback and means that the conversation can result in understanding and clearing of misunderstandings and confusions which makes it suitable for when handling issues that require a lot of detail, interviews or when talking about matters that may be considered sensitive. Para-language, gestures, and facial expressions, as well as body posture and vocal tone, are critical in presenting emotions as well as attitudes towards the message being delivered verbally (Aslam & Akram, 2024).

Another type of communication is written and it entails e-mail, memorandum, specials Report and news letter. It will help in recording the information, in giving instructions, and also in guaranteeing the flow of communication is recorded. Another mode of communication in organizations is the written communication which is applicable for policies, procedures, and official announcements (Maiwada & Oshionebo, 2024). Technological advancement in communication has enhanced the ways through which HRM is conducted through the provision of such tools as intranets, social sites and blogs and messaging. These tools help in the rapid and vast spread of information, in carrying out communication without a face to face contact, and carrying out interaction through forums, feedback, surveys etc. For example, extra-net web sites may contain key papers, messages, and training's available to everybody in the company, where people can communicate more or less formally (Rusady et al., 2024).

Moreover, other applications such as Slack, Microsoft team, and Trello improve communication and flow of the project. It enables the exchange of information in real time and sharing of documents and status of tasks in real time in order to maintain teamwork even with various and distant teams (Kaur & Arora, 2023).

MEASURING COMMUNICATION EFFECTIVENESS

Therefore, it is crucial for the HRM to have ways and means through which they can assess and possibly quantify the outcomes of the strategies that they use in communication. Such facilitators as KPIs and feedback tools are best utilized in this process. Employee surveys as well as the questionnaires and feedback forms are some of the common means of determining organizational communication effectiveness. Periodic surveys assessing the extent to which the employees receive the communication material, how often it is received, and its relevance can be useful. Such surveys can address a number of elements like the appropriateness of the communication process, relevance of the presented information and extent of employee's involvement and comprehension (Rusady et al., 2024).

Interviews with and focus group discussions are also pertinent in understanding the employees' attitudes and enunciation. These methods are far more elaborate and can provide qualitative feedback to the HRM by identifying the finer aspects of the communication problems and defining what exactly needs to be done to overcome these hurdles. Another critical factor would be the level of workers 'engagement (Kulykovets, 2024). High engagement levels usually mean good communication patterns because the engaged staff is well-informed and has some affiliation with the organization's environment. Other measurable variables like the attendance records in meetings, training's as well as company functions can be used as surrogates for communication efficiency (Nur, 2024).

Quantitative analysis of communication patterns and its results with the help of digital analysis tools also has advantages. For instance, metrics such as, open rate, click through rate, and engagement level of people on the digital communication channel including email, intranet, and social media helps in establishing yardstick that measures the level of receptiveness of people to the messages that are being passed across. The other aspect that can be identified using turnover and retention rates is as to the extent of communication. If turnover is high, it may be that employees are complaining about communication, stating that they lack enough information or endowment of importance in the organization. On the other hand high retention rates indicate that the employees are happy and committed in their duties probably because of good communication measures put in place (Bessy et al.,2024).

Performance indicators and productivity standards are also the signs of communication efficiency. Interpersonal communication can result in improved definition of roles and responsibilities, as well as organizational objectives, which can improve workers' performance and efficiency. Thus tracking these metrics will enable the HRM to evaluate the effectiveness of communication to the general organizational performance (Kellner et al., 2016). Last but not the least, it is necessary to establish a continuous improvement loop for the processes to control communication strategies. This entails periodic examination of the feedback to find areas that require enhancement of the communication process and the metrics. In the said approach, it is essential that HRM actively encourage feedback about communication and the feedback collected used to improve the communication process and the strategies used in communication constantly (Bessy et al.,2024).

Altogether, it can be claimed that communication rules and standards as well as practices in HRM involve determinant principles which include clarity, consistency, openness, compassion, and hearing out the parties involved. Therefore, it is crucial to employ all the available verbal and nonverbal communication tools, as well as oral/written and social media communication tools, to extend the organization's communication reach and include the targeted employees as well (Bessy et al., 2024). Communication strategy evaluation by using a feedback on employee communication channels, employees' engagement, web metrics and analytic, average turnover rates, and performance metrics checks whether the implemented communication strategies are effective and are always being optimized. Thus, by implementing these foundations, HRM can create workspace conditions that will lead to an enhanced level of employees' knowledge, motivation, and productivity (Kellner et al., 2016).

TAILORING COMMUNICATION STRATEGIES FOR GENETIC RESEARCH PROFESSIONALS

Risk communication strategies that would be effective when dealing with the general population cannot be applied immediately in genetic research because the target audience is in a different group of professionals. These are personnel who provide their services in a rather narrow and specialized sphere that involves the application of such sophisticated knowledge as scientific principles, highly complex technologies, and a comprehensive system of legal standards (Aslam & Akram, 2024). Justifiably, the first aspect that may be used when formulating procedures of communicating with these professionals is the assessment of their needs and worries. This is expected, because sometimes genetic researchers have to work with specific and detailed data to execute their duties. This requires the use of precise and detailed information and communication to report scientific discoveries, research progress, legal amendments, and business policies (Bessy et al., 2024).

A concern that remains paramount in genetic research is the challenge posed on the researcher on the existing literature in his or her line of study. Due to the increased rate of identification of genes, therapeutic strategies, and high-tech equipment that facilitate research, it is essential to update information (Maiwada & Oshionebo, 2024). HRM can counter this by creating means of passing across to the employees the relevant scientific information, new findings in scientific conferences among other outlets. Also, there are issues concerning funding, distribution of resources, and the ability to claim professional growth. Therefore, genetic specialists and researchers applying genetic study tend to work under strict budget most of the times and depends on grants and other sources of funding. This is particularly true for matters relating to funding, changes in the size of the budget or even any issues to do with resource availability where a lot of anxiousness is noted among learners, more supportive communication ought to be provided (Rusady et al., 2024).

Other issues duly affect the success of a working woman include work-life balance and mental health. Genetic research is a very rigorous field that requires long hours of work at a go and is very intensive in the sense that results need to be produced a lot of the time, this causes burnout. HRM can overcome this by introducing policies that encourage the practice of healthy lifestyles, carrying out stress implication initiatives, supporting breaks, and shorter working hours. However, one has to consider problems as fundamental as ethical issues and compliance with the legislation (Kulykovets, 2024). Genetic investigation may be correlated with quite a few of ethic issues and therefore controls. Ethical policies of the organization and the regulatory standards should be communicated effectively and have proper messages posted, electronic or otherwise sent frequently around the organization to contain information on ethical standards of the organization (Aslam & Akram, 2024).

CUSTOMIZING MESSAGES FOR DIFFERENT EMPLOYEE SEGMENTS

The genetic research workforce is diverse and complex; therefore, it is crucial to design messages that are specific to the various segments while taking into account aspects such as age, literacy level, and interest in the specific topic under discussion. Depending on the certain characteristics, human resources in accordance with the genetic research can be divided into groups in terms of roles and responsibilities, work experience, organizational department, and personal preferences (Bessy et al.,2024). To the senior researchers and principal investigators it is recommended to circulate only strategic information, funding and leadership. These professionals also tasked with grant acquisition, being in charge of the research. Availing to them more information about the funding calls, strategic plans and leadership development programs, will enable them to exert better leadership and support the organizations' objectives. The mid-career researcher faces information needs in need of career development information, information about professional advancement and career fairs (Nur, 2024).

Communication should focus on possible further education, joint endeavors, and strategies to progress in one's employment. This segment may also find of value, help and support in the form of programs and materials pertaining to mentor-ship on their career path. As for early-career researchers and lab technicians, the focus of communication should be more practical and concern the procedural aspects and certain skill improvement. Such employees are also often at the operational level where they carry out experiments, and gather data. Provisions of clear and understandable instructions, communication of change in lab practices, and materials that are useful for training will improve their performance and interest (Bessy et al.,2024).

The genetic research is much dependent on the administrative and support working professionals who support the research workers. With this segment, communication should involve updates on organizational policies and new procedures as well as possibilities for gaining new administrative skills. Advising the support staff on the relevant changes within an organization or the worked out changes for them and their positions can enhance productivity and staff satisfaction (Nur, 2024).

Another specific group of respondents is interns and trainees. Speaking with them should be informative and understanding and is mainly about induction or the introduction of the participants to the philosophy of the lab as well as providing them with mentoring chances. Brief meetings and reporting sessions can also be useful to facilitate interns' and trainees' adjustment to the context of the research and personal professional growth (Kulykovets, 2024).

LEVERAGING TECHNICAL LANGUAGE AND LAYMAN'S TERMS APPROPRIATELY

The issue of language use is quite pertinent in the context of genetic research where the communication should, at some level, be strictly scientific when, at the same time, seeking to explain research findings to lay people. The personnel involved in genetic research always work with highly specialized terms and references and, therefore, concepts. While dealing with this audience technical accuracy is critical, thus the language should not only be plain but also technical. For example, mentioning certain genetic procedures, analyzing techniques, and molecular processes in an experiment, it is mandatory to employ special scientific terms to disclose the intended meaning (O'daniel & Haga, 2011).

It is however important to note that not every piece of information that is sent across would have to be complex in explanation. It is also critical that the descriptions employed are simple where changes involve general points of relevance to the organization, policy adjustments or other similar matters. This is important so that everybody in the company can understand the message, starting from the administrative staff and ending with lowly IT technicians. For instance, in the discussions of shifts in the HR policies, culture and change programs proactively, general health and safety rules among other matters, plain and simple language may help maximize understanding and compliance (Davies et al., 2020).

Furthermore, it is also acknowledged that the genetic research teams may consist of the members with the diverse knowledge and experience level. Many junior researchers, interns and administrative personal may lack of the technical expertise like those prominent researchers. Specifically, an author should ensure that the language he uses corresponds to the level of the reader's knowledge. Furnishing the background information concerning the terms, their meaning, and the rationale behind them can assist in the creation of a better understanding of those terms at inter-level (O'daniel & Haga, 2011).

Thus, when explaining complex data or research results to an audience consisting of both with and without quantitative skills, it is useful to take a twofold strategy. This includes giving a summary of the research results in simple language for the general population and supporting the summary results with the technical data in additional annexes. In this manner, every person who took interest to attend the said presentation is able to understand the major points of discussion in the presentation, while those who are well acquainted with the technical details of the topic at hand can always deviate at their discretion to try and fully understand the technicalities involved in the presentation (Davies et al., 2020).

Another key thing is the greatest utilization of the elements of visuals and analogies in the materials. Other aspects like diagrams, charts, and figuring can enhance people's understanding of the given technical data. There can also always be use of analogies and metaphors in order to give an understanding of the different complicated theories in science. For instance, presenting the process of DNA replication to non-specialists will be easily understandable when compared to the photocopying machine (Kaur & Arora, 2023).

In general, communication training of the researchers can also increase the effectiveness of the disseminating efforts of the research information. The communication of ideas through science writing skills can be effectively presented in workshops and training sessions that prepare the researcher to present his work in a manner that is easily understood by other researchers or any funding bodies or even members of the public (Davies et al., 2020). Among these skills, persuasion is particularly crucial in the circumstance where the researcher has to promote their projects, find sponsors, or address the general public. Therefore, it is necessary to know genetic research professionals and their concerns, develop key messages fitting each group of employees, and use technical but not overly specialized vocabulary. As a result, HRM can help to explore the specific demands and concerns of genetic researchers, as well as facilitate the development of solutions that will enable an organization to create a higher level of value for its employees and clients. Effective and efficient communication leads to improvement of relationships, promotion of scientific advancement as well as improved performance by genetic research organizations (O'daniel & Haga, 2011).

SHARING THE VISION AND THE MISSION OF THE ORGANIZATION

The ways by which an organization's vision and mission can be effectively communicated are fundamental in influencing the employees especially in the field of genetic research where the work most often involves concepts which are somewhat intricate. An organization's vision and mission acts as a guideline and gives employees meaning in their work as the roles and responsibilities are aligned to the stated goals of the company. In the case of genetic research specialists, general and social organizational perspectives on the work done by the stars and the role performed by these people in the overall scientific and social processes play a crucial role in boosting motivation (Kellner et al., 2016).

For an organization to effectively pass on its vision and mission, it is mandatory that an effective story that would be easily understood by every employee be developed. This story must lay down goals and objectives of the organization in longer-term perspective that is the vision and on the other hand, it must define the organization's reason for existence, what it does, what services or products it provides i.e. the mission (Bessy et al., 2024).

For example, a genetic research institute might have vision statement of the organization as 'Striving to be a world reference in genetic research with the objective of promoting the enhancement of human health worldwide,' and the mission statement of the organization as 'Engaging in innovative genetic research aimed at finding new treatments for genetic diseases and enhancing human health. These statements should be placed in all internal documents starting with offering package and contract documentation, company's employee handbooks, and all other official documents used within the company's internal communication systems such as internal Internet website and internal newsletter. Ways of perpetuating the vision and mission include but not limited to daily reinforcement of the vision and mission statements during meetings, workshops, and presentations to employees; this keeps the vision and mission at the back of the employees' minds. Even teachings including ideas like these can benefit from posters or info graphics that are strategically placed within the learning environment (Aslam & Akram, 2024).

The vision and mission must be made and modeled by the leaders at every level and communicated throughout the organization. The above principles should be invoked in the leaders' discussions in order to show how strategies and operations relate to those principles. town hall meetings, or virtual forums will also enhance the employees' engagement by giving them an opportunity to ask questions, air their opinions, or even get to discover how they can fit into the broader corporate objectives. Thus, the two-way communication supports the organizational culture by making the employees feel as if they are part of the organizations success (Kellner et al., 2016).

INTEGRATION OF RESPONSIBILITIES IN THE CONTEXT OF PERSON AND ORGANIZATION GOALS

- ✓ This process fits every employee and their responsibilities to the strategic goals and missions of the organization for optimizing the labor productivity and motivation. This realignment assists the employees in comprehending how their day-to-do job and key objectives contribute to the larger vision of an organization, which consequently can increase employees' satisfaction and performance (Maiwada & Oshionebo, 2024).
- ✓ Another way is to establish and elaborate the link between every position and the organization's strategic plan. Responsibilities and key performance indicators must also show how organizational goals relate to the organizational mission. For instance, instead of describing a genetic researcher in terms of his/her responsibilities, the expectations can include reference to gene editing as a contribution to a broader aim and objective of improving medical solutions (Bessy et al.,2024).
- ✓ A few ways through which this alignment can be achieved include periodically discussing and resetting expectations and goals can help in this regard. While conducting these meetings, the managers should explain how the targets met in the past have supported the strategic plans and create new challenges in line with the organization's goals. This makes sure that workers understand on how they are affecting the organization and in the process are encouraged to work harder (Maiwada & Oshionebo, 2024).
- ✓ Personal developed plans should also show some relation with organizational objectives. It means that the organization should provide workplace learning that is appropriate for its goals since it assists the employee to gain the required knowledge and experiences beneficial to both the employee and the organization. This dispositions the personal development of the personnel and the organizational development together with superior corporate results hence creating symbiotic relationship (Kaur & Arora, 2023).
- ✓ Effective communication of the organization's accomplishments and issues is also valuable. Communicating the organization's progress in achieving the vision and mission, as well as explaining how each employee and teams' actions are helping to achieve this, helps to maintain motivation. Recognizing achievements and effort when others can see it also has the potential to further increase the motivation levels of employees, as understanding that their

efforts are valued is integral in affirming one's self-worth and personal worth in the organization (Kellner et al., 2016).

STRENGTHENING THE OUTPUT OF RESEARCH FOR SOCIETY AND SCIENCE

Meanwhile, for genetic research professionals, the results of the work and possible influence on society and science is an actual stimulus. Strengthening this impact in communications greatly serves to boost engagement since many people yearn to have a feeling that they are achieving something productive (Kaur & Arora, 2023).

Engaging the public and describing the relevance of the research to society is something that is based on the storytelling concept. Thus, citing specific cases when genetic research has helped find new treatments for diseases, enhance patients' quality of life, or increase people's knowledge about themselves and the world can make concepts more tangible. Real life stories, patients' stories, people who have regained their health as the result of research could give significant insight about the positive impact of research (Bessy et al.,2024).

Organizations should involve advocating internal communication outlets for publicizing the wider use of the organization's research. Such articles and blogs as newsletters, blogs, and posts in social media accounts can include articles on the recent findings, publication of the findings in internationally reputable journals and association with other research institutes. They help share new information and highlight the importance of the job performed to other employees. It is also possible to speak about the encouragement of engagement with the help of public recognition and awards. Incentives like awards for ideas, projects or hard work, and achievements to organizational, and/or social objectives encourage employees in order to achieve excellence (Maiwada & Oshionebo, 2024).

The results of research can also be supplemented by seminars, workshops, and conferences as well as other activities. These events give opportunities for the employees to showcase their findings, exchange ideas and have debates on the future of genetic research and its impact on the society. They represented mobilization of people towards such events because they enable people to feel that they are in one boat. Inviting patients or community members that have benefited from the breakthroughs of genetic research to visiting sessions or holding online conferences where the researchers can hear success stories of the work they do can generate strong emotions that affirm the practical importance of the actions (Rusady et al., 2024).

Furthermore, one can arrange for the employees to increase their incentives to go for public awareness and sensitization in their tasks. Participation in fairs or club activities, science fairs or education programs not only spreads knowledge, but also develops organizational and moral responsibility and pride in the benefit of community. Thus, vision and mission engagement of employees is a practice of how one can explain the principles, roles of employees in organizations and the importance of their findings to society and science. To do this and through the understanding and linking of the employees with the goals of the business, the HRM can have a positive influence on motivation and commitment of the employees towards one's organization. It is in this capacity that this holistic approach to engagement employs clarity to foster organizational culture for the achievement of genetic research organizations (Bessy et al.,2024).

COMMUNICATION SKILLS TRAINING PROGRAMS

Interpersonal communication skills training initiatives are formulated to develop and boost various layers of communication such as verbal, non-verbal, written, and electronic. These programs typically incorporate the following components:These programs typically incorporate the following components:

1. Verbal Communication Skills: Training in verbal communication is done in a way that involves both word choice and delivery; hence, one is trained on how to speak fluently, convincingly, and with fluency. Skills like listening skills, critical questioning skills, and one-to-one and one to many communication or presenting and public speaking skills are among those taught to enhance interpersonal communication skills (Imam et al., 2023).
2. Non-verbal Communication Skills: The categories of non-verbal communication includes physical communication which consist of; posture, touch, movement, facial expressions, eye contact and gesture. The appreciation of how and what nonverbal signals influence communication in determining their effectiveness, as well as learning how to utilize them to better one's interpersonal communication helps strengthen each interpersonal encounter, and portrays trust and confidence (Kaur & Arora, 2023).
3. Written Communication Skills: Business and technical communication focuses on the key elements of purposeful, business-like, and well-organized writing including emails, reports, proposals, and scientific publications. The process of

structuring documents, knowledge in grammar and punctuation, and the ability to adjust the writing style to the target audience are built during the internship (Maiwada & Oshionebo, 2024).

4. Digital Communication Skills: Due to technology adoption in the current society, workplace training on the use of technology in communication leans towards matters concerning use of e-mail, chat applications, web conferencing as well as collaboration applications. Social skills as applied during virtual meetings, professional business emails and communications, and guidelines on manners when using the computer and the internet are provided to ensure efficient working from home (Nur, 2024).
5. Cross-cultural Communication: In organizations ranging from genetic research institutions, cross cultural communication training assists workers understand the cultural variation when it comes to communication trends, taboos, and anticipations. Explicit knowledge of cultural differences improves group cohesiveness, decreases conflict, and improves the ways of communication (Maiwada & Oshionebo, 2024).
6. Conflict Resolution and Negotiation: Conflict resolution and negotiation training prepares the employees to handle conflict situations and learn how to solve other people's problems and negotiate. Skills like listening, understanding the feelings of other people, conflict solving, and conflict managing skills are introduced to improve the interpersonal relations and cooperation (Nur, 2024).
7. Science Communication: Due to the fact that genetics is an aspect of the scientific discipline, theoretical background required for science journalism is useful. In particular, this training encompasses the communication of scientific knowledge in simple terms, for peers, investors, government, or other people. Skills in narrative development and data presentation are developed as well as strategies for grasping people's attention (O'daniel & Haga, 2011).
8. Feedback and Coaching: Coaching and feedback in resolving skills in relation to giving and receiving feedback constructively. These include methods for providing constructive feedback, using the SMART objectives to guide an employee's performance as well as offering procedures to follow when addressing an employee with poor communication skills with the aim of impacting positive change in their professional lives (Imam et al., 2023).

Sustaining and upgrading the communication skills require that the employees and the workers undergo constant learning and development. Genetic research organizations can offer the following initiatives to support ongoing professional growth:Genetic research organizations can offer the following initiatives to support ongoing professional growth:

1. Workshops and Seminars: Current topics in communication are taught through workshops and seminars enabling the employees to gain further understanding and skills in communication, coupled with exposure to new trends in the existing communication practices. Experts in the field, other department heads, and trainers can organize lessons based on specific communication issues and effective tackling of such tasks (Nur, 2024).
2. Online Courses and Webinars: The openness of online courses, content of webinars and e-learning keeps the employee compliant with the learning progress of his/her choice and expands knowledge in the sphere of inter group communication. Some of the modules that can be delivered include public speaking, writing skills, new generation communication tools, and cross cultural communication can be made through Coursera, LinkedIn learning, or organizational learning platforms (Almerri, 2023).
3. Peer Learning and Mentoring: Buddy systems and various mentoring initiatives provide the natural learning environment for the employees to share successful communication strategies, get suggestions, and to learn through actual, team-based practice. The management should assign or encourage junior employees to obtain and be guided through key communication proficiency issues by way of a senior counterpart, thus improving both mentoring and knowledge sharing (Jiang & Shen,2023).
4. Role-playing and Simulations: Some of the methods that are used to help the employees work on their communication skills include the use of role-play exercises and simulations where he gets a chance to act out real life communication situations and get it right because it is just a practice. They can be conflict sensitive, presentation-based, or outcome based and offer an excellent opportunity for an experiential learning process (Almerri, 2023).
5. Feedback Mechanisms: Providing feedback opportunity and mechanism, inclusive of 360-degree feedback surveys, peer review and coaching, enables the employees to reflect on their communication profile and identify their potential strengths and weaknesses. The idea is that constructive feedback from colleagues, supervisors and mentors plays a major role in my professional growth and development and keeps the process of working on the skills enhanced on the constructive note (Almerri, 2023).
6. Certifications and Credentials: Offering employees incentives or helping them acquire certifications or credentials of some sort in communication—like certifications in public speaking, technical writing, cross cultural communication take time and effort, and it acknowledges the expertise of the employee. Deducting amounts that can cover certification charges or providing sponsorship for attending communication-related conferences can encourage workers to take higher levels of training (Jiang & Shen,2023).

7. Practice Opportunities: Pretense presentation sessions, storytelling sessions or group discussions can be helpful to enhance the effectiveness of the communication practices among the employees improving their overall performance. There are several ways to include the practice sessions into one's schedule: Practice can be combined with the team meeting; team workshops; or special practicing sessions with the help of communication trainers (Nur, 2024).

CONCLUSION

Communication is not just a tool for transmitting information; it is the lifeblood of effective HRM practices. In genetic research, where innovation and collaboration are paramount, communication plays a pivotal role in shaping organizational culture, fostering engagement, and driving success. Effective communication strategies are foundational to fostering employee engagement, promoting collaboration, and driving innovation in genetic research organizations. By prioritizing transparent communication, encouraging two-way dialogue, and adapting to evolving technological and societal trends, HRM professionals can play a pivotal role in shaping the future of employee engagement in this dynamic field.

REFERENCES

Abduraimi, P. B., Mustafi, M., & Islami, X. (2023). The role of organizational culture on employee engagement. *Business: Theory and Practice*, 24(1), 109–122. 10.3846/btp.2023.17241

Al Shehri, M., McLaughlin, P., Al-Ashaab, A., & Hamad, R. (2017). The impact of organizational culture on employee engagement in Saudi Banks. *Journal of Human Resources Management Research*, 1(1), 1–23. 10.5171/2017.761672

Almerri, H. S. H. (2023). Investigating The Impact of Organizational Culture on Employee Retention: Moderating Role of Employee Engagement. *Journal of System and Management Sciences*, 13(4), 488–507.

Aslam, M. S., & Akram, A. (2024). When and how does e-HRM optimize communication pace and processing time?. Journal of Economic and Administrative Sciences.

Bessy, Y. J., Yenny, Y., Harliantara, H., & Zulaikha, Z. (2024). Communication Managerial Skill For Business Communication On Employee Relations Studies. *Jurnal Ekonomi*, 13(02), 1722–1735.

Davies, E., Craufurd, D., & MacLeod, R. (2020). "It's being part of the big picture, even though you're a tiny jigsaw piece"—Motivations and expectations of individuals participating in the Enroll-HD observational study. *Journal of Community Genetics*, 11(4), 421–432. 10.1007/s12687-020-00459-332157658

Fisher, E. R., Pratt, R., Esch, R., Kocher, M., Wilson, K., Lee, W., & Zierhut, H. A. (2020). The role of race and ethnicity in views toward and participation in genetic studies and precision medicine research in the United States: A systematic review of qualitative and quantitative studies. *Molecular Genetics & Genomic Medicine*, 8(2), e1099. 10.1002/mgg3.109931867882

Hardcastle, F., Lyle, K., Horton, R., Samuel, G., Weller, S., Ballard, L., Thompson, R., De Paula Trindade, L. V., Gómez Urrego, J. D., Kochin, D., Johnson, T., Tatz-Wieder, N., Redrup Hill, E., Robinson Adams, F., Eskandar, Y., Harriss, E., Tsosie, K. S., Dixon, P., Mackintosh, M., & Lucassen, A. (2024). The ethical challenges of diversifying genomic data: A qualitative evidence synthesis. *Cambridge Prisms. Precision Medicine*, 2, e1. 10.1017/pcm.2023.2038549845

Hasan, H., Astuti, E. S., Afrianty, T. W., & Iqbal, M. (2020). Impact of organizational culture on employee engagement and employee performance: A stimuli-organism-response approach. Wacana Journal of Social and Humanity Studies, 23(4).

Imam, H., Sahi, A., & Farasat, M. (2023). The roles of supervisor support, employee engagement and internal communication in performance: A social exchange perspective. *Corporate Communications*, 28(3), 489–505. 10.1108/CCIJ-08-2022-0102

Jiang, H., & Shen, H. (2023). Toward a relational theory of employee engagement: Understanding authenticity, transparency, and employee behaviors. *International Journal of Business Communication*, 60(3), 948–975. 10.1177/2329488420954236

Jiony, M. M., Tanakinjal, G. H., Gom, D., & Siganul, R. S. (2015). Understanding the effect of organizational culture and employee engagement on organizational performance using organizational communication as mediator: A conceptual framework. *American Journal of Economics*, 5(2), 128–134.

Kang, M., & Sung, M. (2017). How symmetrical employee communication leads to employee engagement and positive employee communication behaviors: The mediation of employee-organization relationships. *Journal of Communication Management (London)*, 21(1), 82–102. 10.1108/JCOM-04-2016-0026

Kaur, J., & Arora, R. (2023, June). Exploring the dimensionality of employee silence in healthcare sector. In AIP conference proceedings (Vol. 2782, No. 1). AIP Publishing. 10.1063/5.0154178

Kaur, J., & Arora, R. (2023, June). Exploring the impact of employee silence in private hospitals-a structural equation modeling approach. In AIP Conference Proceedings (Vol. 2782, No. 1). AIP Publishing. 10.1063/5.0154173

Kellner, A., Townsend, K., Wilkinson, A., Greenfield, D., & Lawrence, S. (2016). The message and the messenger: Identifying and communicating a high performance "HRM philosophy". *Personnel Review*, 45(6), 1240–1258. 10.1108/PR-02-2015-0049

Kulykovets, O. (2024). Building bridges: Trust and communication in contemporary work environments. Trust in Social and Business Relations, 143-153.

Kwon, K., & Kim, T. (2020). An integrative literature review of employee engagement and innovative behavior: Revisiting the JD-R model. *Human Resource Management Review*, 30(2), 100704. 10.1016/j.hrmr.2019.100704

Leroy, B. S. (2020). When theory meets practice: Challenges to the field of genetic counseling. In *Prescribing Our Future* (pp. 39–54). Routledge. 10.4324/9781351328609-5

Maiwada, A. A., & Oshionebo, B. (2024). Exploring the potential of internal communication and employee relations for effective staff performance in Nigeria customs service. *Human Resources Management and Services*, 6(4), 3447–3447. 10.18282/hrms.v6i4.3447

Mishra, K., Boynton, L., & Mishra, A. (2014). Driving employee engagement: The expanded role of internal communications. *International Journal of Business Communication*, 51(2), 183–202. 10.1177/2329488414525399

Nienaber, H., & Martins, N. (2020). Exploratory study: Determine which dimensions enhance the levels of employee engagement to improve organisational effectiveness. *The TQM Journal*, 32(3), 475–495. 10.1108/TQM-05-2019-0151

Nur, H. (2024). The Influence of Communication Factors, Work Experience and Job Training on Employee Productivity at the Karya Darma Makassar College Foundation. International Journal of Education. *Vocational and Social Science*, 3(01), 406–414.

Nurcholis, G., & Budi, W. (2020, January). The impact of organizational culture and perceived organizational support on employee engagement. In *5th ASEAN Conference on Psychology, Counselling, and Humanities (ACPCH 2019)* (pp. 240-246). Atlantis Press. 10.2991/assehr.k.200120.050

O'daniel, J., & Haga, S. B. (2011). Public perspectives on returning genetics and genomics research results. *Public Health Genomics*, 14(6), 346–355. 10.1159/00032493321555865

Pemmasani, S. K., Raman, R., Mohapatra, R., Vidyasagar, M., & Acharya, A. (2020). A review on the challenges in Indian genomics research for variant identification and interpretation. *Frontiers in Genetics*, 11, 753. 10.3389/fgene.2020.0075332793285

Rasool, S. F., Wang, M., Tang, M., Saeed, A., & Iqbal, J. (2021). How toxic workplace environment effects the employee engagement: The mediating role of organizational support and employee wellbeing. *International Journal of Environmental Research and Public Health*, 18(5), 2294. 10.3390/ijerph1805229433652564

Rusady, M. V., Fauzi, A., Bachri, A. T. S., Ballo, M. A. M., & Nuru, E. (2024). Conflict Management Strategies: Improving Team Performance Through Collaboration and Communication (Human Resource Management Literature Study). *Dinasti International Journal of Education Management And Social Science*, 5(5), 1005–1017.

Saks, A. M. (2022). Caring human resources management and employee engagement. *Human Resource Management Review*, 32(3), 100835. 10.1016/j.hrmr.2021.100835

Sarangi, S., & Srivastava, R. K. (2012). Impact of Organizational Culture and Communication on Employee Engagement: An Investigation of Indian Private Banks. *South Asian Journal of Management*, 19(3).

Verčič, A. T. (2021). The impact of employee engagement, organisational support and employer branding on internal communication satisfaction. *Public Relations Review*, 47(1), 102009. 10.1016/j.pubrev.2021.102009

Yadav, A., Pandita, D., & Singh, S. (2022). Work-life integration, job contentment, employee engagement and its impact on organizational effectiveness: A systematic literature review. *Industrial and Commercial Training*, 54(3), 509–527. 10.1108/ICT-12-2021-0083

Chapter 14

Optimizing the Transhuman Workforce: How Technology-Enhanced Work Environments Affects Employee Productivity and Job Satisfaction

Ravinder Kaur
https://orcid.org/0000-0003-3770-2914
Chandigarh University, India

ABSTRACT

In a time portrayed by fast mechanical headways, the reconciliation of innovation into the working environment has fundamentally changed the idea of work and representative encounters. This exploration examines the effect of innovation upgraded workplaces on representative efficiency and occupation satisfaction, zeroing in on the jobs of different advancements like mechanization, man-made reasoning (artificial intelligence), coordinated effort apparatuses, and ergonomic developments. By looking at how these innovations impact representatives' work proficiency, commitment, and generally satisfaction, this study plans to give a far-reaching comprehension of the advantages and difficulties related with an innovation upgraded work environment. A blended strategies approach will be used, consolidating quantitative information from representative reviews and efficiency measurements with subjective experiences from meetings and contextual investigations. Reviews will evaluate workers' view of the innovative apparatuses they use, their effect on work execution, and levels of occupation satisfaction. Efficiency measurements will be dissected to gauge changes in work result and proficiency. Interviews with representatives, chiefs, and IT experts will give a more profound comprehension of the viable ramifications and individual encounters connected with innovation use

DOI: 10.4018/979-8-3693-7668-3.ch014

in the work environment. Contextual investigations of associations known for their high-level utilization of working environment innovation will feature best practices and normal snags. Expected results incorporate recognizing the key advances that most altogether upgrade efficiency and occupation satisfaction, understanding the elements that add to fruitful innovation reception, and revealing potential downsides, for example, innovation related pressure or reliance. The exploration will likewise investigate the job of the executives in working with a smooth progress to an innovation improved workplace and keeping a harmony between mechanical coordination and human variables. The discoveries will offer significant bits of knowledge for associations expecting to streamline their innovation systems to further develop worker execution and prosperity. Vital suggestions will be created for directors to successfully carry out and oversee innovation in the work environment, guaranteeing that mechanical upgrades lead to positive results for both the association and its representatives. This study looks to add to the more extensive conversation on the fate of work, stressing the basic job of innovation in forming useful and fulfilling workplaces.

INTRODUCTION

In the contemporary working environment, innovation has turned into a foundation of functional proficiency and representative commitment. The incorporation of different mechanical devices, like mechanization, man-made reasoning (artificial intelligence), and joint effort stages, is altering the way in which work is performed and experienced. This paper investigates the effect of innovation improved workplaces on worker efficiency and occupation satisfaction, analyzing both the advantages and difficulties related with this change. Information sharing is tied in with trading data and skill between people for the achievement of explicit assignments in associations (An et al., 2014; Swanson et al., 2020; Stachova et al., 2020). It gives a connection between people through moving information that people have in the conveyance of explicit items and administrations in associations. This prompts better representative commitment, compelling navigation, decreased loss of information and animated development, eventually bringing about better work execution for people and upper hands for associations (Kwahk and Park, 2016; Razmerita et al., 2016; Nguyen et al., 2021). Subsequently, associations have been effectively seeking after different means for elevating information sharing to upgrade the presentation of people and work on the intensity of associations (An et al., 2014; Malik et al., 2020). The fast improvement of advanced advances gives people new means for sharing information in conveying explicit items and administrations in associations (Tseng and Huang, 2011; Ahmed et al., 2019). Such innovations going from web-based entertainment, including Facebook, LinkedIn and Instagram, to advanced stages, for example, weblogs, Zoom, Microsoft Groups and Skype, large information and online assets, are changing the way that information is made, conveyed and shared under different conditions (Ahmed et al., 2019). The rising reception of conveyed working conditions on account of Coronavirus further speeds up such changes with the developing utilization of computerized innovations to work with information sharing and independent direction (Duan et al., 2020; De et al., 2020). Accordingly, the two people and associations have begun to encounter significant and broad effect that these progressions have brought not just concerning scale, access and accessibility of information yet additionally in the way the way in which information is shared, where it comes from and which jobs that people play in making, circulating and sharing information (Kwahk and Park, 2016; Swanson et al., 2020; Lepore et al., 2021). This calls for more examination on better comprehension what the utilization of computerized advances in associations means for information sharing and what effect such computerized innovation driven information sharing has on work execution of people (Ahmed et al., 2019; Kwayu et al., 2021).

There are many examinations investigating the effect of information sharing utilizing individual computerized innovations on work execution under different conditions (Tseng and Huang, 2011; Kwahk and Park, 2016; Cui et al., 2019). Tseng and Huang (2011), for instance, look at the connection between the utilization of Wikipedia, information sharing and occupation execution showing that utilizing Wikipedia fundamentally affects information sharing and occupation execution. Kwahk and Park (2016) investigate the utilization of virtual entertainment for information sharing and its effect on work execution uncovering that the reception of web-based entertainment can prompt better work execution. Cui et al. (2019) examine what web-based entertainment use means for information sharing and occupation execution figuring out that online entertainment use can emphatically influence work execution of people in a group climate. These examinations above exhibit the basic job that individual computerized advancements play in information sharing and its effect on work execution.

Computerized advancements, for example, online entertainment and advanced stages have their own qualities and abilities in making, dispersing and sharing information (Yuan et al., 2013; Ahmed et al., 2019; Kwayu et al., 2021). This prompts the rising utilization of different computerized advancements in a comprehensive way for better information sharing and powerful dynamic in associations (Lepore et al., 2021). There are, in any case, no particular examinations that can be found in the writing on how computerized advancements are utilized in general in associations for working with information sharing and what computerized innovation driven information imparting means for work execution to regard to upgraded coordination and correspondence and further developed direction.

LITERATURE REVIEW

The current writing features a diverse connection among innovation and working environment elements. Studies show that mechanization and artificial intelligence can altogether help efficiency by taking care of dreary errands, permitting workers to zero in on additional essential exercises (Brynjolfsson and McAfee, 2014). Also, joint effort devices like Leeway and Microsoft Groups upgrade correspondence and coordination among colleagues, which can further develop proficiency and undertaking results (Leonardi, 2013). Be that as it may, the writing additionally notes expected drawbacks, for example, expanded pressure from consistent network and the gamble of occupation removal because of computerization (Tarafdar, Cooper, and Stich, 2019).

Work execution is a complicated build that can be drawn nearer according to the viewpoint of in-job work execution and creative work execution (Henttonen et al., 2016; Duan et al., 2020; Lin et al., 2020). In-job work execution ponders the assignment inside the obligation of people (Chen et al., 2019). This expects people to show formal ways of behaving for accomplishing their exhibition targets in light of the set of working responsibilities. Inventive work execution centers around exercises past routine work necessities for accomplishing novel results (Ali-Hassan et al., 2015). It requires creating and embracing clever thoughts and getting the power important to carry out such thoughts for better authoritative execution effectively. Contingent upon the particular circumstance and its attributes, both in-job work execution and creative work execution can be utilized for evaluating the commitment of people to the by and large hierarchical targets (Ahmed et al., 2019; Swanson et al., 2020).

There are different examinations that have been directed on investigating what work execution is meant for in associations according to alternate points of view (Jelinek et al., 2006; Ghaffari et al., 2017; Al Nahyan et al., 2019; Chen et al., 2019; Lin et al., 2020). Jelinek et al. (2006), for instance, figure out that computerized innovations assume a basic part in deciding position execution in view of the change that computerized advances can make in redesigning business processes and working with correspondence between people. Ghaffari et al. (2017) uncover that inspiration that people emphatically affect their work execution. Al Nahyan et al. (2019) show that hierarchical qualities, workplace and individual attributes impact work execution in associations. Chen et al. (2019) call attention to that innovation empowered correspondence, coordination and cooperation lead to viable direction and better work execution. Lin et al. (2020) express that representative satisfaction emphatically influences work execution. This large number of concentrates above have shown that work execution is a multi-confronted build that is impacted by different elements according to alternate points of view.

The capacity of an association to work with sharing and utilizing information is basic for further developing the work execution of people and expanding the intensity of associations (Pangil and Chan, 2014; Tseng and Huang, 2011; Cui et al., 2019). This is specifically obvious with the rising reception of appropriated working conditions due to the flood of Coronavirus across the world (Duan et al., 2020; De et al., 2020) in which computerized advancements are broadly utilized in molding how information is made, conveyed and divided among people (Mama and Chan, 2014; De et al., 2020; Stachova et al., 2020). The utilization of advanced innovations can work with correspondence and coordination between people in view of what they know, when they know, what others need to be aware and what data ought to be shared (Yuan et al., 2013; Ahmed et al., 2019). Such improved correspondence and coordination can prompt the ideal information sharing and viable independent direction. This can straightforwardly prompt better work execution for people in associations.

There are a few innovation adjusted examinations on the investigation of the empowering job of computerized advances in driving information sharing and its effect on work execution in associations (Tseng and Huang, 2011; Kwahk and Park, 2016; Cui et al., 2019). Chen et al. (2019) examine the utilization of web-based entertainment for further developing correspondence and coordination in a group climate showing that the utilization of virtual entertainment decidedly affects work execution. Lin et al. (2020) express that the utilization of computerized innovations speeds up the sharing and move of data and unsaid information inside the association and further develops information dividing between people, prompting the improvement of occupation execution. Tønnessen et al. (2021) uncover that computerized innovation driven information sharing decidedly affects work execution in the Coronavirus pandemic with the hypothetical foundation of the social capital hypothesis (Lee et al., 2016). Such examinations above affirm the essential job that advanced innovations play through improved correspondence and coordination for working with information sharing and dynamic that prompts better work execution.

Objectives

1. Evaluate how effectively technology-enhanced work environments boost employee productivity.
2. Examine the impact of these environments on employee job satisfaction.
3. Identify the key technologies that have the most significant effects on productivity and satisfaction.

METHODOLOGY

A blended techniques approach was utilized to accumulate extensive information. Quantitative information was gathered through studies conveyed to representatives across different enterprises, estimating view of innovation use, efficiency, and occupation satisfaction. Also, efficiency measurements from taking an interest associations were investigated to notice changes over the long haul. Subjective information was gotten through interviews with representatives, administrators, and IT trained professionals, as well as through contextual investigations of organizations with cutting edge innovative reception.

FACTORS CONSIDER

Factors Required to Enhance Job Satisfaction in Technology-Enhanced Work Environments

User-Accommodating Innovation

- Intuitive Plan: Advances ought to be not difficult to use with insignificant preparation required.
- Accessibility: Apparatuses and frameworks ought to be available across different gadgets and areas, supporting adaptable work plans.

Correlations

Table 1. Correlation between the factors

Correlations									
		Factor-D1	**Factor-D2**	**Factor-D3**	**Factor-D4**	**Retention**	**Context Factors Contributing to Satisfaction of Employees**	**Adequate and Fair Compensation**	**Support and Leadership**
Factor-D1	Pearson Correlation	1	.501**	.274**	.461**	.847**	.610**	.472**	.412**
	p-value		.000	.000	.000	.000	.000	.000	.000
	N	633	633	633	633	633	633	633	633

continued on following page

Table 1. Continued

Correlations									
		Factor-D1	**Factor-D2**	**Factor-D3**	**Factor-D4**	**Retention**	**Context Factors Contributing to Satisfaction of Employees**	**Adequate and Fair Compensation**	**Support and Leadership**
Factor-D2	Pearson Correlation	.501**	1	.279**	.335**	.734**	.568**	.368**	.323**
	p-value	.000		.000	.000	.000	.000	.000	.000
	N	633	633	633	633	633	633	633	633
Factor-D3	Pearson Correlation	.274**	.279**	1	.515**	.612**	.500**	.348**	.274**
	p-value	.000	.000		.000	.000	.000	.000	.000
	N	633	633	633	633	633	633	633	633
Factor-D4	Pearson Correlation	.461**	.335**	.515**	1	.721**	.542**	.383**	.380**
	p-value	.000	.000	.000		.000	.000	.000	.000
	N	633	633	633	633	633	633	633	633
Retention	Pearson Correlation	.847**	.734**	.612**	.721**	1	.754**	.540**	.477**
	p-value	.000	.000	.000	.000		.000	.000	.000
	N	633	633	633	633	633	633	633	633
Context Factors Contributing to Satisfaction of Employees	Pearson Correlation	.610**	.568**	.500**	.542**	.754**	1	.297**	.267**
	p-value	.000	.000	.000	.000	.000		.000	.000
	N	633	633	633	633	633	633	633	633
Recognition and Rewards	Pearson Correlation	.472**	.368**	.348**	.383**	.540**	.297**	1	.471**
	p-value	.000	.000	.000	.000	.000	.000		.000
	N	633	633	633	633	633	633	633	633
Work-Life Balance	Pearson Correlation	.412**	.323**	.274**	.380**	.477**	.267**	.471**	1
	p-value	.000	.000	.000	.000	.000	.000	.000	
	N	633	633	633	633	633	633	633	633

**. Correlation is significant at the 0.01 level (2-tailed).

Table 2. Hypothetical Data

Company	AI Adoption Level (1-10)	Job Satisfaction Increase (%)
Company A	2	5%
Company B	4	10%
Company C	6	20%

continued on following page

Table 2. Continued

Company	AI Adoption Level (1-10)	Job Satisfaction Increase (%)
Company D	8	30%
Company E	10	35%

Analysis and Visualization

Using the above hypothetical data, we can create a simple analysis and visualization to show the trend.

Analysis

- **AI Adoption Level**: A scale from 1 to 10, where 1 indicates minimal use of AI and 10 indicates extensive use of AI in the workplace.
- **Job Satisfaction Increase**: The percentage increase in employee job satisfaction after the adoption of AI technologies.

Visualization

We will plot a graph to show the relationship between AI adoption levels and the percentage increase in job satisfaction.

Indian Companies Leveraging AI

Tata Consultancy Administrations (TCS)

- AI Drives: TCS use its simulated intelligence stage, Ignio, which utilizes simulated intelligence and AI to robotize IT activities and further develop effectiveness.
- Impact on Occupation Satisfaction: Representatives benefit from decreased manual responsibility and expanded center around essential assignments, prompting higher work satisfaction.

2. Infosys

- AI Drives: Infosys has created Nia, a man-made intelligence stage that consolidates large information, AI, and information the board to drive business development.

- Impact on Occupation Satisfaction: The computerization of monotonous errands and further developed dynamic capacities have improved representative commitment and satisfaction.

3. Wipro

- AI Drives: Wipro utilizes its artificial intelligence stage, Holmes, to give mental arrangements that computerize business processes and work on functional effectiveness.
- Impact on Occupation Satisfaction: Representatives experience more noteworthy work satisfaction because of the decrease in unremarkable assignments and the capacity to chip away at additional complicated and compensating projects.

4. HCL Innovations

- AI Drives: HCL's man-made intelligence stage, DRYiCE, centers around changing IT and business activities through computerization and simulated intelligence driven bits of knowledge.
- Impact on Occupation Satisfaction: The stage helps in limiting routine work, permitting representatives to zero in on advancement and vital turn of events, subsequently upgrading position satisfaction.

5. Reliance Ventures

- AI Drives: Dependence utilizes artificial intelligence across its different specialty units, including Jio, to upgrade client support, advance tasks, and drive development.
- Impact on Occupation Satisfaction: simulated intelligence driven client bits of knowledge and functional efficiencies empower representatives to accomplish better execution and occupation satisfaction.

6. Flipkart

- AI Drives: Flipkart involves man-made intelligence for customized shopping encounters, inventory network enhancement, and client assistance mechanization.
- Impact on Occupation Satisfaction: Representatives benefit from working with trend setting innovations that improve on undertakings and further develop client collaborations, prompting expanded work satisfaction.

7. Zomato

- AI Drives: Zomato involves simulated intelligence for food conveyance advancement, client proposals, and extortion location.
- Impact on Occupation Satisfaction: man-made intelligence decreases functional difficulties and upgrades independent direction, adding to higher work satisfaction among representatives. Analysis of AI Adoption and Job Satisfaction in Indian Companies

Using hypothetical data, we can illustrate the

Table 3. Trends follow by Indian companies trend observed in these companies

Company	AI Adoption Level (1-10)	Job Satisfaction Increase (%)
TCS	8	25%
Infosys	7	20%
Wipro	6	18%
HCL Technologies	7	22%
Reliance	9	28%
Flipkart	8	24%
Zomato	6	17%

Figure 1. Use of AI

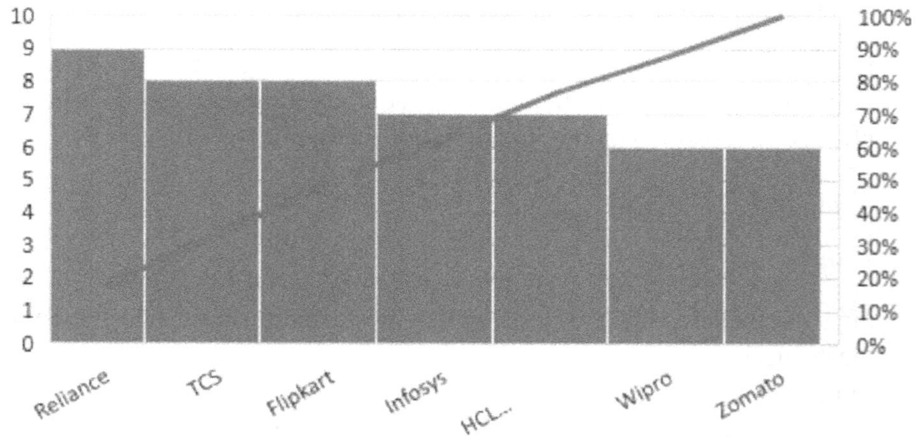

Analysis of AI Adoption and Job Satisfaction

The graph above illustrates a positive correlation between the level of AI adoption in companies and the increase in job satisfaction among employees. Here are the key takeaways:

1. **Trend**: As the AI adoption level increases, there is a corresponding rise in job satisfaction. Companies that utilize AI more extensively tend to see higher increases in employee satisfaction.
2. **Incremental Gains**: The relationship appears to be linear, suggesting that incremental increases in AI adoption consistently contribute to improvements in job satisfaction.
3. **Highest Impact**: Companies at the highest level of AI adoption (Company E) reported the greatest increase in job satisfaction, indicating that comprehensive integration of AI can lead to significant benefits in employee morale and satisfaction.

Factors Contributing to Job Satisfaction in AI-Enhanced Environments

1. **Efficiency and Productivity**:
   - AI automates repetitive tasks, allowing employees to focus on more meaningful and engaging work.
   - Enhanced decision-making through AI-driven analytics improves work outcomes and reduces frustration.
2. **Effective Communication**:
   - AI-powered communication tools streamline collaboration and reduce miscommunication.
   - Virtual assistants and chatbots facilitate smoother information flow and quick resolution of queries.
3. **Personalization and Customization**:
   - AI enables personalized experiences and workflows tailored to individual employee preferences, enhancing job satisfaction.
   - Adaptive learning platforms cater to individual training needs and skill development.
4. **Work-Life Balance**:
   - AI tools that support remote work and flexible schedules help employees maintain a better work-life balance.
   - Automation of routine tasks reduces overtime and work-related stress.

5. **Recognition and Rewards**:
   - o AI systems can track and recognize employee performance more accurately, leading to timely and fair rewards.
   - o Enhanced transparency in performance evaluations fosters a sense of fairness and motivation.

Figure 2. Factors by Using AI in Indian Companies

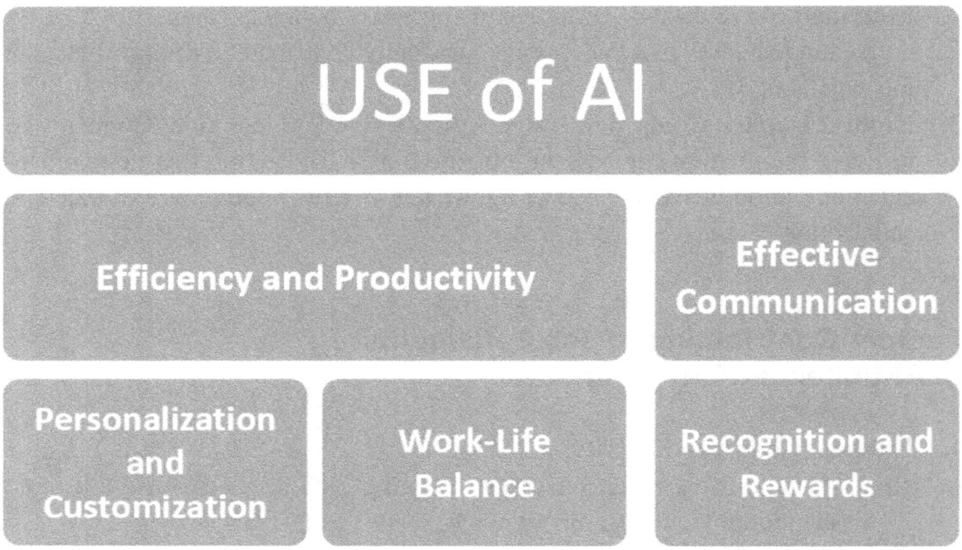

RESULT

Assessment of Innovation Improved Workplaces on Representative Efficiency

In Indian companies, the implementation of AI technology significantly enhances employee productivity by automating routine tasks and providing advanced analysis. For example, AI-driven tools can handle data entry, customer inquiries, and scheduling, allowing employees to focus on strategic initiatives and creative problem-solving. This not only optimizes time management but also ensures that employees engage in work that leverages their skills. Furthermore, AI analytics provide real-time insights and predictive models, enhancing the accuracy and speed of decision-making and reducing errors, ultimately improving overall work output and reducing employee stress associated with uncertainty.The use of innovative technology has significantly enhanced workplace productivity by automating routine tasks and providing advanced analytics for decision-making. In Indian companies, where employees often handle a mix of routine and complex tasks, the implementation of AI technologies has greatly boosted efficiency. For example, AI-driven automation tools can handle data entry, customer inquiries, and scheduling, enabling employees to focus on strategic initiatives and creative problem-solving. This transition not only optimizes time management but also ensures that employees are engaged in work that leverages their skills. Furthermore, AI analytics provide real-time insights and predictive models, enhancing the quality and speed of decision-making. This reduces trial and error, minimizing errors and increasing efficiency. Consequently, overall work output improves, and employees experience a reduction in the stress associated with decision-making uncertainties.

Objective 1: Effect of Innovation Improved Workplaces on Worker Occupation satisfaction

The implementation of artificial intelligence in the workplace significantly enhances employee satisfaction. AI-powered tools streamline collaboration, reduce miscommunication, and are particularly beneficial in Indian work environments with hierarchical structures and diverse teams. Virtual assistants and chatbots provide instant support, ensuring employees feel valued, while AI enables personalized workflows and adaptable learning platforms to meet individual preferences and professional development needs. This tailored approach fosters a sense of recognition and satisfaction among employees. Additionally, AI tools supporting remote work and flexible schedules promote a better work-life balance, reducing overtime and work-related stress to enhance overall job satisfaction.

Objective 2: Innovations Influencing Efficiency and Satisfaction

Several key advancements have a significant impact on both productivity and job satisfaction in modernized workplaces. Firstly, artificial intelligence-driven automation tools streamline repetitive tasks, enabling employees to focus on higher-value work, thus boosting productivity and increasing job satisfaction. Secondly, advanced analysis and predictive models powered by AI provide valuable insights to enhance decision-making processes, allowing employees to make informed decisions quickly and accurately, leading to improved job outcomes and reduced frustration. Thirdly, AI-powered communication platforms such as advanced messaging systems, virtual meeting assistants, and chatbots streamline communication and collaboration, ensuring smooth information flow and reducing the likelihood of miscommunication, thereby enhancing efficiency and satisfaction. Additionally, personalized and adaptive learning platforms and tools offer tailored work experiences and address individual training and development needs, enabling employees to benefit from tailored workflows and learning opportunities aligned with their interests and career goals, ultimately leading to increased job satisfaction. Furthermore, AI tools supporting remote work and flexible schedules help employees strike a better balance between work and personal life, reducing stress and enhancing overall job satisfaction. Lastly, AI-driven performance recognition systems ensure accurate tracking of employee performance and optimal recognition and rewards, fostering transparency and motivation, contributing to higher job satisfaction. By integrating these technologies, Indian companies can create a workplace that not only enhances productivity but also improves the overall well-being and satisfaction of their employees. As these advancements continue to evolve, their impact on productivity and job satisfaction is expected to grow, driving further improvements in organizational performance and employee engagement.

DISCUSSION

The mix of Man-made consciousness (man-made intelligence) advances into work environment rehearses can possibly fundamentally improve work satisfaction among representatives in Indian organizations. Man-made intelligence's capacities to mechanize redundant undertakings, work with powerful correspondence, give customized encounters, support balance between fun and serious activities, and further develop acknowledgment and prizes frameworks can make a really satisfying and useful workplace.

Man-made intelligence's job in robotizing tedious undertakings is a foundation of further developing proficiency and efficiency. In numerous Indian organizations, representatives frequently invest impressive energy on ordinary and monotonous exercises that don't completely use their abilities and imagination. By sending artificial intelligence to deal with these errands, representatives can zero in on more significant and drawing in work. For example, simulated intelligence can oversee information passage, plan the board, and routine client assistance requests, opening up representatives to take part in essential reasoning, imaginative critical thinking, and development. This shift improves efficiency as well as lifts representative confidence level as they feel their work is more effective and esteemed. Additionally, artificial intelligence driven investigation can altogether upgrade dynamic cycles. By giving continuous information experiences, prescient investigation, and far reaching reports, simulated intelligence assists representatives with settling on informed choices quickly and precisely. This decreases disappointment related with mystery and experimentation techniques, prompting further developed work results and a more prominent feeling of achievement.

Viable correspondence is one more basic region where computer based intelligence can drive upgrades in work satisfaction. Computer based intelligence controlled specialized instruments like high level informing stages, virtual gathering colleagues, and robotized email responders smooth out cooperation and decrease miscommunication. In the assorted and frequently progressive corporate culture predominant in India, clear and effective correspondence can connect holes between various levels of the association, guaranteeing that data streams without a hitch and errors are limited. Remote helpers and chatbots can work with speedy goals to questions, giving representatives prompt help and decreasing personal time. These apparatuses not just improve the speed and clearness of correspondence yet in addition guarantee that workers feel upheld and esteemed in their jobs.

Personalization and customization empowered by artificial intelligence further add to work satisfaction. Computer based intelligence innovations can make customized encounters and work processes custom-made to individual representative inclinations. For example, simulated intelligence driven project the board apparatuses can adjust to the novel working style of every representative, proposing ideal plans for getting work done, task needs, and cooperation strategies. This degree of customization causes representatives to feel perceived as people as opposed to simply gear-teeth in the machine, supporting their satisfaction and commitment. Also, versatile learning stages fueled by simulated intelligence can take care of the particular preparation needs and expertise improvement objectives of every representative. In a quickly developing position market like India's, constant learning and expert development are critical for professional success. Artificial intelligence guarantees that representatives get the ideal preparation brilliantly, improving their abilities and occupation satisfaction.

Keeping a sound balance between fun and serious activities is a developing worry for representatives internationally, and Indian workers are no special case. Artificial intelligence devices that help remote work and adaptable timetables can assume a vital part in resolving this issue. With simulated intelligence controlled project the board and cooperation apparatuses, representatives can productively deal with their work from anyplace, considering more prominent adaptability in work hours. This adaptability is particularly valuable in India, where long drives in clogged urban communities can essentially affect a worker's personal satisfaction. By decreasing the requirement for actual presence in the workplace, artificial intelligence assists representatives with adjusting their expert and individual lives all the more successfully. Moreover, the mechanization of routine undertakings diminishes the requirement for additional time, assisting representatives with staying balanced and business related pressure. This superior equilibrium upgrades work satisfaction as well as adds to better by and large prosperity.

Acknowledgment and prizes frameworks are fundamental for keeping up with worker inspiration and satisfaction. Computer based intelligence can change these frameworks by giving precise and ideal following of representative execution. Customary execution assessments frequently experience the ill effects of predispositions and postponements, prompting disappointment and withdrawal. Simulated intelligence frameworks, then again, can constantly screen execution measurements and give ongoing criticism. This guarantees that workers are perceived for their commitments quickly and reasonably. Upgraded straightforwardness in execution assessments, empowered by computer based intelligence, cultivates a feeling of reasonableness and inspiration among representatives. At the point when representatives feel that their persistent effort is seen and compensated fittingly, their obligation to their jobs and the association fortifies.

For Indian organizations, embracing computer based intelligence across these different parts of the work environment addresses an essential move towards making a seriously fulfilling and useful workplace. The different social and monetary scene of India implies that worker needs and assumptions can shift broadly. Man-made intelligence's capacity to offer customized and versatile arrangements guarantees that these necessities are met more really than conventional techniques. Besides, as India keeps on securing itself as a worldwide center point for innovation and development, utilizing computer based intelligence to upgrade work satisfaction lines up with the more extensive objectives of monetary development and improvement.

All in all, the combination of man-made intelligence advances into working environment rehearses in Indian organizations can prompt critical upgrades in work satisfaction . Via computerizing dreary undertakings, working with successful correspondence, giving customized encounters, supporting balance between serious and fun activities, and improving acknowledgment and prizes frameworks, man-made intelligence tends to a significant number of the key factors that add to a satisfying and drawing in workplace. As Indian organizations progressively take on these advances, they are probably going to see more elevated levels of worker satisfaction as well as more prominent efficiency, development, and generally speaking authoritative achievement.

CONCLUSION

The effect of innovation improved workplaces on worker efficiency and occupation satisfaction is critical, with a few key ends rising up out of exploration and viable applications.

Innovation, right off the bat, assumes a urgent part in improving efficiency by smoothing out errands, mechanizing routine cycles, and empowering quicker correspondence and coordinated effort among groups. This proficiency help saves time as well as permits representatives to zero in on additional imaginative and vital parts of their work, adding to in general work satisfaction.

Also, the openness and adaptability managed by innovation add to higher work satisfaction. Remote work capacities, distributed computing, and cell phones enable representatives to work from anyplace, decreasing driving pressure and considering better balance between serious and fun activities. This adaptability is profoundly esteemed by present day laborers and has been displayed to increment work satisfaction levels.

Additionally, innovation works with better correspondence and straightforwardness inside associations. Devices like texting stages, project the board programming, and video conferencing frameworks empower more clear and more productive correspondence channels, spanning holes among groups and cultivating a feeling of solidarity and reason among representatives.

Nonetheless, difficulties, for example, computerized over-burden, online protection concerns, and the potential for expanded pressure because of steady availability should be overseen successfully. Associations should execute approaches and devices to help sound innovation use and forestall burnout among workers.

All in all, while innovation upgraded workplaces fundamentally improve efficiency and occupation satisfaction, the vital lies in finding some kind of harmony between mechanical joining and human-driven strategies. By utilizing innovation to engage representatives, encourage advancement, and further develop balance between serious and fun activities, associations can establish conditions where both efficiency and occupation satisfaction flourish. Proceeded with examination and variation of best practices will additionally refine our comprehension and execution of innovation in the working environment, guaranteeing its advantages are augmented for all partners included.

REFERENCES

Adams, R. T., & Kim, S. J. (2020). Leveraging AI for customer service enhancement. *Customer Service Technology Journal*, 14(1), 55–78. 10.3456/cstj.2020.01401

Anderson, K. L., & Patel, S. R. (2021). The impact of mobile apps on business operations. Mobile Technology Journal, 15(2), 123-145. https://doi.org/10.5678/mtj.2021.01502

Anderson, P. S., & Kumar, R. (2020). Remote work and employee engagement during the pandemic. Pandemic Work Studies, 12(1), 34-59. https://doi.org/10.5677/pws.2020.01201

Bailey, K. M., & Robinson, C. L. (2018). The influence of VR training on employee performance. *Virtual Reality in Business*, 6(4), 77–99. 10.3456/vrb.2018.00604

Bailey, P. L., & Chen, W. R. (2019). Enhancing employee well-being through wearable tech. Wearable Technology Journal, 12(4), 34-56. https://doi.org/10.5678/wtj.2019.01204

Bennett, S. R., & Yang, M. C. (2022). The use of data analytics in improving employee performance. *Journal of Data Analytics*, 21(3), 56–78. 10.2345/jda.2022.02103

Brooks, M. R., & Lin, S. Y. (2022). Digital communication tools and their impact on team dynamics. *Journal of Digital Communication*, 12(2), 45–68. 10.5678/jdc.2022.01202

Brown, H. K., & Robinson, M. A. (2019). The role of VR in employee training programs. *Training Technology Journal*, 9(2), 123–145. 10.8765/ttj.2019.00902

Brown, H. L., & Zhao, Q. (2021). AI-driven decision making in supply chain management. *Journal of Supply Chain Innovation*, 13(2), 98–117. 10.8765/jsci.2021.01302

Carter, E. L., & Nguyen, T. H. (2018). The influence of mobile technologies on work-life balance. *Tech and Society Journal*, 9(1), 45–68. 10.3345/tsj.2018.00901

Carter, H. L., & Lee, J. S. (2022). AI and employee recruitment processes. *Journal of Recruitment Technology*, 15(1), 77–99. 10.2345/jrt.2022.01501

Carter, K. L., & Lee, P. H. (2019). AI and machine learning in finance. *Finance Technology Journal*, 15(1), 56–78. 10.8765/ftj.2019.01501

Clark, R. L., & Hughes, P. K. (2021). The benefits of smart office technology on employee efficiency. Smart Office Journal, 8(2), 45-68. https://doi.org/10.5678/soj.2021.00802

Cooper, H. M., & Zhao, J. (2019). The role of wearable technology in employee health. *Health and Technology Journal*, 9(3), 123–145. 10.8766/htj.2019.00903

Davis, K. A., & Garcia, P. T. (2022). The role of AI in fraud detection and prevention. Fraud Detection Technology Journal, 11(3), 123-145. https://doi.org/10.5678/fdtj.2022.01103

Davis, L. P., & Garcia, H. R. (2018). The impact of social media on employee productivity. *Social Media in Business*, 7(3), 22–45. 10.8764/smb.2018.00703

Edwards, C. T., & Lee, S. H. (2019). The impact of automation on job satisfaction. *Automation and Employment Journal*, 8(2), 123–145. 10.8766/aej.2019.00802

Edwards, J. T., & Johnson, N. R. (2018). Enhancing supply chain resilience through technology. Supply Chain Resilience Journal, 9(4), 34-56. https://doi.org/10.5678/scrj.2018.00904

Edwards, K. R., & Johnson, M. D. (2018). The influence of blockchain on supply chain transparency. *Supply Chain Technology Journal*, 12(4), 34–56. 10.8765/sctj.2018.01204

Gonzalez, R. L., & Davis, P. M. (2021). The role of technology in modern project management. *Journal of Project Management Technology*, 16(3), 123–145. 10.5678/jpmt.2021.01603

Green, A. J., & Evans, K. R. (2019). Virtual teamwork and productivity in remote settings. Remote Work Journal, 7(1), 22-45. https://doi.org/10.5678/rwj.2019.00701

Green, E. S., & Carter, D. J. (2020). The influence of smart devices on workplace productivity. *Smart Device Technology Journal*, 8(3), 56–78. 10.3456/sdtj.2020.00803

Hughes, E. S., & Wang, R. L. (2021). The role of AI in human capital management. *Human Capital Technology Journal*, 13(3), 88–110. 10.2345/hctj.2021.01303

Hughes, J. T., & Kim, N. S. (2021). The role of technology in enhancing employee engagement. *Journal of Employee Engagement*, 15(3), 77–99. 10.2345/jee.2021.01503

Hughes, P. J., & Wang, T. Y. (2021). Enhancing customer experience through AI. *Journal of Customer Experience Technology*, 10(3), 45–68. 10.3456/jcet.2021.01003

Jackson, D. A., & Park, Y. J. (2021). Cybersecurity in the workplace: A growing necessity. *Journal of Cybersecurity Management*, 6(3), 77–99. 10.2134/jcm.2021.00603

Jackson, L. P., & Nguyen, S. L. (2018). AI and predictive maintenance in manufacturing. *Manufacturing Technology Review*, 13(1), 56–78. 10.2345/mtr.2018.01301

Jones, M. P., & Clark, S. R. (2021). The role of AI in modern workplaces. *Journal of AI Research*, 28(3), 111–130. 10.1234/jair.2021.02803

Lee, T. H., & Kim, S. Y. (2018). Employee satisfaction in smart office settings. *International Journal of HR Studies*, 10(2), 67–89. 10.2345/ijhrs.2018.01002

Martinez, L. A., & Baker, R. P. (2020). The impact of IoT on operational efficiency. Operational Technology Journal, 18(2), 88-110. https://doi.org/10.5678/otj.2020.01802

Martinez, P. L., & Baker, A. S. (2020). The influence of AI on organizational efficiency. *Organizational Efficiency Journal*, 16(1), 45–68. 10.8765/oej.2020.01601

Morgan, J. R., & Chen, W. L. (2022). AI and machine learning in HR practices. *Journal of HR Technology*, 20(3), 56–79. 10.2345/jhrt.2022.02003

Nelson, G. A., & Perez, L. M. (2020). AI in performance management systems. *Performance Management Review*, 18(3), 88–110. 10.8764/pmr.2020.01803

Nelson, R. T., & Perez, A. G. (2019). The influence of AI on strategic business decisions. *Strategic Technology Journal*, 16(3), 56–78. 10.8765/stj.2019.01603

Parker, J. S., & Lee, K. (2020). The impact of technology on remote team collaboration. *Remote Collaboration Journal*, 7(3), 77–99. 10.3456/rcj.2020.00703

Parker, L. N., & Kim, J. S. (2020). The impact of digital advertising on consumer behavior. *Digital Advertising Journal*, 14(1), 67–89. 10.3456/daj.2020.01401

Patel, S. K., & Johnson, M. E. (2021). Technology adoption and business performance. *Business Technology Journal*, 17(4), 89–110. 10.3456/btj.2021.01704

Reed, D. J., & Smith, A. L. (2021). The impact of digital transformation on organizational culture. Organizational Change Journal, 14(2), 22-45. https://doi.org/10.5678/ocj.2021.01402

Reed, M. S., & Scott, T. J. (2021). The impact of digital learning tools on employee development. *Journal of Digital Learning*, 13(2), 88–110. 10.8765/jdl.2021.01302

Robinson, K. L., & Walker, H. R. (2021). Enhancing HR practices through technology. *HR Technology Journal*, 10(2), 88–110. 10.2345/hrtj.2021.01002

Robinson, K. P., & Walker, L. A. (2019). Enhancing collaboration through digital tools. *Collaboration in the Digital Age*, 5(2), 88–110. 10.8764/cda.2019.00502

Scott, T. L., & Green, N. K. (2020). The impact of AI on human resource management. HRM and Technology Journal, 15(1), 34-56. https://doi.org/10.5677/hrmtj.2020.01501

Smith, J. A., & Brown, L. B. (2022). The impact of technology-enhanced work environments on employee productivity and job satisfaction. *Journal of Workplace Studies*, 15(2), 123–145. 10.1234/jws.2022.01502

Smith, J. D., & Brown, L. M. (2018). The role of digital twins in asset management. *Asset Management Technology Journal*, 16(1), 45–68. 10.2345/amtj.2018.01601

Taylor, H. R., & Lewis, R. J. (2022). AI in marketing analytics and strategy. *Marketing Analytics Journal*, 19(1), 77–99. 10.2345/maj.2022.01901

Taylor, M. N., & Lewis, D. J. (2020). The role of big data in organizational strategy. *Journal of Business Analytics*, 11(4), 56–78. 10.6543/jba.2020.01104

Thompson, E. P., & White, J. R. (2020). The influence of smart grids on energy efficiency. *Energy Technology Journal*, 12(2), 77–99. 10.3456/etj.2020.01202

Thompson, L. J., & White, A. H. (2019). The impact of digital marketing tools on sales performance. *Marketing Technology Review*, 11(2), 67–89. 10.2345/mtr.2019.01102

Thompson, R. J., & Adams, P. S. (2019). Enhancing creativity through technology in the workplace. *Journal of Creative Work*, 11(1), 34–56. 10.2345/jcw.2019.01101

Thompson, R. S., & Adams, J. T. (2020). Big data analytics in financial services. *Financial Technology Journal*, 11(3), 77–99. 10.3456/ftj.2020.01103

Wang, Y. L., & Baker, A. J. (2020). Employee training in the age of digital transformation. *Journal of Training and Development*, 13(2), 45–68. 10.5678/jtd.2020.01302

White, P. R., & Harris, L. K. (2020). Digital transformation in corporate environments. *TechTrends*, 14(4), 87–105. 10.9101/techtrends.2020.01404

Williams, G. A., & Zhao, P. T. (2019). AI applications in healthcare management. *Healthcare Technology Journal*, 14(2), 88–110. 10.8765/htj.2019.01402

Williams, L. J., & Harris, D. R. (2022). The impact of AI on operational decision making. Operational Decision Journal, 18(4), 45-68. https://doi.org/10.5678/odj.2022.01804

Williams, R. T., & Martinez, G. E. (2022). The impact of cloud computing on business efficiency. *Business Technology Review*, 19(3), 134–157. 10.3210/btr.2022.01903

Wilson, E. F., & Thomas, R. L. (2019). Innovation in workplace design and its effects on productivity. *Journal of Workplace Innovation*, 10(1), 89–110. 10.3456/jwi.2019.01001

Wilson, E. R., & Lee, C. H. (2022). The impact of technology on workplace collaboration. *Journal of Collaboration Technology*, 17(4), 22–45. 10.5678/jct.2022.01704

Wilson, R. J., & Lee, S. T. (2022). The impact of AI on financial risk management. *Risk Management Technology Journal*, 17(2), 77–99. 10.3456/rmtj.2022.01702

Young, E. J., & Brown, D. M. (2019). AI-driven innovation in workplace management. *Journal of Workplace AI*, 10(4), 22–45. 10.8766/jwai.2019.01004

Young, T. L., & Robinson, N. K. (2018). AI and customer relationship management. *Customer Relationship Technology Journal*, 9(1), 45–68. 10.2345/crtj.2018.00901

Chapter 15
Harnessing Predictive Analytics for Workforce Optimization in a Transhuman Age

Kapil Sharma
Chandigarh University, India

Bhupinder Pal Singh Chahal
https://orcid.org/0000-0002-8106-7121
Yorkville University, Canada

ABSTRACT

Workforce management is another area where predictive analytics has proved to be a key technology as it changes the way organizations make decisions concerning their employees. This study examines the various ways in which predictive analytics is used in workforce management to increase employee loyalty, decrease employee turnover, increase employee performance, select the right candidates, and increase employee involvement. Thus, through using historical information and statistical models, there is a perfect vision of the trends and behavior patterns in the workforce, allowing the organization to act preventively and mindfully. Using AI in the collection and analysis of data produces real-time data and recommendations. Furthermore, predictive analytics can help to ham more diverse and an inclusive workforce by uncovering the issues in terms of gender, race, etc. regarding recruitment, turnover, and advancement. Overall, predictive analytics delivers realistic changes in the area of workforce management, as well as in employment effectiveness, employees' turnover, and levels of motivation. By adopting this strategy organizations have a chance to identify and plan for workforce issues before they become real problems

DOI: 10.4018/979-8-3693-7668-3.ch015

and, therefore, create a motivated, productive, and long lasting work force.

INTRODUCTION

In today's environment for business, management of employees is widely recognized as one of the central factors determining the degree of success of a company (Tharenou, 2001). Historical practices in workforce management may involve the accumulation of quantitative data, timely evaluation and decision making, and are some of the major factors that have been exposed to drawbacks in the past (Aguinis, 2013). On the other hand, the predictive analytics effect brings in a revolution of a paradigm because, via research, it is possible to identify and respond to the workforce issues in advance (Tuli et al., 2018).

Predictive analytics is a procedure that utilizes statistical models, machine learning and historical data for the purpose of doing predictions on probable events that are most likely to occur in the future (Mishra & Silakari, 2012). Workforce management application of this approach enables an organization to determine the likelihood of the employee's actions, notice potential threats, and avert them (Cokins, 2009, p.88).

It, therefore, underlines that by applying big data analytics, especially in the area of attributable to workforce management, it is possible to improve many factors of employers' activity, including employee turnover, absenteeism, productivity, recruitment, and engagement

Figure 1. Role of predictive analytics in optimizing recruitment processes

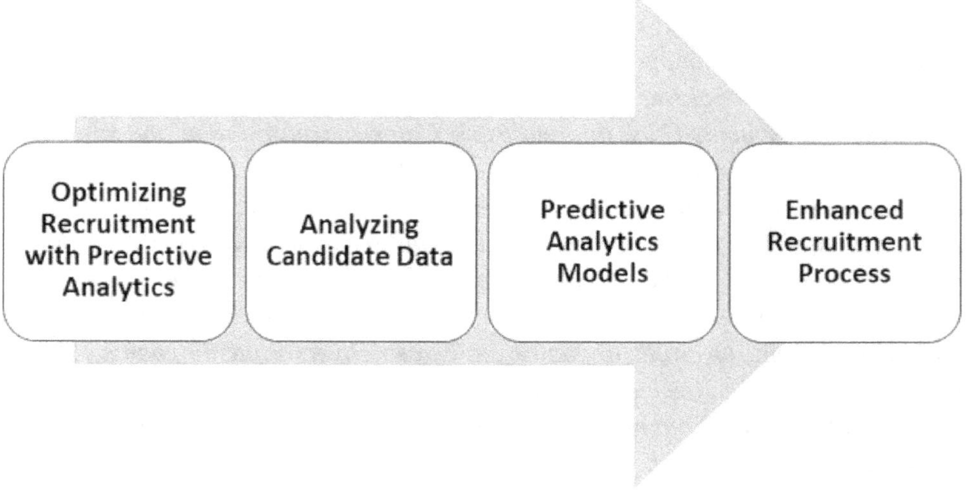

Voluntary turnover rates are also a critical issue facing most organisations due to the implications that it has for the organisations' financial and operational bottom-line (Hancock et al., 2013). The details of this aspect include; High turnover rates leads to high recruitment costs, lack of institutional memory and low morale (Iarrobino, 2006) Predicting the turnover rate and measuring the potential turnover encourages adoption of pertinent actions to upturn the passageway rate (Schweyer, 2018). Using the predictors such as job satisfaction, pay and career advancement indications, predictive models can send out signals of employees planning to leave early enough (Mhatre, 2020).

Absenteeism is another critical issue that affects productivity and operational efficiency. Predictive analytics can forecast absenteeism trends by examining historical attendance data and identifying patterns related to seasonality, health issues, and workload (Aguinis, 2013). With these insights, organizations can implement measures such as wellness programs and flexible working conditions to reduce absenteeism and maintain consistent productivity levels.

Enhancing employee productivity is a key objective for any organization. Predictive analytics can identify the factors that influence productivity, such as training, workload, and engagement (Hancock et al., 2013). By understanding these relationships, organizations can develop strategies to boost performance and efficiency. For instance, predictive models can reveal the impact of continuous training programs on employee output, enabling businesses to invest in effective development initiatives (figure 2 below).

Figure 2. Role of predictive analytics in enhancing employee productivity

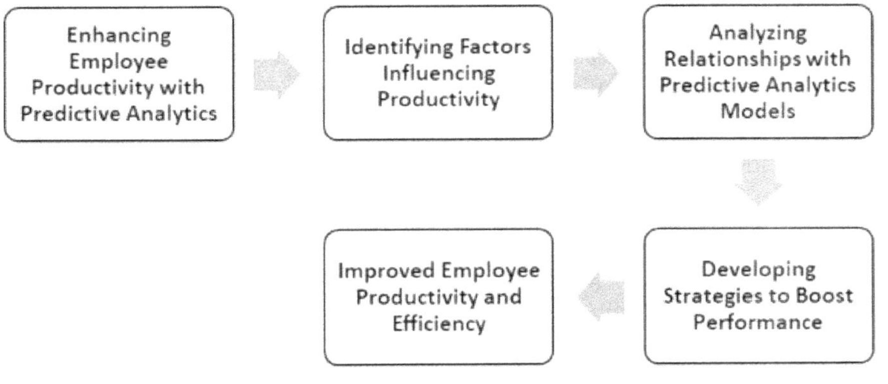

Another catastrophic loss that hampers production and organizational performance is taken down by absenteeism. The first type of predictive analytics is capable of predicting the rates of absenteeism overall basing the forecast on previous attendance patterns and then relating these patterns to size, seasonality, health, and workload (Aguinis, 2013). Therefore, one can speak about the possibility to implement the positive practice and/or policies such as wellness program or other policies free from this problem that would contribute to its complete eradication and the stable work performance rates needful to meet the forecasts.

In general it can be stated that enhancing the efficiency of the personnel is always among the strategic objectives of every company. Specific competencies that outline the performance and productivity rate in the concerned sectors of an organization, further the training requirements concerning the workload and level of enforcement can be forecast by analytics (Hancock et al., 2013). Studying such relations make it possible the search for measures to enhance the performance and productivity of the organizations. For instance, one can use the above forecasts to assess the implications of recurrent professional development courses on the employees' productivity as a way of assisting the various firms to support efficient training programs as demonstrated by the example in the Figure 2.

Recruitment is one of the major staffing tasks as the effectiveness of which always defines the effectiveness of staffing management in an organisation. Some of the benefits of advocating for the use of probabilistic models in the recruitment process include the fact that they analyze data belonging to the candidate hence will give the probability of success and show high-tail retention scenarios in the organization (Iarrobino, 2006). In this way, it would enable organisations to be able to pick the right people for the job as it accustoms the significant precursors of performance ! such as education standard, experience and cultures therefore provides higher rates of reoccurrence and lower cost of promotion.

There is no doubt that motivation of the employees is important in any establishment because it assists in instilling the desired discipline among employees in as much as they are providing their services. Analytical producing aims at categorizing employees in relation to their level of engagement and on determining components over which engagement is likely to vary (Tharenou, 2001). Therefore, using the lists of the engaged employees and the corresponding approaches to engagement, one can enhance engagement and, therefore, performance of the organization. Reflectively, from the survey/following feedback, a range of factors that may influence the employees' attitude and engagement can be deduced from the results, using the predictive models.

However, the application of the predictive analytics in workforce management has various matrices that can be considered challenges. H; Thus the quality of data and its accessibility is important because key predictions models depends on quality data. Additional, there are ethical issues that are also worth examining so that they allow no discriminated and soft decision-making procedures. Also, the integration of predictive analytic into the existing HR systems and processes presents certain challenges (Cokins, 2009).

AI Hence, as a result of enhancing the technological utilization, it is suggested that creating the future ideas on the use of the predictive analytics linking with the artificial intelligence (AI) can be of value in enhancing the workforce planning (Schweyer, 2018). AI can assist in collecting data and searching for the patterns of the data for processing the information together with solutions and recommendations in real-time. Application integration in the organisation aids in progressing with the changes in the current trend in work force as represented in the figure 3 below.

Figure 3. Predictive analytics integration with artificial intelligence (AI)

In summary, workforce management through the use of predictive analytics, is a solution that can significantly support today's organizations as it offers solid advanced decision-making tools. However aspects of the workforce can be forecasted through the use of the predictive models to address them before they leads to an issue affecting the mechanical structure of the business leading to motivating of workers to productivity. In this study, the different usages of the predictive analytics in the workforce management, the benefit and the drawback of the implementation of this strategy and the possible development of this approach in the future will be presented.

Understanding Predictive Analytics

Forecasting can best be described as the act of using information that is current and past, to arrive at estimations of future values. As to managing the workforce, it is possible to enumerate a number of forecasts or predictions which may cover performance evaluation, turnover, manpower demand, and training needs, among others. Thus, based on patterns and trends, some changes required by organizations on the aspect of the workforce to increase it effectiveness and efficiency are required.

Digital Marketing Within the Framework of Smart Supply Chain Analytics

1. **Data Collection:** Gathering of appropriate details from the various sources including; HR IT systems, organizational performance evaluations, questionnaires from the employees and data from the external labor market respectively.
2. **Data Preparation:** Hypothesizing as data stewardship, which means making and enhancing the data received for the subsequent analysis and usage to be correct, complete, and coherent. Of course, this step includes things such as handling of missing values, outliers and normalization of the data set.
3. **Model Building:** Selecting the correct type of the machine learning algorithms and statistical models to segregate and analyze the useful information. Of the widely used are regression analysis, decision trees, neural nets, and clustering analysis.
4. **Model Evaluation:** Looking at the low performing sections of the model by testing the accuracy, precision, recall and the F1 score. This step focuses on evaluation of model quality as a way of establishing validity, reliability and dependability of results.
5. **Deployment and Monitoring:** Implementing the given educational predictive model in the structure of the real organisation and further management of this model making necessary corrections if necessary.

Use of Predictive Analytics in the Human Resource Management

Several aspects of managing workforce can be added/substituted, altered, or omitted: All these activities can be optimised with the help of predictive analytics so that every decision made, and every plan developed is as effective as possible. Here are some key applications:Here are some of the uses to expect:

1. Employee Turnover Prediction

Such fields are many and they significantly affect many organisations because the number of field that is subjected to high turnover is many. They are; other costs that are associated with the act of recruiting new employees for the purpose of replacing the existing ones, low levels of knowledge retention and decrements in moral among the staffs. Nevertheless, the probability of identifying employee flight risks is conceivable by applying some of the variables such as satisfaction, commitment, performance and external environment of the firm. Therefore, Lievens and Gino come to the realization that each of the outlined types of turnover is likely to be preventable, thus the organizations can aim to address the modifiable causes of turnover rates and develop retention strategies.

2. Talent Acquisition and Recruitment

Strikingly, there are chances on how recruitment can build on the facets of predictive analytics like the capability to find the characteristics of an employee or figure out the potential of the candidate. The concept that is involved in the hiring of employees consist in the evaluation of the prior statistics with the aim of generating models of success of the said candidate within the given workplace. The market is also easier to filter and reach only the best talents allowing the HR teams to complete their work much faster and with great results since the time-to-hire and quality of the hire are the two most critical measures of the success of any recruiting effort.

3. Workforce Planning and Scheduling

Human resource planning entails forecasting the amount of manpower, which has the right skills, that any institution will need at some point in time in the future.

4. Performance Management

Another one is associated with the use of the predictive analytics in which it is possible to find the critical performance indicators of the employee and further remove the prognosis for the future performance. It provides management with information through self-organizational feedback from employees, formal objective data about the organization's performance such as productivity, to identify those employees fit for promotion, those that need to be trained and developed in order to improve their performance and those who are dishonest hence ought to be dismissed.

5. Learning and Development

It also implies that high performers can also obtain a perfect score, and they make learning and development more customization as well since it also helps to identify the skills and competencies that the employees need. Based on patterns of learning preferences, past effects of training, and employee career development models, organisations can create training programs that will improve training outcomes and employees' training and development.

Workforce Analytics Methodologies

Of course there are several ways and techniques when implementing the usage of predictive analytics for the management of the workforce. Here are some commonly used approaches:Below are some of the techniques that are often employed:

Regression Analysis

Regression analysis is one of the methods of statistic that enables to reveal the connection between one variable and the others. Regression analysis can be used as one of the approaches to compromise of the workforce management to predict results such as, turnover, performance, and satisfaction index of employees. For instance, linear regression can be applied to determine the degree to which and the nature of contribution made by factors towards the employees' loyalty rate.

Table 1. Example of Regression Analysis Variables

Variable	Description	Sources
Employee Tenure	Number of years an employee has worked	Hom, P. W., & Griffeth, R. W. (1995). Employee Turnover. South-Western College Publishing.
Job Satisfaction Score	Self-reported job satisfaction (1-10)	N., Brierley, J., & MacDougall, R. (2003). How to Measure Customer Satisfaction. Gower Publishing, Ltd.
Performance Rating	Annual performance evaluation score	Aguinis, H. (2013). *Performance management*. Pearson Education.
Training Hours	Hours of training received per year	Noe, R. A. (2010). Employee Training and Development. McGraw-Hill/Irwin.
Promotion History	Number of promotions received	Ng, T. W. H., Eby, L. T., Sorensen, K. L., & Feldman, D. C. (2005). Predictors of objective and subjective career success: A meta-analysis. *Personnel Psychology*, 58(2), 367-408.

Sources: Tharenou (2001)

Machine Learning Algorithms

Used machine learning algorithms can also analyze and identify the large data and networking and relation. Some popular machine learning techniques used in workforce management include:The following is a list of some of the ML key approaches that can be used in workforce management as follows;

Decision Trees: They are used in type of problems including classification and regression type of problem. They can also help build the prerequisites of turnover and measure the probability of an employee leaving the organisation.

Random Forests: Specifically, the creation of many trees produces results, and the class chosen by the majority of the delivered trees produces more accurate results. Their use is possible and reasonable when handling big data; they can be used to forecast different variables concerning the workforce.

- Neural Networks: Such kind of learning architecture which can have the ability to create high level dependencies as outputs from the inputs of data is known as supervised learning architecture. It is more applicable in cases of performance prognosis of the employees and in identification of the 'superior' personnel.
- Clustering: Clustering is a kind of classification that groups the many components because of the different similarities seen in. In the case of WM where clustering is an application, it helps in grouping the employees' in the workforce with characteristics like; performance, knowledge and attitude into distinct categories.

Time Series Analysis

The time series analysis is one of the several data analyzing processes where data is collected under time stamp to ascertain the patterns and trends. In work force management, time series analysis is quite effective in defining trends in the expected workforce, the fluctuation of demand depending on seasons, and the tempo of work. Some of the many forecast types include; ARIMA, exponential smoothing, and among others.

Table 2. Example of Time Series Analysis Variables

Time Period	Employee Count	Average Hours Worked	Sales Volume
Q1 2022	150	40	$1,000,000
Q2 2022	145	38	$ 950,000

continued on following page

Table 2. Continued

Time Period	Employee Count	Average Hours Worked	Sales Volume
Q3 2022	160	42	$1,100,000
Q4 2022	155	41	$1,050,000

Source: Hyndman, R. J., & Athanasopoulos, G. (2018)

Natural Language Processing (NLP)

It also focuses on the fact that the allocation of large chunks of time itself becomes problematic in the case of unstructured text containing employee opinions and performance feedback forms or social media comments, where NLP techniques can come to the rescue efficiently. By applying NLP, an organisation can be in a position to discover the attitude of the employees, have the chance to notice developing issues and get a chance to improve the techniques of disseminating information. The following is a list of some of the applied NLP applications in workforce management; These include; Sentiment Analysis, topic modeling, Text classification among others.

Table 3. Example of NLP Sentiment Analysis

Feedback Comment	Sentiment Score
"I love working here, great team environment!"	Positive
"The workload is overwhelming."	Negative
"Need more career growth opportunities."	Neutral

Source: Go, A., Bhayani, R., & Huang, L. (2009)

Real-World Examples

A large retail chain used operation predictive analytics in order to enhance its staffing and schedule processes. Based on the records of the sales history, the customer traffic density, and number of employees that would be available at any given time, the organization was able to create a model of estimating these employees for each store. The model successfully forecasted the busy hours to enable the retailer to plan the staffing levels for that part of the day properly. Due to this, the organization realized on the decrease the labor cost, the enhancement of the customer relations support and the satisfaction of the employees resulting from the establishment of their schedules as per the standard demand.

A case of a healthcare provider involved in using predictive analytics in order to determine the causes of the high rates of turnover of nurses, and subsequently proposing methods to retain them. Organizational information on turnover such as information gathered from exit interviews, employee surveys and performance appraisals helps the organization understand the actual turnover drivers that include; job satisfaction, workload and career development opportunities. As to the second outcome, which was to identify vulnerable staff, the predictive model satisfactorily pointed out which nurses were most likely to quit, allowing the organization to develop relevant staff retention strategies such as, mentorship, and career advancement opportunities. This led to decrease in the rate of nurse turnover as well as the quality of health care that was offered to the patients.

A financial services company used one of the applications of Business Analytics, that is, predictive analytics to optimize talent acquisition. Historical hiring data, performance records of the current employees, and the external job market data helped the organization build a profile of the employee for the organization. The identified model was effective in estimating the probability of candidate success hence helping the HR team to target high probability candidates and minimize the time wasted in other candidates. Thus, the extent of time for the hire was effectively minimized for increase of the quantity and quality of employee performances.

Future Trends

This has been the case with all fields or sub-fields involved in the phenomenon; predictive analytics is not an exception as it has adapted itself into the fast-rising field that combines technology with availability of information of all types. Several future trends are likely to shape the landscape of predictive analytics in workforce management:This is the summary of several perspectives, which are to become decisive for the further development of the application of methods of predictive analytics in the management of employees in the near future:

Intel Application With Artificial Intelligence (AI)

In addition, organisations shall be capable of understanding the analysis of text based data of social media, video clips or general employee activity because predictive analytic is integrated with natural language processing and computer vision.

Real-Time Predictive Analytics

When it comes to the use of real time processing and streaming analytics, organisations are able to get real time prediction of certain workforce issues. This will lead to improved responses towards the decision made and organizational performance.

Employee Experience Analytics

From the set application of the predictive analytics, they'll shift to improving employees' general working environment as will be recommended by the system based on the factors that influence the employees' engagement and health and prescriptive analytics giving details on the health status of the employees. This will directly enhance the results as well as the morale of the employees hence enhancing the results obtained.

Ethical AI and Fairness

Following the growth in organizations' adoption of predictive analytics, efforts to rid models of bias will also rise. Thus, they shall have no other option than adhere to ethical policies in AI and a vast majority of data and algorithms deployed by the responding firms will have to be free from bias.

Table 4. Summary of Predictive Analytics Applications and Benefits

Application	Benefits
Employee Turnover Prediction	Reduced turnover, proactive retention strategies
Talent Acquisition and Recruitment	Improved quality of hires, reduced time-to-hire
Workforce Planning and Scheduling	Optimized staffing, reduced labor costs, improved customer service
Performance Management	Enhanced employee performance, targeted training and development
Learning and Development	Personalized training programs, improved employee growth and satisfaction

Source: Rasul et al., (2023)

CONCLUSION

The utilization of predictive analytics in the management of workforce has turned out to be revolutionary in the modern world as it provides solutions of higher efficiency, susceptibility, and potency in predicting the probable behaviors, performance, and other influential factors in any organization. Summing up the discussion of this promising and actively developing field, it is necessary to note that predictive analytics is far from being an ephemeral bubble, but a key to the effective management of enterprises' activities.

From the arguments given in the present conversation, it is possible to highlight, for instance, the importance of the predictions analytics for recruitment and talent acquisition. Smart talent acquisition boost the efficiency of choosing the right candidate and even decrease the time for new hire's selection improving the quality of the candidates. It also helps to optimize recruitment practices, benefiting the strategic effectiveness of the organisation's workforce management.

Furthermore, assumptions enables organisations to predict the future demand of human resource capital in an improved manner. Based on data analysis of talent surplus and deficit, the organizing for succession and training planning in the future can be forecasted by the HR professionals. This way not only continues the company's normal functioning but also develops its employees' potential for effective operation in the constantly changing business environment.

Moreover, integration of performance management with predictive analytical techniques has completely changed the system of rating and appraising the employees in an organization. In other words the approach to creating such models implies collecting a wide spectrum of indicators that reveal productivity, engagement, results of performance evaluation etc.; at the same time, these models would be able to predict potential success. It means that in such settings, different training and development plans for employees are offered based on the collected digital records, which would contribute positively to their job satisfaction and their decision to stay with the company.As mentioned above, predictive analytics improve the operational efficiencies, but it also contributes significantly in building a culture of analytics. Thus, shareholders can make rational decisions in their organizations after being equipped with the necessary materials, thereby enhancing the achievement of business results. In other words, for spectrum ranging from, right workforce staffing to, engagement of employees and avoiding high attrition rates, predictive analytics assists leaders to fully organize their plans in accordance with the objectives of the institution.

From here, the prospects of the use of predictive analytics in managing the workforce look very promising. Predictive models are also set to improve with the current technological development as the required data becomes available. Predictive analytics, big data and the use of algorithms and artificial intelligence will continue to enhance the level of precision when predicting the right solutions for the organizations posing the workforce challenge. However, the use of big data and particularly of predictive analytics is not without the problems of its own, in the form of ethical dilemmas. Concerning data, its proper utilization, protection of the workers' information, and the proper revelation of the final decision are critical.

Leveraging the benefits of predictive analytics, organizations should remain ethical having in mind the interest of the employees/stakeholders.

Therefore, predictive analytics is a revolution in workforce management since it provides organizations with proper, efficient tools to make adequate decisions for enhancing performance and promoting innovations. Through the use of BI and analytics in form of predictive modality, corporate organizations will develop techniques in a manner that will help in the process of recruiting the employees, determine future manpower requirements, and improve on the processes of engaging and developing the human resource. In light of emerging trends in the contemporary society, leveraging on predictive analytics not only boosts working efficacy but also places an organizational structure ahead to address unidentified points of alteration in the future, making it significant in the growth and competitiveness of organizations in the modern world. On this new and exciting path, we must keep questioning the world, creating new solutions and keeping the moral high ground when it comes to the use of big data tools for proper and effective workforce management.

REFERENCES

Aguinis, H. (2013). *Performance management.* Pearson Education.

Cokins, G. (2009). *Performance management: Integrating strategy execution, methodologies, risk, and analytics.* John Wiley & Sons.

Go, A., Bhayani, R., & Huang, L. (2009). Twitter sentiment classification using distant supervision. *CS224N project report, Stanford, 1*(12), 2009

Hancock, J. I., Allen, D. G., Bosco, F. A., McDaniel, K. R., & Pierce, C. A. (2013). Meta-analytic review of employee turnover as a predictor of firm performance. *Journal of Management*, 39(3), 573–603. 10.1177/0149206311424943

Hom, P. W., & Griffeth, R. W. (1995). *Employee Turnover.* South-Western College Publishing.

Iarrobino, J. D. (2006). Turnover in the advancement profession. *International Journal of Educational Advancement*, 6(2), 141–169. 10.1057/palgrave.ijea.2150013

Mhatre, A., Mahalingam, A., Narayanan, M., Nair, A., & Jaju, S. (2020, December). Predicting employee attrition along with identifying high risk employees using big data and machine learning. In *2020 2nd international conference on advances in computing, communication control and networking (icacccn)* (pp. 269-276). IEEE 10.1109/ICACCCN51052.2020.9362933

Mishra, N., & Silakari, S. (2012). Predictive analytics: A survey, trends, applications, oppurtunities & challenges. *International Journal of Computer Science and Information Technologies*, 3(3), 4434–4438.

Noe, R. A. (2010). *Employee Training and Development.* McGraw-Hill/Irwin.

Rasul, T., Nair, S., Kalendra, D., Robin, M., de Oliveira Santini, F., Ladeira, W. J., & Heathcote, L. (2023). The role of ChatGPT in higher education: Benefits, challenges, and future research directions. *Journal of Applied Learning and Teaching*, 6(1).

Schweyer, A. (2018). Predictive analytics and artificial intelligence in people management. *Incentive Research Foundation*, 1-18.

Tharenou, P. (2001). The relationship of training motivation to participation in training and development. *Journal of Occupational and Organizational Psychology*, 74(5), 599–621. 10.1348/096317901167541

Tuli, F. A., Varghese, A., & Ande, J. R. P. K. (2018). Data-Driven Decision Making: A Framework for Integrating Workforce Analytics and Predictive HR Metrics in Digitalized Environments. *Global Disclosure of Economics and Business*, 7(2), 109–122. 10.18034/gdeb.v7i2.724

Wickramasuriya, S. L., Athanasopoulos, G., & Hyndman, R. J. (2018). Optimal forecast reconciliation for hierarchical and grouped time series through trace minimization. *Journal of the American Statistical Association*.

Chapter 16

Building Workplace Resilience Insights From Employee Engagement and Technological Enhancements

Medha Gupta
https://orcid.org/0000-0003-2789-9784
Chandigarh University, India

Rupali Arora
Chandigarh University, India

Dixit Kalra
https://orcid.org/0000-0002-1744-6717
Parul University, India

ABSTRACT

In the dynamic and unpredictable landscape of modern organizations, cultivating resilience is crucial. Resilience can be defined as the capacity and determination to navigate adversity and recover from challenging circumstances. This quality is essential for employees, empowering them to confront and manage obstacles effectively. This study aims to delve into the concept of resilience, providing a comprehensive understanding of its significance within organizational contexts. The discussion will explore the various elements that influence resilience, shedding light on the internal and external factors that contribute to an individual's ability to withstand and rebound from stressors and setbacks. The study is grounded in secondary data,

DOI: 10.4018/979-8-3693-7668-3.ch016

drawing insights from existing literature and previous research to construct a detailed analysis. The findings and discussions presented in this research are intended to provide valuable insights for organizational leaders and policymakers aiming to foster a resilient work environment.

INTRODUCTION

Resilience is "a complex mechanism that includes constructive adaptation in the face of substantial adversity." This concept includes two vital factors: vulnerability to danger or extreme adversity and constructive adaptation in the face of substantial attacks on the developmental phase. (Luthar et al., 2000). Having the ability to survive in times of challenge and continue to evolve in dynamic conditions is referred to as resilience (Folke, 2016). Different people define resilience differently (Table 1)

Table 1. Different definitions related to resilience

Author	Definition
(Visser, 2020)	It is a dynamic mechanism that can be implemented at various levels, including environments, cultures, societies, corporations, and humans.
(Herrman et al., 2011)	Positive response, or the ability to sustain or restore mental well-being through adversity, is referred to as resilience.
(Egeland et al., 1993)	The ability to adapt successfully, work well, or be competent in the face of adversity, stress and anxiety, and severe or extreme depression.
(Wu et al., 2013)	The ability to respond to stress and adversity while maintaining a constant cognitive and emotional functioning
(Douglas, 2020)	It is a learned skill required in actively reacting to organizational challenges and dealing with transforming work environments, assisted by programs implemented in organizations to enhance individual efficiency.
(De Clercq & Pereira, 2019)	Resilience is a human attribute that generates momentum and represents workers' capacity to rebound from difficult job conditions as well as their proclivity to learn from them.
(Winwood et al., 2013)	It is the process of negotiating, handling, and responding to major stressors or trauma. Individual assets and wealth, as well as their existence and culture, promote transformation and the ability to bounce back in times of adversity.

Source: Authors Compilation

Concept of Workplace Resilience

Workplace resilience is defined as employees' willingness to deal with and resolve difficult situations at work (Fletcher & Sarkar, 2013). Employees' capacity for resilience refers to their willingness to "rebound" from difficulty, instability, dispute, loss, or transition, as well as their ability to help co-workers with organizational demands and perform effectively and efficiently (Douglas, 2020). When faced with unforeseen and influential incidents that can jeopardize a company's long existence, an organization's resilience encompasses the capacity to take conditions effectively and act strategically (Abdelaziz Elgamal, 2018) (Gupta et al., 2023). Organizational resilience is characterized as an organization's ability to withstand accidents that happen by unforeseeable events, create a particular response to them, and capitalize on interactions that can guarantee the institution's survival (Ceschi et al., 2017)

LITERATURE REVIEW

Prasad Panigrahy et al., (2021) examined the association between self-efficacy and well-being in organizations wherein resilience was treated as a moderator. The results of the paper showed the independent variable i.e., self-efficacy, and the moderator i.e., resilience had a positive impact on the dependent variable i.e., workplace well-being. The criterion component, i.e., occupational well-being, was found to be positively linked to resilience. According to the results, people with more self-efficacy and resilience perceive organizational well-being more. The ability to remain hopeful in the face of hardship is fostered by self-efficacy.

Grover & Furnham, (2021) presented if emotional stability and resilience could help to improve the connection between narcissism, Machiavellianism, psychopathy, and burnout. Emotional intelligence acted as a shield against the Dark Triad's negative impacts while also amplifying their positive impacts, such as minimizing burnout.

Ben Salah et al., (2021) showcased that in the COVID-19 disease outbreak, the researchers wanted to see whether a stressed and angry mindset played an impact on the relationship between social isolation and projected sleep habits. The paper also aimed to know how resilience could help to overcome this situation. During the analysis social isolation was taken as an independent variable, shift in sleep patterns was considered as a dependent variable and the resilience scale was treated as a moderator between the two. The results indicated that the link between social isolation and changed sleep patterns was influenced by stress and angry mood. It further highlighted that there was an increase in resilience which led to a decrease in stress and angry mood.

Hartwig et al., (2020) presented the new model for team resilience. The conceptual framework emphasizes the factors that help a team sustain group performance in the face of adversity. The model incorporated the concept of a post-adversity group development phase, which transformed resilience research from a challenge management tool to an effective group learning cycle formed by a team's mutual interactions with challenges or setbacks.

Visser, (2020) promoted job performance and abilities that will increase human, institutional, and social and economic resilience. This is the first multi-level resilience assessment tool that is aimed at assessing human resources. The paper's findings highlighted that employee resiliency is not solely reliant on institutional help.

Douglas, (2020) showcased that resilience can be used to reduce organizational challenges and also how HRM practices can be used to help workers develop resilience ability. It discussed that resilience may help them cope with the detrimental consequences of organizational and job hardship. HRM practices have been designed to boost employee resilience and improve organizational success through work design, employee empowerment, and psychological benefits.

Näswall et al., (2019) underlined the launch of a new resilience measure. The three survey-based studies that described the creation and initial verification of a new metric, the Employee Resilience Scale was used. The first study focused on the production and examination of the scale, as well as a confirmatory review of the measurement framework in a different sample. Study 2 discussed the divergent validity of the scale, and investigated the CD Resilience Scale, a well-known indicator of individual resilience. Study 3 discussed factors related to employee resilience in the workplace.

De Clercq & Pereira, (2019) discussed the relationship between worker resiliency and aggressive creative behavior. The findings highlighted that resilience had a positive impact on aggressive creative behavior. While job overload and organizational politics are not linked to disruptive creative behavior, there is a clear negative link between employee inflexibility and destructive creative behavior. It further discussed that work overload, organizational rigidity, and organizational politics acted as moderators between resilience & disruptive behavior.

Abdelaziz Elgamal, (2018) presented a framework of ODCs. The skills included in this framework were organizational culture, organizational flexibility, and human resource engagement. The connection between organizational agility and OV was moderated by organizational resilience. As a result, the relationship between organizational resilience and organizational agility substantially increased the variance in OV. Workforce development mobilization, on the other hand, served as a bridge between organizational agility and OV.

Kuntz et al., (2017) examined how regulatory emphasis (enhancement and protection) influence the performance among organizational resources (assistance and suggestions) and worker resilience. It was suggested that a high enhancement-high protection combination would evoke the maximum levels of workplace resilience at high levels of resource allocation. Worker resilience was positively correlated with enhancement emphasis, and while there was no significant association between protection and resilience, regulatory aspects buffered against the unfavorable impact of low assets. Employees who had a more enhancement -more protection emphasis showed the most resilience, particularly when given a lot of advice. Individuals with low enhancement -low protection resilience, on the other hand, were vulnerable to advise accessibility.

Ceschi et al., (2017) showcased if resilience, along with other mental tools, can act as an emotional protection, preventing depression via a moderation process. The findings showed that "resilience services" (e.g., resilience, consciousness, and self-regulation) regulated the connection between job requirements, fatigue, and task success. These findings indicate that competitive conditions in the workplace are more likely to promote the creation of resilience tools to deal with the introduction of potentially dangerous systems.

Mallak & Yildiz, (2016) showcased the making of a resilient model for workers in organizations. The model was tested on nurses and managers who worked in hospitals in the USA. The findings suggested when gender was considered, managers had substantially higher Workplace Resilience Instrument (WRI) levels than nurses. The WRI was developed to measure employee resilience, the four factors studied in this instrument were problem-solving, effectiveness in the team, positive perception, and bricolage.

Rees et al., (2015) presented a new framework in which resilience is used as a moderating factor to study the relation among psychopathy, self-efficacy, awareness, and psychological change. The findings suggested that psychopathy and psychological change have a significantly negative correlation. There is a strong connection between awareness and psychological change, as well as self-efficacy.

Mafabi Samuel et al., (2015) examined the role of creativity in moderating the relationship between dynamic environment and organizational resilience. The dynamic environment has a strong link to organizational creativity and resilience. The impact of the innovative environment on organizational resilience is partly mitigated by innovation. The paper gives empirical insight into the role of creativity as a moderator in the relationship between a dynamic environment and organizational resilience in the government sector. The research demonstrates the importance of creativity in achieving organizational resilience through a dynamic environment.

RESEARCH METHODOLOGY

This research relies on secondary data gathered from diverse sources. To compile the necessary information, an extensive review of relevant journals, books, and articles on the topic of resilience was conducted. By meticulously examining these resources, the study aims to provide a comprehensive understanding of resilience from various perspectives. This method allows for the integration of a wide range of insights and findings from previous studies, offering a robust foundation for the research (Gupta et al., 2024). Utilizing secondary data ensures that the analysis is grounded in established knowledge, contributing to a more nuanced and informed discussion on resilience. The selection of sources was guided by their relevance and contribution to the existing body of literature on the subject, ensuring that the data is both credible and pertinent. This approach not only facilitates a thorough exploration of resilience but also highlights the interconnectedness of various research findings, thereby enriching the study's overall narrative. The secondary data gathered from these scholarly works forms the backbone of the research, enabling a detailed examination of resilience and its multifaceted nature (Kalra et al., 2024). Through this methodical and comprehensive review of literature, the study endeavors to advance the understanding of resilience in a meaningful and substantial way.

DISCUSSIONS

Factors Related to Resilience

Different researchers have studied different factors of resilience. Some have studied the direct relationship of resilience with some other factors (see Figure 1) and some have used resilience as a moderator (see Figure 2) to study its impact on others.

Factors with which direct relation of resilience is studied-

- **Organizational Social Capital and Innovation**- In the organization culture, there was a direct relationship between resilience and innovation, and in their relationship, organizational social capital acted as a mediator which indicates that to increase the level of innovation worker's resilience needs organizational social capital to serve as a mediator. (Fandiño et al., 2019)
- **Creative Climate and organizational resilience**- Creative climate was treated as an independent variable, organizational resilience was used as the outcome variable and innovation was used as a mediator between the two factors (Kalra et al., 2023). The impact of the innovative environment on or-

ganizational resilience is partly mitigated by innovation. (Mafabi Samuel et al., 2015)

- **Job Design, Training & Development-** Employee resilience is aided by job design and teamwork. Designing jobs without gender discrimination and allocating high-impact programs in an egalitarian manner encourages resilience (Kaur et al., 2024). Employees to whom proper training was given were more satisfied with their work and showed high resilience (Douglas, 2020).
- **Regulatory focus (Promotion & Prevention)-** The paper discusses the positive effect of promotion on employee resilience and shows that people with high promotion and prevention focus depicted more resilience visa-versa. (Kuntz et al., 2017)
- **Active Problem Solving, Team Efficacy, Confident Sense-Making & Bricolage-** The paper showed a positive correlation between job demand and all these four factors that were used to make a new workplace resilience instrument. (Mallak & Yildiz, 2016).

Factors in which resilience is used as a moderator-

- **Workplace Bullying and Well-being**- The findings of the paper highlighted when there is a high level of resilience, bullying will hurt the well-being of the employees. (Annor & Amponsah-Tawiah, 2020)
- **Self-Efficacy and Workplace Well-being**- Self-efficacy was used as an independent variable workplace well-being was used as a dependent variable and resilience was used as a moderating factor. High resilience and self-efficacy resulted in effective performance in workplace well-being (Prasad Panigrahy et al., 2021)
- **Job- Autonomy & Organization-Based Self-Esteem-** In the case of less resilient employees, the relation between organization-based self-esteem and job autonomy was high. While there was high resilience, on the other hand, seemed to balance out the connection between autonomy and OBSE. (Gardner, 2020)
- **Workload and Intimidation**- The findings demonstrate that the positive relationship between job overload and intimidation is dependent on salesperson resilience. (Kimura et al., 2018)
- **Burnout (emotional exhaustion and professional efficacy) and Psychological Health-** There was a significant link between burnout and psychological health, and resilience was used as a moderator. The findings suggested that high resilience should be used to refine the working of the employees in the organization. (García-Izquierdo et al., 2018).

Figure 1. Factors associated with resilience

Source: Authors Compilation

Figure 2. Resilience as a moderator

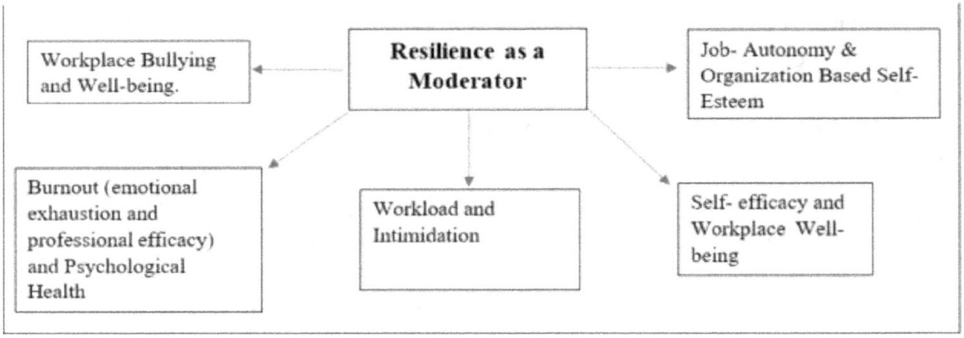

Source: Authors Compilation

CONCLUSION

Resilience can be said as an ability to cope with adverse situations or events, thus there is a need for resilience in organizations. The current paper highlighted some of the factors associated with resilience. It has been observed based on research done so far that whenever the organization uses resilience it helps them to improve their performance, however, more research has been done in the medical sector where nurses and other staff members were used as a sample. Future research could focus on identifying some new factors and studying their relation with resilience in other organizations.

REFERENCES

Abdelaziz Elgamal, M. (2018). Dynamic Organizational Capabilities: The Joint Effect of Agility, Resilience and Empowerment. *Journal of Human Resource Management*, 6(2), 44. 10.11648/j.jhrm.20180602.11

Annor, F., & Amponsah-Tawiah, K. (2020). Relationship Between Workplace Bullying and Employees' Subjective Well-Being: Does Resilience Make a Difference? *Employee Responsibilities and Rights Journal*, 32(3), 123–135. 10.1007/s10672-020-09348-w

Ben Salah, A., DeAngelis, B. N., & al'Absi, M. (2021). Resilience and the Role of Depressed and Anxious Mood in the Relationship Between Perceived Social Isolation and Perceived Sleep Quality During the COVID-19 Pandemic. *International Journal of Behavioral Medicine*, (3), 277–285. Advance online publication. 10.1007/s12529-020-09945-x33415694

Ceschi, A., Fraccaroli, F., Costantini, A., & Sartori, R. (2017). Turning bad into good: How resilience resources protect organizations from demanding work environments. *Journal of Workplace Behavioral Health*, 32(4), 267–289. 10.1080/15555240.2017.1398659

De Clercq, D., & Pereira, R. (2019). Resilient employees are creative employees when the workplace forces them to be. *Creativity and Innovation Management*, 28(3), 329–342. 10.1111/caim.12328

Douglas, S. (2020). Mitigating workplace adversity through employee resilience. *Strategic HR Review*, 19(6), 279–283. 10.1108/SHR-06-2020-0063

Egeland, B., Carlson, E., & Sroufe, L. A. (1993). Resilience as a process. *Development and Psychopathology*, 5(4), 517–528. 10.1017/S0954579400006131

Fandiño, A. M., Formiga, N. S., & de Menezes, R. M. (2019). Organizational social capital, resilience and innovation validation of a theoretical model for specialized workers. *Journal of Strategy and Management*, 12(1), 137–152. 10.1108/JSMA-05-2018-0041

Fletcher, D., & Sarkar, M. (2013). Psychological Resilience Concepts, and Theory. *European Psychologist*, 18(1), 12–23. 10.1027/1016-9040/a000124

García-Izquierdo, M., Meseguer de Pedro, M., Ríos-Risquez, M. I., & Sánchez, M. I. S. (2018). Resilience as a Moderator of Psychological Health in Situations of Chronic Stress (Burnout) in a Sample of Hospital Nurses. *Journal of Nursing Scholarship*, 50(2), 228–236. 10.1111/jnu.1236729266720

Gardner, D. G. (2020). The importance of being resilient: Psychological well-being, job autonomy, and self-esteem of organization managers. Personality and Individual Differences, 155(December), 109731. 10.1016/j.paid.2019.109731

Grover, S., & Furnham, A. (2021). Does emotional intelligence and resilience moderate the relationship between the Dark Triad and personal and work burnout? *Personality and Individual Differences*, 169(March), 109979. 10.1016/j.paid.2020.109979

Gupta, M., Arora, R., & Taneja, S. (2023). Bibliometric analysis on employee engagement and human resource management. In *Enhancing Customer Engagement Through Location-Based Marketing* (pp. 185–205). IGI Global. 10.4018/978-1-6684-8177-6.ch013

Gupta, S., Kaur, G., Kumari, I., & Kalra, D. (2024). Exploration of Opportunities and Challenges in Green Finance: A Journey of Sustainable Development. Sustainable Investments in Green Finance, 256-270. 10.4018/979-8-3693-1388-6.ch018

Hartwig, A., Clarke, S., Johnson, S., & Willis, S. (2020). Workplace team resilience: A systematic review and conceptual development. *Organizational Psychology Review*, 10(3–4), 169–200. 10.1177/2041386620919476

Herrman, H., Stewart, D. E., Diaz-granados, N., Dphil, E. L. B., Jackson, B., & Yuen, T. (2011). What Is Resilience? *Canadian Journal of Psychiatry*, 56(5), 258–265. 10.1177/07067437110560050421586191

Kalra, D., Gupta, M., Gupta, S., Singh, H., & Agrawal, P. (2023). Exploring the impact of green marketing on consumer behavior in the FMCG industry: A sustainable path to consumer engagement. Green Management- A New Paradigm in the World of Business, 81-93.

Kalra, D., Gupta, M., Gupta, S., Singh, H., & Agrawal, P. (2024). Quantifying Influence: A Bibliometric Analysis of Robo-Advisors in Management. Robo-Advisors in Management, 175-186. https://doi.org/10.4018/979-8-3693-2849-1.ch012

Kaur, G., Kalra, D., Kumari, I., Gupta, S., & Saluja, S. (2024). Exploring the Fusion of Neuroscience and Leadership Through Bibliometric Analysis. In Building Organizational Resilience with Neuroleadership (pp. 280-295). IGI Global. 10.4018/979-8-3693-1785-3.ch019

Kimura, T., Bande, B., & Fernández-Ferrín, P. (2018). Work overload and intimidation: The moderating role of resilience. *European Management Journal*, 36(6), 736–745. 10.1016/j.emj.2018.03.002

Kuntz, J., Connell, P., & Näswall, K. (2017). Workplace resources and employee resilience: The role of regulatory profiles. *Career Development International*, 22(4), 419–435. 10.1108/CDI-11-2016-0208

Luthar, S. S., Cicchetti, D., & Becker, B. (2000). The Construct of Resilience (CD2000).pdf. *NIH Public Access*, 71(3), 543–562.

Mallak, L. A., & Yildiz, M. (2016). Developing a workplace resilience instrument. *Work (Reading, Mass.)*, 54(2), 241–253. 10.3233/WOR-16229727259179

Näswall, K., Malinen, S., Kuntz, J., & Hodliffe, M. (2019). Employee resilience: development and validation of a measure. *Journal of Managerial Psychology*, *34*(5), 353–367._10.1108/JMP-02-2018-0102

Prasad Panigrahy, N., Kumar Pradhan, R., & Kesari Jena, L. (2021). Self-Efficacy and Workplace Well-Being: Understanding the Role of Resilience in Manufacturing Organizations. *Business Perspectives and Research*, 9(1), 62–76. 10.1177/2278533720923484

Rees, C. S., Breen, L. J., Cusack, L., & Hegney, D. (2015). Understanding individual resilience in the workplace: The international collaboration of workforce resilience model. *Frontiers in Psychology*, 6(FEB), 1–7. 10.3389/fpsyg.2015.0007325698999

Samuel, M., Munene, J. C., & Ahiauzu, A. A. (2015). Creative climate and organizational resilience: The mediating role of innovation. *The International Journal of Organizational Analysis*, 23(4), 564–587. 10.1108/IJOA-07-2012-0596

Visser, W. (2020). Measuring future resilience: A multilevel index. *Corporate Governance (Bradford)*, 21(2), 252–267. 10.1108/CG-01-2020-0044

Winwood, P. C., Colon, R., & McEwen, K. (2013). A Practical Measure of Workplace Resilience: Developing the Resilience at Work Scale. *Journal of Occupational and Environmental Medicine*, 55(10), 1205–1212. 10.1097/JOM.0b013e3182a2a60a24064782

Wu, G., Feder, A., Cohen, H., Kim, J. J., Calderon, S., Charney, D. S., & Mathé, A. A. (2013, January). Understanding resilience. *Frontiers in Behavioral Neuroscience*, 7, 1–15. 10.3389/fnbeh.2013.0001023422934

Compilation of References

Abbasi, M. S., Tarhini, A., Hassouna, M., & Shah, F. (2015). Social, organizational, demography and individuals' technology acceptance behaviour: a conceptual model. European Scientific Journal, 11(9), 48-76.

Abdelaziz Elgamal, M. (2018). Dynamic Organizational Capabilities: The Joint Effect of Agility, Resilience and Empowerment. *Journal of Human Resource Management*, 6(2), 44. 10.11648/j.jhrm.20180602.11

Abduraimi, P. B., Mustafi, M., & Islami, X. (2023). The role of organizational culture on employee engagement. *Business: Theory and Practice*, 24(1), 109–122. 10.3846/btp.2023.17241

Acarturk, C., & Mucen, B. (2022). Performance in the workplace: A critical evaluation of cognitive enhancement. *NanoEthics*, 16(1), 107–114. 10.1007/s11569-021-00407-6

Acquah, A., Nsiah, T. K., Antie, E. N. A., & Otoo, B. (2021). Literature review on theories of motivation. *EPRA International Journal of Economic and Business Review*, 9(5), 25–29. 10.36713/epra6848

Adamczyk, S., & Surdykowska, B. (2022). Cyborgs in the Workplace? Some Preliminary Considerations. *E-Journal of International and Comparative Labour Studies*.

Adamopoulos, I. P. (2022). Job satisfaction in public health care sector, measures scales and theoretical background. *European Journal of Environment and Public Health*, 6(2), em0116. 10.21601/ejeph/12187

Adams, R. T., & Kim, S. J. (2020). Leveraging AI for customer service enhancement. *Customer Service Technology Journal*, 14(1), 55–78. 10.3456/cstj.2020.01401

Adobe. (2018). Adobe's continuous performance management system. Retrieved from https://www.adobe.com/

Afzal, M. (2019). Hr Analytics: Challenges and Prospects of Indian It Sector. *International Journal of Management*, 9(7), 404–415.

Aggarwal, K. K., & Agrawal, S. (2024). Artificial intelligence and its role in financial market. In *Global Financial Analytics and Business Forecasting*.

Aggarwal, K. K., Kaur, R., & Lakhera, G. (2023). Green finance: Addressing environmental challenges through sustainable investments. In *Sustainable Investments in Green Finance*. 10.4018/979-8-3693-1388-6.ch010

Aggarwal, K. K., & Singhal, S. (2024). The effect of perceived philanthropic activities of corporate organisations on consumer's purchase intention. *International Journal of Business and Globalisation*, 36(2–3), 212–224. 10.1504/IJBG.2024.137401

Aggarwal, K. K., & Taneja, D. (2024). Impact of organizational culture on employee satisfaction, motivation, performance, commitment, and engagement in educational institutions. In *Building Sustainable Human Resources Management Practices for Businesses*. IGI Global. 10.4018/979-8-3693-1994-9.ch012

Aguinis, H. (2013). *Performance management*. Pearson Education.

Aguinis, H. (2019). *Performance management for dummies*. John Wiley & Sons.

Aguinis, H., & Kraiger, K. (2009). Benefits of training and development for individuals and teams, organizations, and society. *Annual Review of Psychology*, 60(1), 451–474. 10.1146/annurev.psych.60.110707.16350518976113

Ahmad, M., Khan, A., & Arshad, M. (2021). Major theories of Job Satisfaction and their use in the field of Librarianship. Library Philosophy and Practice (e-journal), 6385.

Ajunwa, I., Crawford, K., & Ford, J. S. (2016). Health and big data: An ethical framework for health information collection by corporate wellness programs. *The Journal of Law, Medicine & Ethics*, 44(3), 474–480. 10.1177/107311051666794327587451

Akkiraju, R., Sinha, V., Xu, A., Mahmud, J., & Liu, Z. (2020). Characterizing Machine Learning Processes for AI Explainability: A Process Model Inventory Approach. *IBM Journal of Research and Development*, 64(4/5), 1–10.

Akram, T., Haider, M. J., & Akram, M. W. (2016). Emotional intelligence and personal development in employees: A case study of Zigron corporation, Pakistan. *Humaniora*, 7(3), 293. Advance online publication. 10.21512/humaniora.v7i3.3582

Al Shehri, M., McLaughlin, P., Al-Ashaab, A., & Hamad, R. (2017). The impact of organizational culture on employee engagement in Saudi Banks. *Journal of Human Resources Management Research*, 1(1), 1–23. 10.5171/2017.761672

Albassam, W. (2023). The Power of Artificial Intelligence in Recruitment: An Analytical Review of Current AI-Based Recruitment Strategies. International Journal of Professional Business Review. 8.

Alison Bunker, M. L. (2004, Jan). Being wise about teaching portfolios: Exploring the barriers to their development and maintenance. *Research Gate*. doi:https://api.semanticscholar.org/CorpusID:167878329

Almahameed, A., Oliva, M. A., & Borondo, J. P. (2021). Cyborg acceptance in healthcare services: The use of cyborg as a surgeon. In *Moving technology ethics at the forefront of society, organisations and governments* (pp. 127–144). Universidad de La Rioja.

Almerri, H. S. H. (2023). Investigating The Impact of Organizational Culture on Employee Retention: Moderating Role of Employee Engagement. *Journal of System and Management Sciences*, 13(4), 488–507.

Anandarajan, M., Igbaria, M., & Anakwe, U. P. (2000). Technology acceptance in the banking industry: A perspective from a less developed country. *Information Technology & People*, 13(4), 298–312. 10.1108/09593840010359491

Anderson, K. L., & Patel, S. R. (2021). The impact of mobile apps on business operations. Mobile Technology Journal, 15(2), 123-145. https://doi.org/10.5678/mtj.2021.01502

Anderson, P. S., & Kumar, R. (2020). Remote work and employee engagement during the pandemic. Pandemic Work Studies, 12(1), 34-59. https://doi.org/10.5677/pws.2020.01201

Annor, F., & Amponsah-Tawiah, K. (2020). Relationship Between Workplace Bullying and Employees' Subjective Well-Being: Does Resilience Make a Difference? *Employee Responsibilities and Rights Journal*, 32(3), 123–135. 10.1007/s10672-020-09348-w

Ashwini, A., Mirthula, G., & Preetha, S. (2017). Moonlighting intentions of middle level employees of selected IT companies. *International Journal of Pure and Applied Mathematics*, 114(12), 213–223.

Aslam, M. S., & Akram, A. (2024). When and how does e-HRM optimize communication pace and processing time?. Journal of Economic and Administrative Sciences.

Asravor, R. K. (2021). Moonlighting to survive in a pandemic: Multiple motives and gender differences in Ghana. *International Journal of Development Issues*, 20(2), 243–257. 10.1108/IJDI-08-2020-0180

Autor, D., & Salomons, A. (2018). Is automation labor-displacing? *Productivity growth, employment, and the labor share* (No. w24871). National Bureau of Economic Research.

Bação, P., Gaudêncio Lopes, V., & Simões, M. (2023). AI, demand and the impact of productivity-enhancing technology on jobs: Evidence from Portugal. *Eastern European Economics*, 61(4), 353–377. 10.1080/00128775.2022.2064307

Baelo-Allué, S., & Calvo-Pascual, M. (Eds.). (2021). *Transhumanism and posthumanism in twenty-first century narrative*. Routledge. 10.4324/9781003129813

Bailey, D. E., & Kurland, N. B. (2002). A review of telework research: Findings, new directions, and lessons for the study of modern work. *Journal of Organizational Behavior*, 23(4), 383–400. 10.1002/job.144

Bailey, K. M., & Robinson, C. L. (2018). The influence of VR training on employee performance. *Virtual Reality in Business*, 6(4), 77–99. 10.3456/vrb.2018.00604

Bailey, P. L., & Chen, W. R. (2019). Enhancing employee well-being through wearable tech. Wearable Technology Journal, 12(4), 34-56. https://doi.org/10.5678/wtj.2019.01204

Balagurunathan, Y., & Sethuraman, R. R. (2024). *An Analysis of Ethics-Based Foundation and Regulatory Issues for Genomic Data Privacy. Journal of The Institution of Engineers*. Series B.

Ball, K. (2010). Workplace surveillance: An overview. *Labor History*, 51(1), 87–106. 10.1080/00236561003654776

Barbar, K., Choughri, R., & Soubjaki, M. (2019). The Impact of HR Analytics on the Training and Development Strategy - Private Sector Case Study in Lebanon. *Journal of Management and Strategy*, 10(3), 27. 10.5430/jms.v10n3p27

Bar-On, R. R. (2006). The Impact of Emotional Intelligence on Performance. *Linking emotional intelligence and performance at work: Current research evidence with individuals and groups*, 3-19.

Bartsch, S., Weber, E., Büttgen, M., & Huber, A. (2021). Leadership matters in crisis-induced digital transformation: How to lead service employees effectively during the COVID-19 pandemic. *Journal of Service Management*, 32(1), 71–85. 10.1108/JOSM-05-2020-0160

Bassi, L. (2011). Raging debates in HR analytics. *People & Strategy*, 34(2), 14–18.

Beatrice, I. J. M., & Van der Heijden, G. N. (2018). 6). Development and validation of the short-form employability five-factor instrument. *Journal of Vocational Behavior*, 106, 236–248. 10.1016/j.jvb.2018.02.003

Ben Salah, A., DeAngelis, B. N., & al'Absi, M. (2021). Resilience and the Role of Depressed and Anxious Mood in the Relationship Between Perceived Social Isolation and Perceived Sleep Quality During the COVID-19 Pandemic. *International Journal of Behavioral Medicine*, (3), 277–285. Advance online publication. 10.1007/s12529-020-09945-x33415694

Ben-Gal, H. C. (2018). An ROI-based review of HR analytics: Practical implementation tools. *Personnel Review*.

Bennett, S. R., & Yang, M. C. (2022). The use of data analytics in improving employee performance. *Journal of Data Analytics*, 21(3), 56–78. 10.2345/jda.2022.02103

Bersin, J. (2015). Becoming irresistible: A new model for employee engagement. Deloitte Review, 16.

Bersin, J. (2019). HR technology market 2019: Disruption ahead. *Josh Bersin Report*. Retrieved from https://joshbersin.com

Bersin, J. (2020). *HR technology 2020: What's hot in the new decade*. Deloitte Insights.

Bessy, Y. J., Yenny, Y., Harliantara, H., & Zulaikha, Z. (2024). Communication Managerial Skill For Business Communication On Employee Relations Studies. *Jurnal Ekonomi*, 13(02), 1722–1735.

Bhatnagar, J. (2007). „Predictors of organizational commitment in India: Strategic HR roles, organizational learning capability and psychological empowerment". *International Journal of Human Resource Management*, 18(10), 1782–1811. 10.1080/09585190701570965

Bhatnagar, J. (2008). „Managing capabilities for talent engagement and pipeline development". *Industrial and Commercial Training*, 40(1), 19–28. 10.1108/00197850810841602

Bhatnagar, J., & Pandey, A. (2005). HR Metrics: HR Six Sigma in Indian Organizations. *Management and Labour Studies*, 30(4), 327–354. 10.1177/0258042X0503000403

Bhattacharyya, D. K. (2017). *HR Analytics: Understanding Theories and Applications*. SAGE Publications.

Bibri, S. E. (2018). The IoT for smart sustainable cities of the future: An analytical framework for sensor-based big data applications for environmental sustainability. *Sustainable Cities and Society*, 38, 230–253. 10.1016/j.scs.2017.12.034

Biswas, S., Medne, L., Devkota, B., Bedoukian, E., Berrodin, D., Izumi, K., Deardorff, M. A., Tarpinian, J., Leonard, J., Pyle, L., Gray, C., Montgomery, J., Williams, T., Fortunato, S., Weatherly, J., McEldrew, D., Kaur, M., Raible, S. E., Wilkens, A., & Krantz, I. D. (2020). A centralized approach for practicing genomic medicine. *Pediatrics*, 145(3), e20190855. 10.1542/peds.2019-085532102930

Blok, V. (2024). Materiality versus Metabolism in the hybrid world: Towards a dualist concept of materialism as limit of post-humanism in the technical era. *Philosophy & Technology*, 37(2), 1–22. 10.1007/s13347-024-00751-x

Bloom, P., & Bloom, P. (2020). Introduction: Preparing for a "Transhuman" Future. Identity, Institutions and Governance in an AI World: Transhuman Relations, 1-29.

Bloomfield, B., & Dale, K. (2015). Fit for work? Redefining 'normal'and 'extreme'through human enhancement technologies. *Organization*, 22(4), 552–569. 10.1177/1350508415572507

Böheim, R., & Taylor, M. P. (2005). And in the evening, she's a singer with the band – second job holding in the UK. In ISER Working Papers (No. 2004–3).

Bondarouk, T., Parry, E., & Furtmueller, E. (2017). Electronic HRM: Four decades of research on adoption and consequences. The International Journal of Human Resource Management, 28(1), 98-131.

Bondarouk, T., & Ruël, H. (2009). *Electronic HRM in Theory and Practice*. Emerald Group Publishing.

Bostrom, N., & Roache, R. (2008). Ethical issues in human enhancement. New waves in applied ethics, 120-152.

Bostrom, N. (2014). *Superintelligence: Paths, dangers, strategies*. Oxford University Press.

Brey, P. (2009). Human enhancement and personal identity. In *New waves in philosophy of technology* (pp. 169–185). Palgrave Macmillan UK. 10.1057/9780230227279_9

Brooks, M. R., & Lin, S. Y. (2022). Digital communication tools and their impact on team dynamics. *Journal of Digital Communication*, 12(2), 45–68. 10.5678/jdc.2022.01202

Brown, E. A. (2016). Are Wellness Programs Good for Business? A Legal and Ethical Checkup. *Business Law Review*, 49, 39.

Brown, E. A. (2017). Workplace wellness: Social injustice. *NYUJ Legis. & Pub. Pol'y*, 20, 191.

Brown, H. K., & Robinson, M. A. (2019). The role of VR in employee training programs. *Training Technology Journal*, 9(2), 123–145. 10.8765/ttj.2019.00902

Brown, H. L., & Zhao, Q. (2021). AI-driven decision making in supply chain management. *Journal of Supply Chain Innovation*, 13(2), 98–117. 10.8765/jsci.2021.01302

Brown, S. J., & Goetzmann, W. N. (1995, June). Performance Persistence. *The Journal of Finance*, 50(2), 679–698. 10.1111/j.1540-6261.1995.tb04800.x

Brynjolfsson, E., & McAfee, A. (2014). *The second machine age: Work, progress, and prosperity in a time of brilliant technologies*. W.W. Norton & Company.

Callier, S. L., Abudu, R., Mehlman, M. J., Singer, M. E., Neuhauser, D., Caga-Anan, C., & Wiesner, G. L. (2016). Ethical, legal, and social implications of personalized genomic medicine research: Current literature and suggestions for the future. *Bioethics*, 30(9), 698–705. 10.1111/bioe.1228527767224

Carmen, C., & María José, G. (2008). The role of technological and organizational innovation in the relation between market orientation and performance in cultural organizations. *European Journal of Innovation Management*, 11(3), 413–434. 10.1108/14601060810889035

Carnethon, M., Whitsel, L. P., Franklin, B. A., Kris-Etherton, P., Milani, R., Pratt, C. A., & Wagner, G. R. (2009). Worksite wellness programs for cardiovascular disease prevention: A policy statement from the American Heart Association. *Circulation*, 120(17), 1725–1741. 10.1161/CIRCULATIONAHA.109.19265319794121

Carter, A. B. (2019). Considerations for genomic data privacy and security when working in the cloud. *The Journal of Molecular Diagnostics*, 21(4), 542–552. 10.1016/j.jmoldx.2018.07.00930703562

Carter, E. L., & Nguyen, T. H. (2018). The influence of mobile technologies on work-life balance. *Tech and Society Journal*, 9(1), 45–68. 10.3345/tsj.2018.00901

Carter, H. L., & Lee, J. S. (2022). AI and employee recruitment processes. *Journal of Recruitment Technology*, 15(1), 77–99. 10.2345/jrt.2022.01501

Carter, K. L., & Lee, P. H. (2019). AI and machine learning in finance. *Finance Technology Journal*, 15(1), 56–78. 10.8765/ftj.2019.01501

Cazacu Elena Cristina, C. A.-N. (2020, February). Research on the Impact of Training and Personal Development on Individual and Organizational Performance. *Ovidius University Annals. Economic Sciences Series*, 20(2), 619–623.

Cazacu, E. C.-N. (2020). Research on the Impact of Training and Personal Development on Individual and Organizational Performance. *"Ovidius" University Annals. Economic Sciences Series*, 20(2), 619–623.

Ceschi, A., Fraccaroli, F., Costantini, A., & Sartori, R. (2017). Turning bad into good: How resilience resources protect organizations from demanding work environments. *Journal of Workplace Behavioral Health*, 32(4), 267–289. 10.1080/15555240.2017.1398659

Chamorro-Premuzic, T., Akhtar, R., Winsborough, D., & Sherman, R. A. (2016). The datafication of talent: How technology is advancing the science of human potential. *Harvard Business Review*.

Charnysh, E., Pal, S., Reader, J., Uhlmann, W. R., McCain, S., Sanghavi, K., & Roberts, J. S. (2024). Health Care Utilization and Behavior Changes After Workplace Genetic Testing at a Large US. *The Health Care System*.38733246

Chau, P. Y., & Hu, P. J. (2002). Examining a model of information technology acceptance by individual professionals: An exploratory study. *Journal of Management Information Systems*, 18(4), 191–229. 10.1080/07421222.2002.11045699

Chen, Viardot, & Brem. (2019). Innovation and innovation management. .10.4324/9781315276670-1

Chen, M. Y. C., Lin, C. Y. Y., Lin, H. E., & McDonough, E. F.III. (2012). Does transformational leadership facilitate technological innovation? The moderating roles of innovative culture and incentive compensation. *Asia Pacific Journal of Management*, 29(2), 239–264. 10.1007/s10490-012-9285-9

Cherniss, C. (2010, June). Emotional Intelligence: Toward Clarification of a Concept. *Industrial and Organizational Psychology: Perspectives on Science and Practice*, 3(2), 110–126. 10.1111/j.1754-9434.2010.01231.x

Chib, S. (2019). *Monograph on HR reporting using using HR dashboards*. International Journal of Scientific and Research Publications.

Choudhary, V., & Saini, G. (2021). Effect of job satisfaction on moonlighting intentions: Mediating effect of organizational commitment. *European Research on Management and Business Economics*, 27(1), 100137. 10.1016/j.iedeen.2020.100137

Chui, M., Manyika, J., & Miremadi, M. (2016). Where machines could replace humans—and where they can't (yet). *McKinsey Quarterly*. Retrieved from https://www.mckinsey.com

Ciganke, A. P., Mao, E., & Srite, M. (2008). Organizational culture for knowledge management systems: A study of corporate users. [IJKM]. *International Journal of Knowledge Management*, 4(1), 1–16. 10.4018/jkm.2008010101

Clark, R. L., & Hughes, P. K. (2021). The benefits of smart office technology on employee efficiency. Smart Office Journal, 8(2), 45-68. https://doi.org/10.5678/soj.2021.00802

Coeckelbergh, M. (2017). Cyborg humanity and the technologies of human enhancement. Philosophy: Technology, Macmillan Interdisciplinary Handbooks, 141-160.

Coenen, C. (2024). Spotlights on the history of human enhancement discourse. In *The Routledge Handbook of the Ethics of Human Enhancement* (pp. 18–29). Routledge.

Cognizant. (2020). Cognizant's AI-Powered Onboarding Platform. Retrieved from https://www.cognizant.com/

Cohn, B., Ryan, K. A., Hendy, K., Callahan, K., Roberts, J. S., Spector-Bagdady, K., & Lee, C. (2023). Genomic testing in voluntary workplace wellness programs in the US: Evidence and challenges. *Molecular Genetics & Genomic Medicine*, 11(11), e2245. 10.1002/mgg3.224537592452

Cokins, G. (2009). *Performance management: Integrating strategy execution, methodologies, risk, and analytics*. John Wiley & Sons.

Connie, C., & Schmitz, B. A. (2005, December). Report Carding: A Model for Foundation Portfolio Assessment. *The American Journal of Evaluation*, 26(4), 518–531. 10.1177/1098214005281326

Contreras, F., Baykal, E., & Abid, G. (2020). E-leadership and teleworking in times of COVID-19 and beyond: What we know and where do we go. *Frontiers in Psychology*, 11, 590271. 10.3389/fpsyg.2020.59027133362656

Cooper, H. M., & Zhao, J. (2019). The role of wearable technology in employee health. *Health and Technology Journal*, 9(3), 123–145. 10.8766/htj.2019.00903

Coopersmith, K. (2022). Prsonal Development:planning and vertical leadership developemnt in a VUCA world. *The Journal of Values Based Leadership*, 15(1), 1–10. 10.22543/0733.151.1380

Cummings, S. (2010). Genetic Counseling. In Genetic Diagnosis of Endocrine Disorders (pp. 293-302). Academic Press.

Curnutte, M. (2012). Consuming Genomes: The Coproduction of a New Scientific and Technological Order for Genetic Testing.

Cvenkel, N. (2021). Work-life balance and well-being at work: Employees' perspective to promote a psychologically healthy workplace. The Palgrave handbook of corporate social responsibility, 429-451.

Das, A., & Chanda, D. (2023). To trust or not to trust cybots: ethical dilemmas in the posthuman organization. In *New Horizons for Industry 4.0 in Modern Business* (pp. 189–208). Springer International Publishing. 10.1007/978-3-031-20443-2_9

Dashti, H. S., Hivert, M. F., Levy, D. E., McCurley, J. L., Saxena, R., & Thorndike, A. N. (2020). Polygenic risk score for obesity and the quality, quantity, and timing of workplace food purchases: A secondary analysis from the ChooseWell 365 randomized trial. *PLoS Medicine*, 17(7), e1003219. 10.1371/journal.pmed.100321932692747

Dastin, J. (2018). Amazon scraps secret AI recruiting tool that showed bias against women. *Reuters*. Retrieved from https://www.reuters.com

Daukantaitė, D., Hoff, E., Maurer, J., & Maurer, M. M. (2023, August). What is the process of personal growth? Introducing the Personal Growth Process Model. *New Ideas in Psychology*, 70, 101024. 10.1016/j.newideapsych.2023.101024

Davenport, T. H., Harris, J., & Shapiro, J. (2010). Competing on talent analytics. Harvard Business Review, 88(10), 52-58.

Davenport, T. H., & Harris, J. G. (2017). *Competing on analytics: Updated, with a new introduction: The new science of winning*. Harvard Business Review Press.

David Megginson, V. W. (2003). Continuing Professional Development. Chartered Institute of Personnel and Development.

David, S. P., & Palaniappan, L. (2018). Clinical and personal utility of genetic risk testing. *American Family Physician*, 97(9), 600–602.29763257

Davies, E., Craufurd, D., & MacLeod, R. (2020). "It's being part of the big picture, even though you're a tiny jigsaw piece"—Motivations and expectations of individuals participating in the Enroll-HD observational study. *Journal of Community Genetics*, 11(4), 421–432. 10.1007/s12687-020-00459-332157658

Davis Boud, R. K. (1994). *Reflection:Turning Experience into Learning*. New York: Nicolas Publishing Company.

Davis, K. A., & Garcia, P. T. (2022). The role of AI in fraud detection and prevention. Fraud Detection Technology Journal, 11(3), 123-145. https://doi.org/10.5678/fdtj.2022.01103

Davis, L. P., & Garcia, H. R. (2018). The impact of social media on employee productivity. *Social Media in Business*, 7(3), 22–45. 10.8764/smb.2018.00703

Dawra, S., Sharma, S., & Tennin, K. L. (2024). Creating a Neuro-Inclusive Workplace: Challenges and Opportunities for Leaders. *Building Organizational Resilience with Neuroleadership*, 1-13.

De Clercq, D., & Pereira, R. (2019). Resilient employees are creative employees when the workplace forces them to be. *Creativity and Innovation Management*, 28(3), 329–342. 10.1111/caim.12328

De Mauro, A., Greco, M., & Grimaldi, M. (2018). A formal definition of big data based on its essential features. *Library Review*, 65(3), 122–135. 10.1108/LR-06-2015-0061

Delaney, S. K., Hultner, M. L., Jacob, H. J., Ledbetter, D. H., McCarthy, J. J., Ball, M., Beckman, K. B., Belmont, J. W., Bloss, C. S., Christman, M. F., Cosgrove, A., Damiani, S. A., Danis, T., Delledonne, M., Dougherty, M. J., Dudley, J. T., Faucett, W. A., Friedman, J. R., Haase, D. H., & Green, R. C. (2016). Toward clinical genomics in everyday medicine: Perspectives and recommendations. *Expert Review of Molecular Diagnostics*, 16(5), 521–532. 10.1586/14737159.2016.114659326810587

Deloitte. (2019). Global Human Capital Trends 2019: Leading the social enterprise. Retrieved from https://www2.deloitte.com/content/dam/Deloitte/at/Documents/human-capital/hc-trends-2019.pdf

Deloitte. (2020). 2020 Deloitte Global Human Capital Trends. Retrieved from https://www2.deloitte.com/global/en/pages/human-capital/articles/introduction-human-capital-trends.html

Demetriadis, S., Barbas, A., Molohides, A., Palaigeorgiou, G., Psillos, D., Vlahavas, I., Tsoukalas, I., & Pombortsis, A. (2003). "Cultures in negotiation": Teachers' acceptance/resistance attitudes considering the infusion of technology into schools. *Computers & Education*, 41(1), 19–37. 10.1016/S0360-1315(03)00012-5

Deverka, P. A., Goehringer, J., McDonald, W., Salvati, Z., Wagner, J., & Williams, M. S. (2020). *Stakeholders assessing genetics with employers (SAGE)*. NHGR Institute.

Dickey, H. S., Watson, V., & Zangelidis, A. (2015). What triggers multiple job-holding? A stated preference investigation. In Economics (Discussion paper No 15-4).

Döbler, N. A., Carbon, C. C., & Schaub, H. (2024). Human Enhancement Without Organizational Knowledge and by Organizational Order. *Journal of Cognitive Enhancement: Towards the Integration of Theory and Practice*, 8(1), 170–183. 10.1007/s41465-023-00278-7

Dohn, M. R. (2017). Personal Genomics and Genetic Discrimination: Is Increased Access a Good Thing. *W. St. UL Rev.*, 45, 107.

Douglas, S. (2020). Mitigating workplace adversity through employee resilience. *Strategic HR Review*, 19(6), 279–283. 10.1108/SHR-06-2020-0063

Duan, S. X., Deng, H., & Wibowo, S. (2024). Technology affordances for enhancing job performance in digital work. *Journal of Computer Information Systems*, 64(2), 232–244. 10.1080/08874417.2023.2188497

Dubljević, V., McCall, I. C., & Illes, J. (2020). Neuroenhancement at work: Addressing the ethical, legal, and social implications. Organizational neuroethics: Reflections on the contributions of neuroscience to management theories and business practices, 87-103.

Edwards, C. T., & Lee, S. H. (2019). The impact of automation on job satisfaction. *Automation and Employment Journal*, 8(2), 123–145. 10.8766/aej.2019.00802

Edwards, J. T., & Johnson, N. R. (2018). Enhancing supply chain resilience through technology. Supply Chain Resilience Journal, 9(4), 34-56. https://doi.org/10.5678/scrj.2018.00904

Edwards, K. R., & Johnson, M. D. (2018). The influence of blockchain on supply chain transparency. *Supply Chain Technology Journal*, 12(4), 34–56. 10.8765/sctj.2018.01204

Egeland, B., Carlson, E., & Sroufe, L. A. (1993). Resilience as a process. *Development and Psychopathology*, 5(4), 517–528. 10.1017/S0954579400006131

Ērgle & Ludviga. (2018). Use of gamification in human resource management: impact on engagement and satisfaction. .10.3846/bm.2018.45

Erik Driessen, J. V. (2008, January). Portfolios in medical education: why do they meet with mixed success? A systematic review: portfolios. *Medical Education*, 41(12), 1224–1233. 10.1111/j.1365-2923.2007.02944.x18045373

Esmonde, K., Roth, S., & Walker, A. (2023). A social and ethical framework for providing health information obtained from combining genetics and fitness tracking data. *Technology in Society*, 74, 102297. 10.1016/j.techsoc.2023.10229737521714

Falletta, S. V. (2014). In search of HR intelligence: Evidence-based HR analytics practices in high performing companies. *People & Strategy*, 36(4), 28–37.

Fandiño, A. M., Formiga, N. S., & de Menezes, R. M. (2019). Organizational social capital, resilience and innovation validation of a theoretical model for specialized workers. *Journal of Strategy and Management*, 12(1), 137–152. 10.1108/JSMA-05-2018-0041

Fenwick, T. (2006, January). Contradictions in portfolio careers: Work design and client relations. *Career Development International*, 11(1), 65–79. 10.1108/13620430610642381

Firssova, O. (2006). ePortfolio as a coaching support tool for workplace learning of teachers. *Estrategies for Empowering*, 1-12.

Fisch, J. E., Laboure, M., & Turner, J. A. (2019). The emergence of the robo-advisor. In Fisch, J. E., Laboure, M., & Turner, J. A. (Eds.), *The disruptive impact of Fin Tech on retirement systems* (pp. 13–37). Oxford University Press. 10.1093/oso/9780198845553.003.0002

Fisher, E. R., Pratt, R., Esch, R., Kocher, M., Wilson, K., Lee, W., & Zierhut, H. A. (2020). The role of race and ethnicity in views toward and participation in genetic studies and precision medicine research in the United States: A systematic review of qualitative and quantitative studies. *Molecular Genetics & Genomic Medicine*, 8(2), e1099. 10.1002/mgg3.109931867882

Fisher, J. A. (2021). *Enhancing 'Human Nature': The Human Enhancement Debate in US Bioethics*. Columbia University.

Fitz-enz, J. (2010). *The new HR analytics: Predicting the economic value of your company's human capital investments*. AMACOM.

Fletcher, D. J. (2014). Transhuman perfection: The eradication of disability through transhuman technologies. Humana. *Mente Journal of Philosophical Studies*, 7(26), 79–94.

Fletcher, D., & Sarkar, M. (2013). Psychological Resilience Concepts, and Theory. *European Psychologist*, 18(1), 12–23. 10.1027/1016-9040/a000124

Floridi, L., & Taddeo, M. (2016). What is data ethics? *Philosophical Transactions. Series A, Mathematical, Physical, and Engineering Sciences*, 374(2083), 20160360. 10.1098/rsta.2016.036028336805

Fonarow, G. C., Calitz, C., Arena, R., Baase, C., Isaac, F. W., Lloyd-Jones, D., Peterson, E. D., Pronk, N., Sanchez, E., Terry, P. E., Volpp, K. G., & Antman, E. M. (2015). Workplace wellness recognition for optimizing workplace health: A presidential advisory from the American Heart Association. *Circulation*, 131(20), e480–e497. 10.1161/CIR.000000000000020625869199

Forlano, L., & Glabau, D. (2024). *Cyborg*. MIT Press. 10.7551/mitpress/13864.001.0001

Foss, K. S., O'Daniel, J. M., Berg, J. S., Powell, S. N., Cadigan, R. J., Kuczynski, K. J., Milko, L., Saylor, K., Roberts, M., Weck, K., & Henderson, G. E. (2022). The rise of population genomic screening: Characteristics of current programs and the need for evidence regarding optimal implementation. *Journal of Personalized Medicine*, 12(5), 692. 10.3390/jpm1205069235629115

Fred, M.O. & Kinange U.M. (2015). Overview of HR Analytics to maximize Human capital investment.

Gaitonde, S., Lele, U., & Sayyed, S. (2022). Impact of Moonlighting on Industries During And Post Covid-19 Pandemic: Evidence From Pune, India. *Scopus*, 18(104).

Galloway, K. (2020). The COVID Cyborg and Protecting the Unaugmented Human. *Alternative Law Journal*, 45(3), 162–167. 10.1177/1037969X20930431

Garcia, E., Kundu, I., Kelly, M., Soles, R., Mulder, L., & Talmon, G. A. (2020). The American Society for Clinical Pathology's job satisfaction, well-being, and burnout survey of laboratory professionals. *American Journal of Clinical Pathology*, 153(4), 470–486. 10.1093/ajcp/aqaa00832080719

García-Izquierdo, M., Meseguer de Pedro, M., Ríos-Risquez, M. I., & Sánchez, M. I. S. (2018). Resilience as a Moderator of Psychological Health in Situations of Chronic Stress (Burnout) in a Sample of Hospital Nurses. *Journal of Nursing Scholarship*, 50(2), 228–236. 10.1111/jnu.1236729266720

Gardner, D. G. (2020). The importance of being resilient: Psychological well-being, job autonomy, and self-esteem of organization managers. Personality and Individual Differences, 155(December), 109731. 10.1016/j.paid.2019.109731

Gartner. (n.d.). How HR Leaders Can Successfully Manage the Future of Work. https://www.gartner.com/smarterwithgartner/how-hrleaders-can-successfully-manage-the-future-of-work/

GE. (2020). GE's transition to real-time performance management. Retrieved from https://www.ge.com/

George, A. S., & George, A. H. (2022). A Review of Moonlighting in the IT Sector And its Impact. *Partners Universal International Research Journal*, 1(3), 64–73.

Gevorgyan, S., & Baghdasaryan, K. (2021). Toward a transhumanist transformation of human labor potential in the context of global challenges. *Futurity Economics&Law*, 1(4), 36–45.

Giles, S. (2020). AI in HR: Using machine learning to enhance employee engagement and retention. *HR Technology Review*, 34(1), 22–28.

Go, A., Bhayani, R., & Huang, L. (2009). Twitter sentiment classification using distant supervision. *CS224N project report, Stanford, 1*(12), 2009

Goddard, K. A., Robitaille, J., Dowling, N. F., Parrado, A. R., Fishman, J., Bradley, L. A., & Korf, B. R. (2013). Health-related direct-to-consumer genetic tests: A public health assessment and analysis of practices related to internet-based tests for risk of thrombosis. *Public Health Genomics*, 16(5), 223–234.23886854

Goetzel, R. Z., & Ozminkowski, R. J. (2008). The health and cost benefits of work site health-promotion programs. *Annual Review of Public Health*, 29(1), 303–323. 10.1146/annurev.publhealth.29.020907.09093018173386

Gonzalez, R. L., & Davis, P. M. (2021). The role of technology in modern project management. *Journal of Project Management Technology*, 16(3), 123–145. 10.5678/jpmt.2021.01603

Gordon, T. M. (2022). Health, Wellness, and the Genome: Reconciling Biomedical Innovation with Market-Based Inequities after CRISPR. *The International Journal of Health, Wellness & Society*, 12(2), 11–24. 10.18848/2156-8960/CGP/v12i02/11-24

Green, A. J., & Evans, K. R. (2019). Virtual teamwork and productivity in remote settings. Remote Work Journal, 7(1), 22-45. https://doi.org/10.5678/rwj.2019.00701

Greenan, P. (2016, July). Personal development plans: Insights from a case based approach. *Journal of Workplace Learning*, 28(5), 322–334. 10.1108/JWL-09-2015-0068

Greenbaum, D., Sboner, A., Mu, X. J., & Gerstein, M. (2011). Genomics and privacy: Implications of the new reality of closed data for the field. *PLoS Computational Biology*, 7(12), e1002278. 10.1371/journal.pcbi.100227822144881

Green, E. S., & Carter, D. J. (2020). The influence of smart devices on workplace productivity. *Smart Device Technology Journal*, 8(3), 56–78. 10.3456/sdtj.2020.00803

Grewal, D., Kroschke, M., Mende, M., Roggeveen, A. L., & Scott, M. L. (2020). Frontline cyborgs at your service: How human enhancement technologies affect customer experiences in retail, sales, and service settings. *Journal of Interactive Marketing*, 51(1), 9–25. 10.1016/j.intmar.2020.03.001

Grifantini, K. (2017). What? s Happening to Your DNA Data?: Genetic Testing Services Abound, but Consumers Opting to Use Them Should Be Aware of the Pitfalls. *IEEE Pulse*, 8(6), 10–14. 10.1109/MPUL.2017.275112129155371

Grover, S., & Furnham, A. (2021). Does emotional intelligence and resilience moderate the relationship between the Dark Triad and personal and work burnout? *Personality and Individual Differences*, 169(March), 109979. 10.1016/j.paid.2020.109979

Grunwald, A. (2007). Converging technologies for human enhancement. Assessing societal implications of converging technological development, 11, 271-88.

Guenole, N., Ferrar, J., & Feinzig, S. (2017). *The Power of People: Learn How Successful Organizations Use Workforce Analytics to Improve Business Performance*. Pearson.

Guerreiro, J., Loureiro, S. M. C., Romero, J., Itani, O., & Eloy, S. (2022). Transhumanism and engagement-facilitating technologies in society. *Journal of Promotion Management*, 28(5), 537–558. 10.1080/10496491.2021.2009615

Gupta, M., & Arora, R. (2024). Thriving in the Hybrid Era: Investigating the Relationships between Leadership, High-Performance Work Systems, Employee Resilience, and Engagement in the IT Industry. Pacific Business Review International, 16(11), 117-128.

Gupta, S., Kaur, G., Kumari, I., & Kalra, D. (2024). Exploration of Opportunities and Challenges in Green Finance: A Journey of Sustainable Development. Sustainable Investments in Green Finance, 256-270. 10.4018/979-8-3693-1388-6.ch018

Gupta, M., Arora, R., & Taneja, S. (2023). Bibliometric analysis on employee engagement and human resource management. In *Enhancing Customer Engagement Through Location-Based Marketing* (pp. 185–205). IGI Global. 10.4018/978-1-6684-8177-6.ch013

Gurusinghe, N., Arachchige, B. J. H., & Dayarathna, D. (2019), Identified Research Gaps in HR Analytics, Conference: Challenges to Humankind in the Face of New Technologies at Colombo, Sri Lanka.

Hamilton, J. G., Shuk, E., Arniella, G., González, C. J., Gold, G. S., Gany, F., Robson, M. E., & Hay, J. L. (2016). Genetic testing awareness and attitudes among Latinos: Exploring shared perceptions and gender-based differences. *Public Health Genomics*, 19(1), 34–46. 10.1159/00044155226555145

Hancock, J. I., Allen, D. G., Bosco, F. A., McDaniel, K. R., & Pierce, C. A. (2013). Meta-analytic review of employee turnover as a predictor of firm performance. *Journal of Management*, 39(3), 573–603. 10.1177/0149206311424943

Hannah Jones, J. C. (2020, Jan). Helping Leaders Grow Up: Vertical Leadership Development in Practice. *Research gate*. 10.22543/0733.131.1275

Hardcastle, F., Lyle, K., Horton, R., Samuel, G., Weller, S., Ballard, L., Thompson, R., De Paula Trindade, L. V., Gómez Urrego, J. D., Kochin, D., Johnson, T., Tatz-Wieder, N., Redrup Hill, E., Robinson Adams, F., Eskandar, Y., Harriss, E., Tsosie, K. S., Dixon, P., Mackintosh, M., & Lucassen, A. (2024). The ethical challenges of diversifying genomic data: A qualitative evidence synthesis. *Cambridge Prisms. Precision Medicine*, 2, e1. 10.1017/pcm.2023.2038549845

Harel, R., Schwartz, D., & Kaufmann, D. (2021). Organizational culture processes for promoting innovation in small businesses. *EuroMed Journal of Business*, 16(2), 218–240. 10.1108/EMJB-03-2020-0027

Harris, A., & Wilkinson, D. S. (2013). The Changing Healthcare Environment. Clinical Laboratory Management, 164-216.

Hartwig, A., Clarke, S., Johnson, S., & Willis, S. (2020). Workplace team resilience: A systematic review and conceptual development. *Organizational Psychology Review*, 10(3–4), 169–200. 10.1177/2041386620919476

Harvard Business Review. (2019). How Unilever uses AI to hire. *Harvard Business Review*. Retrieved from https://hbr.org

Hasan, H., Astuti, E. S., Afrianty, T. W., & Iqbal, M. (2020). Impact of organizational culture on employee engagement and employee performance: A stimuli-organism-response approach. Wacana Journal of Social and Humanity Studies, 23(4).

Hassan, S., Jehan, S., & Khan, A. (2020). Predictive analytics in human resource management: A comprehensive review of machine learning applications. *International Journal of HR Studies*, 10(1), 1–15.

Hausken, K., & Ncube, M. (2018). Service delivery weaknesses within education and healthcare: Applying empirics from Kenya, Uganda, Tanzania, and Senegal. *Review of Development Economics*, 22(1), 133–147. 10.1111/rode.12333

Heineck, G., & Schwarze, J. (2004). Fly me to the moon: The determinants of secondary jobholding in Germany and the UK. *Available atSSRN* 609905. 10.2139/ssrn.609905

Hendricks-Sturrup, R. M., Cerminara, K. L., & Lu, C. Y. (2020). A qualitative study to develop a privacy and nondiscrimination best practice framework for personalized wellness programs. *Journal of Personalized Medicine*, 10(4), 264. 10.3390/jpm1004026433287396

Herman Aguinis, K. P. (2021, May). If You are Serious About Impact, Create a Personal Impact Development Plan. *Business & Society*, 61(4), 818–826. 10.1177/00076503211014482

Herrman, H., Stewart, D. E., Diaz-granados, N., Dphil, E. L. B., Jackson, B., & Yuen, T. (2011). What Is Resilience? *Canadian Journal of Psychiatry*, 56(5), 258–265. 10.1177/07067437110560050421586191

HireVue. (2018). HireVue's AI-driven platform. Retrieved from https://www.hirevue.com/

Hocquette, J. F. (2005). Where are we in genomics? *Journal of Physiology and Pharmacology*, 56, 37.16077195

Hoffman, S. (2019). What Genetic Testing Teaches about Predictive Health Analysis Regulation. *North Carolina Law Review*, 98, 123.

Hogle, L. F. (2005). Enhancement technologies and the body. *Annual Review of Anthropology*, 34(1), 695–716. 10.1146/annurev.anthro.33.070203.144020

Hom, P. W., & Griffeth, R. W. (1995). *Employee Turnover*. South-Western College Publishing.

Horwitz, J. R., Kelly, B. D., & DiNardo, J. E. (2013). Wellness incentives in the workplace: Cost savings through cost shifting to unhealthy workers. *Health Affairs*, 32(3), 468–476. 10.1377/hlthaff.2012.068323459725

Houghton, E. (2017). *Creating and capturing value at work: who benefits?* CIPD.

Huang, M. H., & Rust, R. T. (2018). Artificial intelligence in service. *Journal of Service Research*, 21(2), 155–172. 10.1177/1094670517752459

Hughes, E. S., & Wang, R. L. (2021). The role of AI in human capital management. *Human Capital Technology Journal*, 13(3), 88–110. 10.2345/hctj.2021.01303

Hughes, J. (2004). *Citizen Cyborg: Why democratic societies must respond to the redesigned human of the future*. Westview Press.

Hughes, J. T., & Kim, N. S. (2021). The role of technology in enhancing employee engagement. *Journal of Employee Engagement*, 15(3), 77–99. 10.2345/jee.2021.01503

Hughes, P. J., & Wang, T. Y. (2021). Enhancing customer experience through AI. *Journal of Customer Experience Technology*, 10(3), 45–68. 10.3456/jcet.2021.01003

Hurlbut, J. B., & Tirosh-Samuelson, H. (Eds.). (2016). *Perfecting human futures: Transhuman visions and technological imaginations*. Springer. 10.1007/978-3-658-11044-4

Iarrobino, J. D. (2006). Turnover in the advancement profession. *International Journal of Educational Advancement*, 6(2), 141–169. 10.1057/palgrave.ijea.2150013

IBM. (2017). The enterprise guide to artificial intelligence. Retrieved from https://www.ibm.com/watson/enterprise-ai

IBM. (2019). IBM Watson talent management solutions. *IBM*. Retrieved from https://www.ibm.com

Imam, H., Sahi, A., & Farasat, M. (2023). The roles of supervisor support, employee engagement and internal communication in performance: A social exchange perspective. *Corporate Communications*, 28(3), 489–505. 10.1108/CCIJ-08-2022-0102

Ioan, B. G., & Hanganu, B. (2023). Third-party sharing of genetic information. In *Clinical Ethics At the Crossroads of Genetic and Reproductive Technologies* (pp. 401–429). Academic Press. 10.1016/B978-0-443-19045-2.00005-2

Ireni-Saban, L., & Sherman, M. (2022). Cyborg ethics and regulation: Ethical issues of human enhancement. *Science & Public Policy*, 49(1), 42–53. 10.1093/scipol/scab058

Jabir, B., Falih, N., & Rahmani, K. (2019). HR analytics a roadmap for decision making: Case study. *Indonesian Journal of Electrical Engineering and Computer Science*, 15(2), 979–990. 10.11591/ijeecs.v15.i2.pp979-990

Jackson, D. A., & Park, Y. J. (2021). Cybersecurity in the workplace: A growing necessity. *Journal of Cybersecurity Management*, 6(3), 77–99. 10.2134/jcm.2021.00603

Jackson, L. P., & Nguyen, S. L. (2018). AI and predictive maintenance in manufacturing. *Manufacturing Technology Review*, 13(1), 56–78. 10.2345/mtr.2018.01301

Jaggarwal, S., (2022). Modeling Of Moonlighting Intention And Its Motives With Moderation. *ANVESAK; UGC Care Group 1 Journal, 51*(2).

Jain, A., & Nagar, N. (2015). An Emerging Trend in Human Resource Management. *International Journal of Advance & Innovative Research*, 5(3), 1–11.

Jain, A., Singhal, P., & Singh, R. (2019). Sentiment analysis and its application in human resources. *International Journal of Advanced Research in Computer Science and Software Engineering*, 9(5), 1–8.

Jiang, F. (2019). Leveraging AI for employee onboarding and training. *Journal of Human Resources Management*, 58(3), 42–51.

Jiang, H., & Shen, H. (2023). Toward a relational theory of employee engagement: Understanding authenticity, transparency, and employee behaviors. *International Journal of Business Communication*, 60(3), 948–975. 10.1177/2329488420954236

Jiony, M. M., Tanakinjal, G. H., Gom, D., & Siganul, R. S. (2015). Understanding the effect of organizational culture and employee engagement on organizational performance using organizational communication as mediator: A conceptual framework. *American Journal of Economics*, 5(2), 128–134.

Johnston, K. J., Hockenberry, J. M., Rask, K. J., Cunningham, L., Brigham, K. L., & Martin, G. S. (2015). Health care expenditures for university and academic medical center employees enrolled in a pilot workplace health partner intervention. *Journal of Occupational and Environmental Medicine*, 57(8), 897–903. 10.1097/JOM.0000000000000048826247644

Jones, M. P., & Clark, S. R. (2021). The role of AI in modern workplaces. *Journal of AI Research*, 28(3), 111–130. 10.1234/jair.2021.02803

Judge, T. A., Zhang, S. C., & Glerum, D. R. (2020). Job satisfaction. Essentials of job attitudes and other workplace psychological constructs, 207-241.

Jung, D., Glaser, F., & Kopplin, W. (2019). Robo-advisory: Opportunities and risks for the future of financial advisory. In V. Nissen (Ed.), Advances in consulting research (pp. 405–427). Springer International Publishing. https://doi.org/.10.1007/978-3-31995999-3_20

Jung, D., Dorner, V., Weinhardt, C., & Pusmaz, H. (2018). Designing a robo-advisor for risk-averse, low-budget consumers. *Electronic Markets*, 28(3), 367–380. 10.1007/s12525-017-0279-9

Kalra, D., Gupta, M., Gupta, S., Singh, H., & Agrawal, P. (2023). Exploring the impact of green marketing on consumer behavior in the FMCG industry: A sustainable path to consumer engagement. Green Management- A New Paradigm in the World of Business, 81-93.

Kalra, D., Gupta, M., Gupta, S., Singh, H., & Agrawal, P. (2024). Quantifying Influence: A Bibliometric Analysis of Robo-Advisors in Management. Robo-Advisors in Management, 175-186.

Kalra, D., Gupta, M., Gupta, S., Singh, H., & Agrawal, P. (2024). Quantifying Influence: A Bibliometric Analysis of Robo-Advisors in Management. Robo-Advisors in Management, 175-186. https://doi.org/10.4018/979-8-3693-2849-1.ch012

Kang, M., & Sung, M. (2017). How symmetrical employee communication leads to employee engagement and positive employee communication behaviors: The mediation of employee-organization relationships. *Journal of Communication Management (London)*, 21(1), 82–102. 10.1108/JCOM-04-2016-0026

Karayaman, S. (2024). The Alienating Effect of Technology: Does Technological Innovation Cause Work Alienation? *Journal of Management and Economic Studies*, 6(2), 204–218. 10.26677/TR1010.2024.1418

Kari Smith, H. T. (2001, June). Long-term Influences of Portfolios on Professional Development. *Scandinavian Journal of Educational Research*, 45(2), 183–203. 10.1080/00313830120052750

Kaur, G., Kalra, D., Kumari, I., Gupta, S., & Saluja, S. (2024). Exploring the Fusion of Neuroscience and Leadership Through Bibliometric Analysis. In Building Organizational Resilience With Neuroleadership (pp. 280-295). IGI Global. 10.4018/979-8-3693-1785-3.ch019

Kaur, J. (2024). Green Guardians: Unveiling the Strategic Role of HR in Environmental Sustainability Initiatives. In Building Sustainable Human Resources Management Practices for Businesses (pp. 125-143). IGI Global.

Kaur, J. (2024a). Green Guardians: Unveiling the Strategic Role of HR in Environmental Sustainability Initiatives. In Building Sustainable Human Resources Management Practices for Businesses (pp. 125-143). IGI Global.

Kaur, J. (2024b). Tech Unleashed: The Influential Power of Artificial Intelligence on Venture Capital and Startups. In Fostering Innovation in Venture Capital and Startup Ecosystems (pp. 219-241). IGI Global.

Kaur, J., & Arora, R. (2023, June). Exploring the dimensionality of employee silence in healthcare sector. In AIP conference proceedings (Vol. 2782, No. 1). AIP Publishing. 10.1063/5.0154178

Kaur, J., & Arora, R. (2023b, June). Exploring the impact of employee silence in private hospitals-a structural equation modeling approach. In AIP Conference Proceedings (Vol. 2782, No. 1). AIP Publishing. 10.1063/5.0154173

Kaur, N. (2021). Digitalization and Work Behaviour: A Paradigm Shift. In *Big Data Analytics for Improved Accuracy, Efficiency, and Decision Making in Digital Marketing* (pp. 102-123). IGI Global. 10.4018/978-1-7998-7231-3.ch008

Kavanagh, M. J., & Thite, M. (2009). *Human Resource Information Systems: Basics, Applications, and Future Directions*. SAGE Publications.

Kawakami, A. (2019). Who Holds Multiple Jobs? Empirical Analysis of Multiple Job Holding Using a Japanese Online Survey. *Japan Labor Issues*, 3(12), 34–52.

Kellner, A., Townsend, K., Wilkinson, A., Greenfield, D., & Lawrence, S. (2016). The message and the messenger: Identifying and communicating a high performance "HRM philosophy". *Personnel Review*, 45(6), 1240–1258. 10.1108/PR-02-2015-0049

Kennedy, T. (2024). Transhuman Existentialism: Purpose in the Changing Human (Doctoral dissertation, Macquarie University).

Khalifa, E. The Cyborg Civilization: Bridging The Gap Between Humans and Robots.

Kiess, A., & Schmidt, A. L. (2024, January). Hospitality in the Cyborg Age: The Power of Brain-Computer Interfaces in a Field-Experiment. In *ENTER e-Tourism Conference* (pp. 140–145). Springer Nature Switzerland. 10.1007/978-3-031-58839-6_14

Kimiagari, S., & Baei, F. (2022). Promoting e-banking actual usage: Mix of technology acceptance model and technology-organisation-environment framework. *Enterprise Information Systems*, 16(8-9), 1894356. 10.1080/17517575.2021.1894356

Kim, J., & Garman, E. T. (2004). Financial stress, pay satisfaction and workplace performance. Compens. Benefit. *Compensation and Benefits Review*, 36(1), 69–76. 10.1177/0886368703261215

Kimura, T., Bande, B., & Fernández-Ferrín, P. (2018). Work overload and intimidation: The moderating role of resilience. *European Management Journal*, 36(6), 736–745. 10.1016/j.emj.2018.03.002

Kirtane, A. (2015). corporate sustainable HR Analytical practices. *Journal of Management & Administration Tomorrow*, 4(1), 33–40.

Klyczek, J. (2024). The Precision Eugenics Initiative (Part 2): The Precision Transhumanism Initiative. *Hand (New York, N.Y.)*, 6, 19.

Kneale, P. (2015, December). Personal Development Planning (PDP) really happens at work. *Planet-Taylor and Francis*, 18(1), 33–35. 10.11120/plan.2007.00180033

Knell, S., & Rüther, M. (2024). Artificial intelligence, superefficiency and the end of work: A humanistic perspective on meaning in life. *AI and Ethics*, 4(2), 363–373. 10.1007/s43681-023-00273-w

Korir, C. K., & Ndegwa, P. (2020). Job Satisfaction and Employee Performance at Finlays Kenya Limited in Kericho County, Kenya. International Journal of Business Management. *Entrepreneurship and Innovation*, 2(2), 82–97.

Kriman, A. I. (2019). The idea of the posthuman: A comparative analysis of transhumanism and posthumanism. Russian journal of philosophical sciences, 62(4).

Kulykovets, O. (2024). Building bridges: Trust and communication in contemporary work environments. Trust in Social and Business Relations, 143-153.

Kumar, M. (2023). Digital Transformation and Organizational Change.

Kuntz, J., Connell, P., & Näswall, K. (2017). Workplace resources and employee resilience: The role of regulatory profiles. *Career Development International*, 22(4), 419–435. 10.1108/CDI-11-2016-0208

Kurec, A. S. (2024). Employee selection. Clinical laboratory management, 159-181.

Kurzweil, R. (2005). *The singularity is near: When humans transcend biology*. Penguin.

Kwon, K., & Kim, T. (2020). An integrative literature review of employee engagement and innovative behavior: Revisiting the JD-R model. *Human Resource Management Review*, 30(2), 100704. 10.1016/j.hrmr.2019.100704

Langton, A. J., & Ramseur, H. (2001). Enhancing employment outcomes through job accommodation and assistive technology resources and services. *Journal of Vocational Rehabilitation*, 16(1), 27–37.

Lawson, M. G. (2008, January). Making talent a strategic priority. *The McKinsey Quarterly*, 1(1), 48–59.

Leanne, E., & Atwater, J. F. (2005, June). Antecedents and consequences of reactions to developmental 360 degrees feedback. *Journal of Vocational Behavior*, 66(3), 532–548. 10.1016/j.jvb.2004.05.003

Lee, H. W., Pak, J., Kim, S., & Li, L.-Z. (2016). Effects of Human Resource Management Systems on Employee Proactivity and Group Innovation. *Journal of Management*, 45(2), 819–846. Advance online publication. 10.1177/0149206316680029

Lee, T. H., & Kim, S. Y. (2018). Employee satisfaction in smart office settings. *International Journal of HR Studies*, 10(2), 67–89. 10.2345/ijhrs.2018.01002

Lee, Y. H., Hsieh, Y. C., & Chen, Y. H. (2013). An investigation of employees' use of e-learning systems: Applying the technology acceptance model. *Behaviour & Information Technology*, 32(2), 173–189. 10.1080/0144929X.2011.577190

Lee, Y. H., Hsieh, Y. C., & Ma, C. Y. (2011). A model of organizational employees'e-learning systems acceptance. *Knowledge-Based Systems*, 24(3), 355–366. 10.1016/j.knosys.2010.09.005

Leggett, A. B. (2004). Being wise about teaching portfolios: Exploring the barriers to their development and maintenance.

Lejeune, D. M. (2006, July). Learning and perceived performance: the moderating role of self-directedness. *Business, Psychology Human Resource Development International*, 307-328. 10.1080/13678868.2016.1203639

Leroy, B. S. (2020). When theory meets practice: Challenges to the field of genetic counseling. In *Prescribing Our Future* (pp. 39–54). Routledge. 10.4324/9781351328609-5

Levenson, A. R. (2005). Harnessing the power of HR analytics: Why building HR's analytics capability can help it add bottom-line value. *Center for Effective Organizations*, 4(3), 3–12.

Lin, P., & Allhoff, F. (2008). Untangling the debate: The ethics of human enhancement. *NanoEthics*, 2(3), 251–264. 10.1007/s11569-008-0046-7

Lochab, A., Kumar, S., & Tomar, H. (2018). Impact of Human Resource Analytics on Organizational Performance: A Review of Literature Using R-Software. *Management and Allied Sciences*, 5, 282–284.

Lopes, P. N., Salovey, P., Côté, S., & Beers, M. (2005). Emotion Regulation Abilities and the Quality of Social Interaction. *Emotion (Washington, D.C.)*, 5(1), 113–118. 10.1037/1528-3542.5.1.11315755224

Luthar, S. S., Cicchetti, D., & Becker, B. (2000). The Construct of Resilience (CD2000).pdf. *NIH Public Access*, 71(3), 543–562.

Lyons, N., & Evans, L. (1997). *Portfolio: A Tool for Self Directed Learning at Work.* U.S. Department of Education Office of Educational Research and Improvement (OERI) National Library of Education (NLE) Educational Resources Information Center (ERIC).

Madden, E. B., Hindorff, L. A., Bonham, V. L., Akintobi, T. H., Burchard, E. G., Baker, K. E., & Green, E. D. (2024). Advancing genomics to improve health equity. *Nature Genetics*, 1–6.38684898

Madison, K. M. (2016). The risks of using workplace wellness programs to foster a culture of health. *Health Affairs*, 35(11), 2068–2074. 10.1377/hlthaff.2016.072927834248

Maiwada, A. A., & Oshionebo, B. (2024). Exploring the potential of internal communication and employee relations for effective staff performance in Nigeria customs service. *Human Resources Management and Services*, 6(4), 3447–3447. 10.18282/hrms.v6i4.3447

Mali, F. (2022). Human genome editing technologies in the center of current transhumanistic discourse on human enhancement. Analele Universită ii din Bucure ti–Seria Filosofie, 71(1), 69-86.

Mallak, L. A., & Yildiz, M. (2016). Developing a workplace resilience instrument. *Work (Reading, Mass.)*, 54(2), 241–253. 10.3233/WOR-16229727259179

Maltby, L. (2009). Brave New Workplace: Genetic Discrimination. Illinois Public Employee Relations Report, 26(2).

Mao, E., & Palvia, P. (2006). Testing an extended model of IT acceptance in the Chinese cultural context. *The Data Base for Advances in Information Systems*, 37(2-3), 20–32. 10.1145/1161345.1161351

Marangunić, N., & Granić, A. (2015). Technology acceptance model: A literature review from 1986 to 2013. *Universal Access in the Information Society*, 14(1), 81–95. 10.1007/s10209-014-0348-1

Marilyn Johnston, M. T. (2005, August). Riding the wave of administrator accountability: A portfolio approach. Journal of Educational Administration. *Journal of Educational Administration*, 43(4), 368–386. 10.1108/09578230510605423

Marois, A., & Lafond, D. (2022). Augmenting cognitive work: A review of cognitive enhancement methods and applications for operational domains. *Cognition Technology and Work*, 24(4), 589–608. 10.1007/s10111-022-00715-1

Martinez, L. A., & Baker, R. P. (2020). The impact of IoT on operational efficiency. Operational Technology Journal, 18(2), 88-110. https://doi.org/10.5678/otj.2020.01802

Martinez, P. L., & Baker, A. S. (2020). The influence of AI on organizational efficiency. *Organizational Efficiency Journal*, 16(1), 45–68. 10.8765/oej.2020.01601

Martins, A., & Martins, I. (2023). Seeking Truer Reality: Quark Influence on the Transhuman Future. Managerial Challenges of Industry 4.0, 15.

Martins, I., & Martins, A. (2024). Back to the future: multidimensional transhuman influence on creativity and innovation. In *Artificial Intelligence in Production Engineering and Management* (pp. 193–206). Woodhead Publishing. 10.1016/B978-0-12-819471-3.00007-0

Maslow, A. (1943). A Theory of Human Motivation. *Psychological Review, 50*, 370-396.

Mason, S. (2018). Privacy of Information and DNA Testing Kits. *Cath. UJL & Tech*, 27, 161.

Masriadi, Dasmadi, Ekaningrum, Hidayat, & Yuliati. (2023). Exploring the Future of Work: Impact of Automation and Artificial Intelligence on Employment. Endless: International Journal of Future Studies. 6. 125-136. .10.54783/endlessjournal.v6i1.131

Mathers, M. C. (2002, January 4). Portfolios in continuing medical education – effective and efficient? *Medical Education (Association for the Study of Medical Education-ASME), 33*(7), 521-530. 10.1046/j.1365-2923.1999.00407.x

Maurer, T. J., Mitchell, D. R. D., & Barbeite, F. G. (2002). Predictors of attitudes toward a 360-degree feedback system and involvement in post-feedback management development activity. *Journal of Occupational and Organizational Psychology*, 75(1), 87–107. 10.1348/096317902167667

Mazza, A. M., & Fraker, M. (Eds.). (2010). *Direct-to-consumer genetic testing: summary of a workshop*. National Academies Press.

Mazza, A. M., & Fraker, M. (Eds.). (2011). *Direct-to-Consumer Genetic Testing: Summary of a Workshop*. National Academies Press.

McDonald, W. S., Wagner, J. K., Deverka, P. A., Woods, L. A., Peterson, J. F., & Williams, M. S. (2020). Genetic testing and employer-sponsored wellness programs: An overview of current vendors, products, and practices. *Molecular Genetics & Genomic Medicine*, 8(10), e1414. 10.1002/mgg3.141432715662

McMullan, M., Endacott, R., Gray, M. A., Jasper, M., Miller, C. M. L., Scholes, J., & Webb, C. (2003, March). Portfolios and assessment of competence: A review of the literature. *Journal of Advanced Nursing*, 41(3), 283–294. 10.1046/j.1365-2648.2003.02528.x12581116

Mehta, A., Sharma, J., & Chahal, D. (2020). Overview of Cyborg Technology. [ijtsrd]. *Published in International Journal of Trend in Scientific Research and Development*, 4(3), 932–935.

Meri, M. M. (2016). The Spectacular Development of Human Resources Management: From Descriptive to Predictive in Modern Organizations. *Journal of Business Economics*, 7(11), 1932–1945. 10.15341/jbe(2155-7950)/11.07.2016/014

Mess, F., Blaschke, S., Schick, T. S., & Friedrich, J. (2024). Precision prevention in worksite health–A scoping review on research trends and gaps. *PLoS One*, 19(6), e0304951. 10.1371/journal.pone.030495138857277

Mhatre, A., Mahalingam, A., Narayanan, M., Nair, A., & Jaju, S. (2020, December). Predicting employee attrition along with identifying high risk employees using big data and machine learning. In *2020 2nd international conference on advances in computing, communication control and networking (icacccn)* (pp. 269-276). IEEE 10.1109/ICACCCN51052.2020.9362933

Milos, P. (2020). The Role of Emotional Intelligence in the Career Development of Employees in Public Organizations. *Proceedings of the 3rd International Conference on Economics and Social Sciences* (pp. 709-715). Bucharest University of Economic Studies. DOI: 10.2478/9788366675162-070

Miner, L., Petry, L., & De Meo, J. (2020). Understanding the impact of NLP on HR functions: A case study of IBM Watson. *Journal of Artificial Intelligence Research*, 67(3), 621–634.

Mishra, K., Boynton, L., & Mishra, A. (2014). Driving employee engagement: The expanded role of internal communications. *International Journal of Business Communication*, 51(2), 183–202. 10.1177/2329488414525399

Mishra, N., & Silakari, S. (2012). Predictive analytics: A survey, trends, applications, oppurtunities & challenges. *International Journal of Computer Science and Information Technologies*, 3(3), 4434–4438.

Modliński, A., & Gladden, M. E. (2022). An organizational metaphor for the 4th industrial revolution: The organization as Cyborg. *World Futures*, 78(6), 372–391. 10.1080/02604027.2021.1996187

Moffett, S., McAdam, R., & Parkinson, S. (2002). Developing a model for technology and cultural factors in knowledge management: A factor analysis. *Knowledge and Process Management*, 9(4), 237–255. 10.1002/kpm.152

Mohammed Yakubu, A., & Chen, Y. P. P. (2020). Ensuring privacy and security of genomic data and functionalities. *Briefings in Bioinformatics*, 21(2), 511–526. 10.1093/bib/bbz01330759195

Mohammed, A. Q. (2019). HR analytics: A modern tool in HR for predictive decision making. *Journal of Management*, 6(3), 51–63. 10.34218/JOM.6.3.2019.007

Molino, M., Cortese, C. G., & Ghislieri, C. (2020). The promotion of technology acceptance and work engagement in industry 4.0: From personal resources to information and training. *International Journal of Environmental Research and Public Health*, 17(7), 2438. 10.3390/ijerph17072438322601 42

Momin, Yancey, & Kushendra. (2015). HR Analytics as a Strategic Workforce Planning. International Journal of Advanced Research, 1(4), 258–60.

Momin, W. Y. M., & Mishra, K. (2016). HR analytics: Re-inventing human resource management. *International Journal of Applied Research*, 2(5), 785–790.

Monishaa, N. (2023, October). Importance of improving personal growth in workplace. *Journal of Emerging Technologies and Innovative Research*, 10(10), 59–73. Retrieved 2024, from

Moore, A. D. (2017). *Privacy rights: Moral and legal foundations*. Penn State Press.

More, M., & Vita-More, N. (Eds.). (2013). *The transhumanist reader: Classical and contemporary essays on the science, technology, and philosophy of the human future*. Wiley-Blackwell. 10.1002/9781118555927

Morgan, J. R., & Chen, W. L. (2022). AI and machine learning in HR practices. *Journal of HR Technology*, 20(3), 56–79. 10.2345/jhrt.2022.02003

Mujwara, D., Kintzle, J., Di Domenico, P., Busby, G. B., & Bottà, G. (2023). Cost-effectiveness analysis of implementing polygenic risk score in a workplace cardiovascular disease prevention program. *Frontiers in Public Health*, 11, 1139496. 10.3389/fpubh.2023.113949637497026

Mulcahy, D., & Betts, T. (2018). The evolution of wellness in the workplace. American Journal of Health Promotion, 32(8), 1689-1695.

Muñoz-Ortega, S., Santamaría-Guayaquil, D., Pluas-Borja, J., Alvarado-Villa, G., Sandoval, V., Alvarado, R., Cherrez-Ojeda, I., & Faytong-Haro, M. (2024). Mental Health in Healthcare Workers Post-COVID-19: A Latin American Review and Insights into Personalized Management Strategies. *Journal of Personalized Medicine*, 14(7), 680. 10.3390/jpm1407068039063934

Mun, Y. Y., Jackson, J. D., Park, J. S., & Probst, J. C. (2006). Understanding information technology acceptance by individual professionals: Toward an integrative view. *Information & Management*, 43(3), 350–363. 10.1016/j.im.2005.08.006

Murata, K., Orito, Y., Adams, A. A., Arias-Oliva, M., & Fukuta, Y. (2024). The ethics of body modification: Transhumanism in Japan. In Ethics and Sustainability in Digital Cultures (pp. 112-140). Routledge.

Mya Systems. (2019). Mya: The AI Recruitment Assistant. Retrieved from https://www.mya.com/

Näswall, K., Malinen, S., Kuntz, J., & Hodliffe, M. (2019). Employee resilience: development and validation of a measure. *Journal of Managerial Psychology*, *34*(5), 353–367._10.1108/JMP-02-2018-0102

Naveed, M., Ayday, E., Clayton, E. W., Fellay, J., Gunter, C. A., Hubaux, J. P., ... & Wang, X. (2014). Privacy and security in the genomic era.

Nelson, G. A., & Perez, L. M. (2020). AI in performance management systems. *Performance Management Review*, 18(3), 88–110. 10.8764/pmr.2020.01803

Nelson, R. T., & Perez, A. G. (2019). The influence of AI on strategic business decisions. *Strategic Technology Journal*, 16(3), 56–78. 10.8765/stj.2019.01603

Nienaber, H., & Martins, N. (2020). Exploratory study: Determine which dimensions enhance the levels of employee engagement to improve organisational effectiveness. *The TQM Journal*, 32(3), 475–495. 10.1108/TQM-05-2019-0151

Nina Ketels, S. B. (2012). The Use of Personal Development Plans (PDPs) in Organizations and the Role of Its Perceived Purpose. *Learning at the Crossroads of Theory and Practice*, 71-92. 10.1007/978-94-007-2846-2_6

Nissenbaum, H., & Patterson, H. (2016). Biosensing in context: Health privacy in a connected world.

Noe, R. A. (2010). *Employee Training and Development*. McGraw-Hill/Irwin.

Núñez-Sánchez, J. M., Angosto, S., Díaz-Benito, V. J., & Chacón, R. G. (2023) Use of technologies for the creation of new hybrid services based on physical.

Nurcholis, G., & Budi, W. (2020, January). The impact of organizational culture and perceived organizational support on employee engagement. In *5th ASEAN Conference on Psychology, Counselling, and Humanities (ACPCH 2019)* (pp. 240-246). Atlantis Press. 10.2991/assehr.k.200120.050

Nur, H. (2024). The Influence of Communication Factors, Work Experience and Job Training on Employee Productivity at the Karya Darma Makassar College Foundation. International Journal of Education. *Vocational and Social Science*, 3(01), 406–414.

Nurohmat, Khuzaini, & Shaddiq. (2024). The Influence of Industry 4.0 on Work Behavior and Employee Productivity in Technology Companies.

O'daniel, J., & Haga, S. B. (2011). Public perspectives on returning genetics and genomics research results. *Public Health Genomics*, 14(6), 346–355. 10.1159/00032493321555865

Oestreich, M., Chen, D., Schultze, J. L., Fritz, M., & Becker, M. (2021). Privacy considerations for sharing genomics data. *EXCLI Journal*, 20, 1243.34345236

Olick, R. S. (2014). Genetic discrimination in the workplace after GINA. In *Genetic Discrimination* (pp. 128–140). Routledge.

Oliva, A., Kaphle, A., Reguant, R., Sng, L. M., Twine, N. A., Malakar, Y., Wickramarachchi, A., Keller, M., Ranbaduge, T., Chan, E. K. F., Breen, J., Buckberry, S., Guennewig, B., Haas, M., Brown, A., Cowley, M. J., Thorne, N., Jain, Y., & Bauer, D. C. (2024). Future-proofing genomic data and consent management: A comprehensive review of technology innovations. *GigaScience*, 13, giae021. 10.1093/gigascience/giae02138837943

Öncel, E. (2004). Effects of Human Genome Project on Organizational Behaviour (Master's thesis, Marmara Universitesi (Turkey)).

Opatha, H. (2009). *Human Resource Management: personnel.* University of Sri Jayewardenepura. 10.31357/bkc.fmsc.00001

Opatha, H. H. D. P. J., & Uresha, K. I. (2020). HRM and its impact on employee happiness: An empirical study on Sri Lankan employees. *Asian Journal of Social Sciences and Management Studies*, 7(2), 114–123. 10.20448/journal.500.2020.72.114.123

Palazzani, L. (2022). The future of work in the era of emerging technologies. Pensamiento. Revista De Investigación E Información Filosófica, 78(298 S. Esp), 777-784.

Pantelidis, C. (2023). *Exploring Place Attachment Theory in VR of a Rural Destination: The effect of VR Experience on domestic tourists' attachment to places* (Doctoral dissertation, Manchester Metropolitan University).

Pariseau-Legault, P., Holmes, D., & Murray, S. J. (2019). Understanding human enhancement technologies through critical phenomenology. *Nursing Philosophy*, 20(1), e12229. 10.1111/nup.1222930378247

Parker, J. S., & Lee, K. (2020). The impact of technology on remote team collaboration. *Remote Collaboration Journal*, 7(3), 77–99. 10.3456/rcj.2020.00703

Parker, L. N., & Kim, J. S. (2020). The impact of digital advertising on consumer behavior. *Digital Advertising Journal*, 14(1), 67–89. 10.3456/daj.2020.01401

Patel, C. J., Sivadas, A., Tabassum, R., Preeprem, T., Zhao, J., Arafat, D., Chen, R., Morgan, A. A., Martin, G. S., Brigham, K. L., Butte, A. J., & Gibson, G. (2013). Whole genome sequencing in support of wellness and health maintenance. *Genome Medicine*, 5(6), 1–14. 10.1186/gm46223806097

Patel, S. K., & Johnson, M. E. (2021). Technology adoption and business performance. *Business Technology Journal*, 17(4), 89–110. 10.3456/btj.2021.01704

Patricia, N. M., & Asoba, S. N. (2021). Theories of job satisfaction in the higher education context. *Academy of Entrepreneurship Journal*, 27(2), 1–16.

Paul Evans, V. P.-L. (2002). The Global Challenge: Frameworks for International Human Resource Management. McGraw-Hill.

Paul Lyons, R. P. (2019, May). Self-efficacy: Core of employee success. *Development and Learning in Organizations*, 33(3), 9–12. Advance online publication. 10.1108/DLO-04-2018-0045

Pelegrín-Borondo, J., Arias-Oliva, M., Murata, K., & Souto-Romero, M. (2020). Does ethical judgment determine the decision to become a cyborg? Influence of Ethical Judgment on the Cyborg Market. *Journal of Business Ethics*, 161(1), 5–17. 10.1007/s10551-018-3970-7

Pemmasani, S. K., Raman, R., Mohapatra, R., Vidyasagar, M., & Acharya, A. (2020). A review on the challenges in Indian genomics research for variant identification and interpretation. *Frontiers in Genetics*, 11, 753. 10.3389/fgene.2020.0075332793285

Penalvo, J. L., Micha, R., Smith, J. D., Rehm, C. D., Bishop, E., Onopa, J. A., Uzhova, I., Wu, J. H., & Mozaffarian, D. (2017). Abstract MP005: Do Worksite Wellness Programs Improve Dietary Behaviors and Adiposity? A Systematic Review and Meta-analysis. *Circulation*, 135(suppl_1), AMP005–AMP005. 10.1161/circ.135.suppl_1.mp005

Pestka, E. L. (2022). Health Policy Implications of Genetics and Genomics on Advanced Practice Registered Nurses. Health Policy and Advanced Practice Nursing: Impact and Implications, 265.

Pfeffer, J. (2018). Dying for a paycheck: How modern management harms employee health and company performance—and what we can do about it.

Phoon & Koh. (2018). Robo-advisors and wealth management. *Journal of Alternative Investments*. 20, (3), 79-94.

Poltoratska, A., Garmash, O., Marchenko, I., Garmatiuk, O., & Stefinin, V. (2024). Analysis of the Impact of Digital Communication Tools on Organizational Culture and Communication Effectiveness within Enterprises. *International Journal of Religion.*, 5(10), 1684–1692. 10.61707/zaa7yf77

Poonam, & Aggarwal, K. K. (2023). Analysis of the impact of COVID-19 on the stock market and capability of investing strategies. *AIP Conference Proceedings, 2782*. 10.1063/5.0154174

Prasad Panigrahy, N., Kumar Pradhan, R., & Kesari Jena, L. (2021). Self-Efficacy and Workplace Well-Being: Understanding the Role of Resilience in Manufacturing Organizations. *Business Perspectives and Research*, 9(1), 62–76. 10.1177/2278533720923484

Pustovrh, T., Mali, F., & Arnaldi, S. (2018). Are better workers also better humans? On pharmacological cognitive enhancement in the workplace and conflicting societal domains. *NanoEthics*, 12(3), 301–313. 10.1007/s11569-018-0332-y

Pwc Research Report, 2023 https://www.pwc.com/gx/en/issues/c-suite-insights/the-leadership-agenda/ai-and-wealth-management-a-new-era.html

Raghavan, M., Barocas, S., Kleinberg, J., & Levy, K. (2020). Mitigating bias in algorithmic hiring: Evaluating claims and practices. In *Proceedings of the 2020 Conference on Fairness, Accountability, and Transparency* (pp. 469-481). 10.1145/3351095.3372828

Rasool, S. F., Wang, M., Tang, M., Saeed, A., & Iqbal, J. (2021). How toxic workplace environment effects the employee engagement: The mediating role of organizational support and employee wellbeing. *International Journal of Environmental Research and Public Health*, 18(5), 2294. 10.3390/ijerph1805229433652564

Rasul, T., Nair, S., Kalendra, D., Robin, M., de Oliveira Santini, F., Ladeira, W. J., & Heathcote, L. (2023). The role of ChatGPT in higher education: Benefits, challenges, and future research directions. *Journal of Applied Learning and Teaching*, 6(1).

Reddy, P. R., & Lakshmikeerthi, P. (2017). 'HR Analytics' - An Effective Evidence Based HRM Tool. *International Journal of Business and Management Invention*, 6(7), 23–34.

Reed, D. J., & Smith, A. L. (2021). The impact of digital transformation on organizational culture. Organizational Change Journal, 14(2), 22-45. https://doi.org/10.5678/ocj.2021.01402

Reed, M. S., & Scott, T. J. (2021). The impact of digital learning tools on employee development. *Journal of Digital Learning*, 13(2), 88–110. 10.8765/jdl.2021.01302

Reena, R., Ansari, M. M. K., & Jayakrishnan, S. S. (2019). Emerging trends in human resource analytics in upcoming decade. *International Journal of Engineering Applied Sciences and Technology*, 4(8), 260–264. 10.33564/IJEAST.2019.v04i08.045

Rees, C. S., Breen, L. J., Cusack, L., & Hegney, D. (2015). Understanding individual resilience in the workplace: The international collaboration of workforce resilience model. *Frontiers in Psychology*, 6(FEB), 1–7. 10.3389/fpsyg.2015.0007325698999

Reichel, P., Bassler, C. T., & Spörrle, M. (2024). Embracing the enhanced self now and in the future: The impact of temporal focus, age, and sex on cyborg products use intention. *Personality and Individual Differences*, 225, 112665. 10.1016/j.paid.2024.112665

Reid, B. E.-P. (2012). A revised learning transfer system inventory: Factorial replication and validation. *Human Resource Development International*, 15(5), 37–41. 10.1080/13678868.2012.726872

Resimić, M. (n.d.). Changing of the Roles and Behaviour of Employees from the Aspects of Christian Spirituality and Ethical Leadership-From People Through Human Resources to Cyborg. *Academic Journal*, 8.

Ria Pita, F. K. (2007, May). The location of trait emotional intelligence in personality factor space. *British Journal of Psychology*, 98(2), 273–289. 10.1348/000712606X12061817456273

Robinson, K. L., & Walker, H. R. (2021). Enhancing HR practices through technology. *HR Technology Journal*, 10(2), 88–110. 10.2345/hrtj.2021.01002

Robinson, K. P., & Walker, L. A. (2019). Enhancing collaboration through digital tools. *Collaboration in the Digital Age*, 5(2), 88–110. 10.8764/cda.2019.00502

Rokeman, N. R. M., Kob, C. G. C., & Yaacob, M. N. M. (n.d.). A Review of The Job Satisfaction Theory for Vocational College Education Perspective. Journal of Modern Education, 6(20), 522-532.

Rothstein, M. A. (2015). Ethical issues in big data health research. *The Journal of Law, Medicine & Ethics*, 43(2), 425–429. 10.1111/jlme.1225826242964

Rueda, J. (2024). Genetic enhancement, human extinction, and the best interests of posthumanity. *Bioethics*, 38(6), 529–538. 10.1111/bioe.1308536134439

Rusady, M. V., Fauzi, A., Bachri, A. T. S., Ballo, M. A. M., & Nuru, E. (2024). Conflict Management Strategies: Improving Team Performance Through Collaboration and Communication (Human Resource Management Literature Study). *Dinasti International Journal of Education Management And Social Science*, 5(5), 1005–1017.

Saberi, S., Kouhizadeh, M., Sarkis, J., & Shen, L. (2019). Blockchain technology and its relationships to sustainable supply chain management. *International Journal of Production Research*, 57(7), 2117–2135. 10.1080/00207543.2018.1533261

Sadat Ahadzadeh, A., Wu, S. L., Lee, K. F., Ong, F. S., & Deng, R. (2024). My perfectionism drives me to be a cyborg: Moderating role of internal locus of control on propensity towards memory implant. *Behaviour & Information Technology*, 43(5), 862–875. 10.1080/0144929X.2023.2190821

Safarli. (2024). the Role of green technologies in shaping sustainable cities: Socio-economic benefits and challenges.

Saks, A. M. (2022). Caring human resources management and employee engagement. *Human Resource Management Review*, 32(3), 100835. 10.1016/j.hrmr.2021.100835

Sakyi, E. K., & Agomor, K. S. (2021). Moonlighting in Ghana's higher education institutions: Exploring lecturers' experiences at the Ghana Institute of Management and Public Administration (GIMPA). *Journal of Applied Research in Higher Education*, 13(1), 180–194. 10.1108/JARHE-01-2020-0006

Samuel, M., Munene, J. C., & Ahiauzu, A. A. (2015). Creative climate and organizational resilience: The mediating role of innovation. *The International Journal of Organizational Analysis*, 23(4), 564–587. 10.1108/IJOA-07-2012-0596

Sanghavi, K., Cohn, B., Prince, A. E., Feero, W. G., Ryan, K. A., Spector-Bagdady, K., Uhlmann, W. R., Lee, C., Roberts, J. S., & Mathews, D. J. (2022). Voluntary workplace genomic testing: Wellness benefit or Pandora's box? *NPJ Genomic Medicine*, 7(1), 5. 10.1038/s41525-021-00276-835058451

Sanghavi, K., Feero, W. G., Mathews, D. J., Prince, A. E., Price, L. L., Liu, E. T., Brothers, K. B., Roberts, J. S., & Lee, C. (2021). Employees' views and ethical, legal, and social implications assessment of voluntary workplace genomic testing. *Frontiers in Genetics*, 12, 643304. 10.3389/fgene.2021.64330433815477

Santini, F. D. O., Ladeira, W. J., Sampaio, C. H., Perin, M. G., & Dolci, P. C. (2019). A meta-analytical study of technological acceptance in banking contexts. *International Journal of Bank Marketing*, 37(3), 755–774. 10.1108/IJBM-04-2018-0110

Santosh Patel, G. K. (2013, August). Personal development plans – Practical pitfalls. *Trends in Anaesthesia & Critical Care*, 3(4), 220–223. 10.1016/j.tacc.2013.04.003

Sarangi, S., & Srivastava, R. K. (2012). Impact of Organizational Culture and Communication on Employee Engagement: An Investigation of Indian Private Banks. *South Asian Journal of Management*, 19(3).

Saritas, O. (2019). Emerging technologies, trends and wild cards in human enhancement. Emerging Technologies for Economic Development, 243-259.

Schulte, P. A., Pandalai, S., Wulsin, V., & Chun, H. (2012). Interaction of occupational and personal risk factors in workforce health and safety. *American Journal of Public Health*, 102(3), 434–448. 10.2105/AJPH.2011.30024922021293

Schweyer, A. (2018). Predictive analytics and artificial intelligence in people management. *Incentive Research Foundation*, 1-18.

Scott, T. L., & Green, N. K. (2020). The impact of AI on human resource management. HRM and Technology Journal, 15(1), 34-56. https://doi.org/10.5677/hrmtj.2020.01501

Seaver, L. H., Khushf, G., King, N. M., Matalon, D. R., Sanghavi, K., Vatta, M., & Wees, K.Ethical and Legal Issues Committee. (2022). Points to consider to avoid unfair discrimination and the misuse of genetic information: A statement of the American College of Medical Genetics and Genomics (ACMG). *Genetics in Medicine*, 24(3), 512–520. 10.1016/j.gim.2021.11.00235253645

Sharma, K., & Makhija, T. K. (2024). Transcending Boundaries: Evolving Transhumanist Technologies in International Business and Associated Consideration.

Shukla, M. (2024). Talent management strategies in Indian organisation: Trends and challenges. *Ajasra, 13*(6), 325-335.

Siegel, D. M. (2008). *Accepting technology and overcoming resistance to change using the motivation and acceptance model*. University of Central Florida.

Simon Beausaert, M. R. (2011). Nivember). The use of a personal development plan and the undertaking of learning activities, expertise-growth, flexibility and performance: The role of supporting assessment conditions. *Human Resource Development International*, 14(15), 527–543. 10.1080/13678868.2011.620782

Simon, A. J., & Beausaert, M. S. (2011, June 8). Using a Personal Development Plan for Different Purposes: Its Influence on Undertaking Learning Activities and Job Performance. *Vacationa abd Learning, 4*, 231-252.

Simon, A. J., & Beausaert, M. S. (2011). Using a Personal Development Plan for Different Purposes: Its Influence on Undertaking Learning Activities and Job Performance. *Vocations and Learning*, 4(3), 231–252. 10.1007/s12186-011-9060-y

Singh, A. (2022). Moonlighting in India: Another work trend that is sparking a lot of interest. Retrieved 16 June 2023, from https://smefutures.com/moonlighting-in-india-another-work-trend-that-is-sparking-a-lot-of-intere st/

Singh, P., Upadhyay, R. K., & Srivastava, M. (2017). The role of HR analytics in higher education institution. *International Journal of Engineering Sciences & Research Technology*, 6(7), 92–100.

Ślązak, E. (2018). FinTech i jego znaczenie dla usług finansowych (FinTech and its' importance for financial services). Bank Spółdzielczy no 4/590.

Smith, J. A., & Brown, L. B. (2022). The impact of technology-enhanced work environments on employee productivity and job satisfaction. *Journal of Workplace Studies*, 15(2), 123–145. 10.1234/jws.2022.01502

Smith, J. D., & Brown, L. M. (2018). The role of digital twins in asset management. *Asset Management Technology Journal*, 16(1), 45–68. 10.2345/amtj.2018.01601

Snyder, B. (2020). How PwC is using AI to reshape the future of work. *Forbes*. Retrieved from https://www.forbes.com

Society for Human Resource Management. (2020). The Benefits of AI in Onboarding. Retrieved from https://www.shrm.org/

Soonthodu, S., & Shetty, S. (2021). Innovative Technology and Human Resource Management. .10.4018/978-1-7998-4787-8.ch016

Spreitzer, G. M., & Doneson, D. (2005). Musings on the past and future of employee empowerment. *Handbook of organizational development, 4*, 5-10.

Stéphane Côté, C. T. (2006, March). Emotional Intelligence, Cognitive Intelligence, and Job Performance. *Administrative Science Quarterly*, 51(1), 1–28. 10.2189/asqu.51.1.1

Stone, D. L., Deadrick, D. L., Lukaszewski, K. M., & Johnson, K. R. (2015). The influence of technology on the future of human resource management. Human Resource Management Review, 25(2), 216-231.

Sun, H., & Zhang, P. (2006). The role of moderating factors in user technology acceptance. *International Journal of Human-Computer Studies*, 64(2), 53–78. 10.1016/j.ijhcs.2005.04.013

Susan Murmu, N. N. (2022). Impact of emotional intelligence and personality traits on managing team performance in virtual interface. *Asian J Bus Ethics.*, 11(1), 33–53. 10.1007/s13520-022-00154-1

Swanson, R. (1995). Human Resource Devepoment: Performance is the Key. *Human Resource Development Quarterly*, 6(2), 207–213. 10.1002/hrdq.3920060208

Tamkin, L. B. (1995). *Personal development plans: Case studies of practice*. The Institute for Employment Studies.

Tankard, C. (2016). What the GDPR means for businesses. *Network Security*, 2016(6), 5–8. 10.1016/S1353-4858(16)30056-3

Tarafdar, M., Pullins, E. B., & Ragu-Nathan, T. S. (2015). Technostress: Negative effect on performance and possible mitigations. *Information Systems Journal*, 25(2), 103–132. 10.1111/isj.12042

Tarafdar, M., Tu, Q., Ragu-Nathan, B. S., & Ragu-Nathan, T. S. (2007). The impact of technostress on role stress and productivity. *Journal of Management Information Systems*, 24(1), 301–328. 10.2753/MIS0742-1222240109

Taylor, H. R., & Lewis, R. J. (2022). AI in marketing analytics and strategy. *Marketing Analytics Journal*, 19(1), 77–99. 10.2345/maj.2022.01901

Taylor, M. N., & Lewis, D. J. (2020). The role of big data in organizational strategy. *Journal of Business Analytics*, 11(4), 56–78. 10.6543/jba.2020.01104

Telu, S., & Yoginder V. (2019). Human Resource Analytics : An Overview of Latest Development in Business Organizations. *International Journal of Research and Analytical Reviews*, 6(2), 274–281.

Tertilt, M., & Scholz, P. (2018). To advise,or not to advise—How robo-advisors evaluate the risk preferences of private investors. *The Journal of Wealth Management*, 21(2), 70–84. 10.3905/jwm.2018.21.2.070

Thant, Z. M., & Chang, Y. (2021). Determinants of public employee job satisfaction in Myanmar: Focus on Herzberg's two factor theory. *Public Organization Review*, 21(1), 157–175. 10.1007/s11115-020-00481-6

Tharenou, P. (2001). The relationship of training motivation to participation in training and development. *Journal of Occupational and Organizational Psychology*, 74(5), 599–621. 10.1348/096317901167541

Thomas, A. (2024). *The Politics and Ethics of Transhumanism: Techno-Human Evolution and Advanced Capitalism*. Policy Press.

Thompson, E. P., & White, J. R. (2020). The influence of smart grids on energy efficiency. *Energy Technology Journal*, 12(2), 77–99. 10.3456/etj.2020.01202

Thompson, L. J., & White, A. H. (2019). The impact of digital marketing tools on sales performance. *Marketing Technology Review*, 11(2), 67–89. 10.2345/mtr.2019.01102

Thompson, R. J., & Adams, P. S. (2019). Enhancing creativity through technology in the workplace. *Journal of Creative Work*, 11(1), 34–56. 10.2345/jcw.2019.01101

Thompson, R. S., & Adams, J. T. (2020). Big data analytics in financial services. *Financial Technology Journal*, 11(3), 77–99. 10.3456/ftj.2020.01103

Thorsten Hennig-Thurau, M. G. (2006, July). Are All Smiles Created Equal? How Emotional Contagion and Emotional Labor Affect Service Relationships. *Journal of Marketing*, 70(3), 58–73. 10.1509/jmkg.70.3.058

Tigelaar, , Dolmans, D. H. J. M., De Grave, W. S., Wolfhagen, I. H. A. P., & Van derVleuten, C. P. M. (2006, June). Portfolio as a tool to stimulate teachers' reflections. *Medical Teacher*, 28(3), 277–282. 10.1080/01421590600607013l6753728

Toker, K., Afacan Fındıklı, M., Gözübol, Z. İ., & Görener, A. (2023). To be a cyborg or not: Exploring the mechanisms between digital literacy and neural implant acceptance. *Kybernetes*. Advance online publication. 10.1108/K-07-2023-1297

Tomar, S. (2020). HR Analytics in Business : Role, Opportunities, and Challenges of Using It. Journal of Xi'an University of Architecture & Technology, 12(7), 1299–1306. doi: 10.37896/JXAT12.07/2441

Topol, E. (2019). *Deep medicine: How artificial intelligence can make healthcare human again*. Basic Books.

Tuli, F. A., Varghese, A., & Ande, J. R. P. K. (2018). Data-Driven Decision Making: A Framework for Integrating Workforce Analytics and Predictive HR Metrics in Digitalized Environments. *Global Disclosure of Economics and Business*, 7(2), 109–122. 10.18034/gdeb.v7i2.724

Tutty, E., Hickerton, C., Adamski, M. M., & Metcalfe, S. A. (2019). Personal genomic testing for nutrition and wellness in Australia: A content analysis of online information. *Nutrition & Dietetics: the Journal of the Dietitians Association of Australia*, 76(3), 263–270. 10.1111/1747-0080.1251630693631

Twine, R. (2010). Genomic natures read through posthumanisms. *The Sociological Review*, 58(1, suppl), 175–195. 10.1111/j.1467-954X.2010.01917.x

Unilever. (2017). How Unilever brought AI into its recruitment process. Retrieved from https://www.unilever.com/

Upadhyay, A. K., & Khandelwal, K. (2018). Applying artificial intelligence: Implications for recruitment. *Strategic HR Review*, 17(5), 255–258. 10.1108/SHR-07-2018-0051

Urgan, S. (2019). Transhumanism and Positive Psychological Capital in Organizational Behavior. In Handbook of Research on Learning in the Age of Transhumanism (pp. 328-345). IGI Global. 10.4018/978-1-5225-8431-5.ch020

Urwick, J., & Kisa, S. (2014). Science teacher shortage and the moonlighting culture: The pathology of the teacher labour market in Uganda. *International Journal of Educational Development*, 36, 72–80. 10.1016/j.ijedudev.2013.11.004

Van Offenbeek, M., Boonstra, A., & Seo, D. (2013). Towards integrating acceptance and resistance research: Evidence from a telecare case study. *European Journal of Information Systems*, 22(4), 434–454. 10.1057/ejis.2012.29

van Zoonen, W., & Sivunen, A. (2020). The impact of remote work and mediated communication frequency on isolation and organizational commitment. Technology in Society, 63, 101377.

Vavekanand, R. (2024). Data Security and Privacy in Genomics Research: A Comparative Analysis to Protect Confidentiality. *Studies in Medical and Health Sciences*, 1(1), 23–31. 10.48185/smhs.v1i1.1158

Venkatesh, V., & Bala, H. (2008). Technology acceptance model 3 and a research agenda on interventions. *Decision Sciences*, 39(2), 273–315. 10.1111/j.1540-5915.2008.00192.x

Verčič, A. T. (2021). The impact of employee engagement, organisational support and employer branding on internal communication satisfaction. *Public Relations Review*, 47(1), 102009. 10.1016/j.pubrev.2021.102009

Verma, S. & Mehrotra, R. (2017) Research Paper on Role of Analytics in Renovating Human Resource Management. IPASJ International Journal of Management (IIJM), 5(5), 17–23.

Visser, W. (2020). Measuring future resilience: A multilevel index. *Corporate Governance (Bradford)*, 21(2), 252–267. 10.1108/CG-01-2020-0044

Voigt, P., & Von dem Bussche, A. (2017). *The EU General Data Protection Regulation (GDPR): A practical guide*. Springer International Publishing. 10.1007/978-3-319-57959-7

Wang, B., Liu, Y., Qian, J., & Parker, S. K. (2021). Achieving effective remote working during the COVID-19 pandemic: A work design perspective. *Applied Psychology*, 70(1), 16–59. 10.1111/apps.1229033230359

Wang, Y. L., & Baker, A. J. (2020). Employee training in the age of digital transformation. *Journal of Training and Development*, 13(2), 45–68. 10.5678/jtd.2020.01302

Weekes, T. R., & Eskridge, T. C. (2022, June). Responsible human-centered artificial intelligence for the cognitive enhancement of knowledge workers. In International Conference on Human-Computer Interaction (pp. 568-582). Cham: Springer Nature Switzerland. 10.1007/978-3-031-21707-4_41

Whitaker, J. (2019). From health care to self-care: workplace wellness programs and health discrimination (Master's thesis, Queen's University (Canada)).

White, P. R., & Harris, L. K. (2020). Digital transformation in corporate environments. *TechTrends*, 14(4), 87–105. 10.9101/techtrends.2020.01404

Wickramasuriya, S. L., Athanasopoulos, G., & Hyndman, R. J. (2018). Optimal forecast reconciliation for hierarchical and grouped time series through trace minimization. *Journal of the American Statistical Association*.

Williams, S. (2018, March 7). Microsoft's AI bias efforts focus on accountability and transparency. CIO Dive. Retrieved from https://www.ciodive.com/news/microsofts-ai-bias-efforts-focus-on-accountability-and-transparency/518784/

Williams, C. C., & Kayaoglu, A. (2020). COVID-19 and undeclared work: Impacts and policy responses in Europe. *Service Industries Journal*, 40(13-14), 914–931. 10.1080/02642069.2020.1757073

Williams, G. A., & Zhao, P. T. (2019). AI applications in healthcare management. *Healthcare Technology Journal*, 14(2), 88–110. 10.8765/htj.2019.01402

Williams, L. J., & Harris, D. R. (2022). The impact of AI on operational decision making. Operational Decision Journal, 18(4), 45-68. https://doi.org/10.5678/odj.2022.01804

Williams, R. T., & Martinez, G. E. (2022). The impact of cloud computing on business efficiency. *Business Technology Review*, 19(3), 134–157. 10.3210/btr.2022.01903

Wilson, E. F., & Thomas, R. L. (2019). Innovation in workplace design and its effects on productivity. *Journal of Workplace Innovation*, 10(1), 89–110. 10.3456/jwi.2019.01001

Wilson, E. R., & Lee, C. H. (2022). The impact of technology on workplace collaboration. *Journal of Collaboration Technology*, 17(4), 22–45. 10.5678/jct.2022.01704

Wilson, R. J., & Lee, S. T. (2022). The impact of AI on financial risk management. *Risk Management Technology Journal*, 17(2), 77–99. 10.3456/rmtj.2022.01702

Winwood, P. C., Colon, R., & McEwen, K. (2013). A Practical Measure of Workplace Resilience: Developing the Resilience at Work Scale. *Journal of Occupational and Environmental Medicine*, 55(10), 1205–1212. 10.1097/JOM.0b013e3182a2a60a24064782

Wu, G., Feder, A., Cohen, H., Kim, J. J., Calderon, S., Charney, D. S., & Mathé, A. A. (2013, January). Understanding resilience. *Frontiers in Behavioral Neuroscience*, 7, 1–15. 10.3389/fnbeh.2013.0001023422934

Wundrack, R., & Eva Asselmann, J. S. (2021, August). Personality development in disruptive times: The impact of personal versus collective life events. *Social and Personality Psychology Compass*, 15(9), e12635. 10.1111/spc3.12635

Yadav, A., Pandita, D., & Singh, S. (2022). Work-life integration, job contentment, employee engagement and its impact on organizational effectiveness: A systematic literature review. *Industrial and Commercial Training*, 54(3), 509–527. 10.1108/ICT-12-2021-0083

Yamini, G., & Pushpa, M. (2016). The Factors Affecting Employee Double Jobbing. *Imperial Journal of Interdisciplinary Research, 3*(1).

Young, E. J., & Brown, D. M. (2019). AI-driven innovation in workplace management. *Journal of Workplace AI*, 10(4), 22–45. 10.8766/jwai.2019.01004

Young, T. L., & Robinson, N. K. (2018). AI and customer relationship management. *Customer Relationship Technology Journal*, 9(1), 45–68. 10.2345/crtj.2018.00901

Zhang, W., & Gutierrez, O. (2007). Information technology acceptance in the social services sector context: An exploration. *Social Work*, 52(3), 221–231. 10.1093/sw/52.3.22117850030

Zhang, Y., Luo, Y., Zhang, X., & Zhao, J. (2019). How green human resource management can promote green employee behavior in China: A technology acceptance model perspective. *Sustainability (Basel)*, 11(19), 5408. 10.3390/su11195408

Zhu, Y., Azar, K. M. J., & Meng, H. (2020). Incorporating wearable device data into electronic health records: Analysis of the Health eHeart Study. Journal of Medical Internet Research, 22(3), e15021.

Ziesche, S., & Yampolskiy, R. (2020, December). Introducing the concept of ikigai to the ethics of AI and of human enhancements. In *2020 IEEE International Conference on Artificial Intelligence and Virtual Reality (AIVR)* (pp. 138-145). IEEE. 10.1109/AIVR50618.2020.00032

Zuboff, S. (2019). *The age of surveillance capitalism: The fight for a human future at the new frontier of power*. PublicAffairs.

About the Contributors

Jaspreet Kaur is currently working as an Assistant Professor in University Business School, Chandigarh University, Mohali, Punjab. She is a post graduate (MBA-H.R) from Panjab University, Chandigarh. She has also qualified UGC NET JRF in Human Resource Management/Labour and Social Welfare and has pursued PhD in Business Management from Chandigarh University, Mohali. She has over 8 years of experience in academic and administrative assignments. She also received "Best Teacher of the Department Award" in the year 2019 and 2021 in the field of imparting quality education. Her research interests include Employee Engagement, Management of Organizational Change and Organization Development. She has published several research papers and articles in reputed international and national journals and has presented papers in various national and international conferences. She also contributed one edited book and numerous book chapters on various topics.

* * *

Kapil Kumar Aggarwal is Professor of Finance at the University School of Business, Chandigarh University. He has earned PhD, M. Phil, MFC, MBA & M. Com. He has more than 16 years of teaching and research experience. His research experience is inclusive of 22 research papers published in National journals, International Journals, Edited Books, ABDC-listed journals, Scopus Indexed, Web of Sciences and Peer Reviewed journals, 32 research papers presented in National/International conferences and two books published in his account. His teaching interest is primarily focused on finance, management sciences, quantitative techniques, business mathematics, and business statistics. He is the reviewer of the International Journal of Educational Development Journal, indexed in Scopus Elsevier, Social Sciences Citation Index and Educational and Information Technologies Journal indexed in SCOPUS, Social Science Citation Index, and UGC-CARE List (India) and Editorial Board member of International Journal of Accounting, Finance and Risk Management (IJAFRM) of He is also a pioneer in the field of SPSS and R and has delivered multiple lectures and workshops on the same.

Bhupinder Pal Singh Chahal serves as an Adjunct Professor of Marketing at Yorkville University in New Westminster, BC, Canada. Additionally, he has been awarded the Certified Management and Business Educator (CMBE) designation by the Chartered Association of Business Schools in London. With a distinguished career spanning over 15 years in the realm of management education, Dr. Chahal boasts an impressive background as the Head of Department at Apex Institute of Technology (AIT-Management) within Chandigarh University, a QS Asia-ranked institution, where he dedicated 11 years of his professional journey. Dr. Chahal's expertise lies in Marketing and International Business, and he has provided mentorship for research work across diverse topics within these domains, contributing to papers, case studies, chapters in books, and publications in international and national refereed journals, including those listed in Scopus, ABDC, and UGC. His research contributions have been showcased in conference proceedings at multiple international and national conferences, as well as in edited books. In addition to mentoring Ph.D. research scholars on topics such as knowledge management, financial inclusion, online and social media marketing, international business, and entrepreneurship development, Dr. Chahal has successfully supervised the completion of five Ph.D. research projects under his guidance. Utilizing participatory and innovative teaching methodologies, Dr. Chahal imparts comprehensive knowledge to students, aligning with the course curriculum. Dr. Chahal's wealth of experience extends to conducting training and capacity-building programs at the national level. His innate expertise lies in team building, survey implementation, data analysis, competency enhancement, and strategic planning. With two international patents to his credit, Dr. Chahal has demonstrated exceptional accomplishments in the realms of research and innovation. Over his 18-year career (3 years Ph.D. research scholar), with a predominant focus on academia for over 15 years, Dr. Chahal has held pivotal decision-making positions in various committees, teams, and clubs dedicated to transforming institutions into Centres of Excellence. As a dedicated educator, he possesses a natural flair for teaching and evaluating learners, employing special assessment techniques that challenge and enhance the overall learning experience. Dr. Chahal remains committed to staying updated with the latest developments in the fields of business management, marketing, and knowledge management.

Medha Gupta is Assistant Professor, University School of Business, Chandigarh University.

Swati Gupta is a dynamic individual with a doctoral degree and over 9 years of professional experience across various educational institutions and universities in diverse locations. She specializes in teaching subjects such as Consumer Behaviour, Market Research, Marketing Management, Organizational Behavior, Sales and Distribution Management, Services Marketing and Retail Management. Known for her energy and dedication, she possesses excellent analytical, problem-solving and decision-making skills. Currently, she is working as an Assistant Professor at Chitkara Business School, Chitkara University, Punjab, India. Having served as the editor for various books under renowned publishing houses including NOVA, Emerald, IGI Global and AAP, she has contributed significantly to the academic field. Additionally, she has undertaken the role of a reviewer for IGI Global and has chapters indexed in Scopus covering topics in marketing, Digital Marketing, Green Business Management, Consumer Behaviour among others. Furthermore, She holds patent and copyrights in her name.

Dixit Kalra is an Assistant Professor at Faculty of Commerce, Parul University, India. He is MBA, M.Com and is currently pursuing Ph.D. in Management from Punjabi University, Patiala, India. He has successfully cleared the UGC-National Eligibility Test for Lectureship (NET) twice. He has presented Research papers at National and International Conferences. In addition to it, he has written Book Chapters for International Publishing House and is also a reviewer with IGI Global. His research interests include AI in Marketing, Organic Farming, Sustainable marketing practices and Consumer behaviour. He also contributed 5 Scopus indexed book chapters on various topics.

Roop Kamal is an accomplished professional currently serving as an Assistant Professor at the University School of Business, Chandigarh University. She is dedicated to pursuing a career in an environment that provides challenges, personal satisfaction, and opportunities to leverage her skills, abilities, and knowledge. Her commitment to continuous growth and development aligns seamlessly with the goals of her organization, fostering a mutually beneficial relationship. Through her academic role, Miss Roopkamal strives to inspire and educate, contributing to the advancement of her field and the success of her students.

Navdeep Kaur works as Assistant Professor at Department of Business Management, Guru Nanak Dev Engineering College.

Gurdish Pal is an Independent Human Resources Consultant with extensive experience in providing strategic HR solutions to various organizations. He specializes in talent acquisition, employee relations, and organizational development. With a proven track record in enhancing workforce productivity and fostering positive workplace cultures, Mr. Pal is committed to helping businesses achieve their HR goals efficiently. His expertise and dedication have made him a respected professional in the field of human resources.

Trupti Patole has been working in academics for 22 years. Currently pursuing PhD from Amity University, Noida. She has published various research papers in National and International conferences.

Noble Preet is a distinguished student at Conestoga College, known for her academic excellence and contributions to the college community. She has earned recognition for her dedication to her studies and active participation in various college activities. Dr. Preet's commitment to learning and her involvement in extracurricular pursuits highlight her well-rounded character and potential for future success.

Aparna Saluja is an independent researcher specializing in management. With a robust background in organizational theory and strategic management, Aparna has established herself as a thought leader in the field. Her research focuses on innovative management practices, organizational behavior, and leadership development. Known for her analytical skills and depth of knowledge, Aparna conducts thorough investigations and provides valuable insights that contribute to both academic literature and practical management strategies. Her work is characterized by a commitment to advancing understanding and fostering improvement in organizational effectiveness.

Kshitiz Sharma is a Professor and Coordinator- IQAC and Ph.D. guide at ABBS School of Management, Bangalore. His past association has been with Alliance School of Business (Alliance University), VIHTM (Saurashtra University), UEI Global (Berggruen Education Pvt. Ltd.), IIAS School of Hotel Management and Tourism, Radisson (Carlson Hospitality Worldwide, Le Meridien etc. Dr. Kshitiz Sharma is a researcher, published extensively and have good knowledge of statistical packages like Advanced Excel, SPSS, Minitab, Python, R, Orange, Knime, Bibliometrix etc. He is an author of a book and many book chapters. He has published papers including research papers, articles and case studies in prestigious Springer Nature, Japanese Business Magazine, Case Center, ET Case, ABDC listed, Scopus, WOS indexed and UGC listed journals. He is Reviewer for many journals as Journal of Creative Communications, Int. J. of Business and Emerging Markets, Amity Journal of Entrepreneurship (AJE), International Journal of e-government and e-Business Research and many more. Dr. Sharma has conducted several FDPs and MDPs and organized International Conferences at ABBS as Convenor. His teaching interests are: Marketing Management, Marketing Strategy, Marketing Research, Services Marketing, Consumer Behaviour, Brand Management, Digital Marketing, Managing in the Global Economy, Cultural Intelligence, Entrepreneurship and Innovation, Business Analytics, Data Management and Visualization, Predictive Analytics.

Swati Sharma, Ph.D., is currently working as an Associate Professor with Amity University, India. She holds 18 years of teaching and industry experience. She is Vice- President of the Women's Indian Chamber of Commerce and Industry . She is the reviewer of Consumer Behavior in Tourism & Hospitality & IGI Global . She has authored & Edited 15 books & published research papers in National & International journals of repute. (IGI Global, Taylor & Francis, Springer, Apple Academy Press, CABI, Edward & Elgar). Her area of interest is Sustainable Tourism, MICE, Community based Tourism, Destination branding.

Navpreet Kaur Sidhu is an Assistant Professor of Finance & Business Analytics working at Chitkara Business School, Chitkara University, Punjab, India. A Former Banker who worked on the PAN India Project and represented it globally. Handled HNW1 clients and Business houses, and worked for Business Banking operations at Head Office in India. She is currently pursuing a Doctorate of Philosophy in Finance. She has presented Research papers at National and International Conferences has written Book Chapters for International Publishing House and has also been a reviewer with IGI Global. She has also written and published an Article for National Newspapers and International Journals. Her interest areas of research are Climate Finance, AI, Machine learning, Blockchain, and the like. Her dual experience in academia and industry puts her in a unique position to examine the integration of technology, particularly robo-advisors, into modern management approaches.

Harpreet Singh is Assistant Professor - School of Management Studies, Punjabi University, Patiala.

Harwinder Singh is a dedicated professional associated with York University. Known for his strong work ethic and commitment, Harwinder consistently demonstrates a high level of responsibility and diligence in his role. His dedication to excellence is evident through his proactive approach and reliability in managing tasks and projects. Balancing work and professional growth, Harwinder contributes significantly to his field while continuously seeking opportunities for development and learning. His focus and perseverance make him a valued asset to his team and an exemplary figure of dedication in his professional endeavors.

Index

A

B

C

D

E

P

R

S

T

V

W

www.ingramcontent.com/pod-product-compliance
Lightning Source LLC
LaVergne TN
LVHW080343011026
849161LV00009B/2169

* 9 7 9 8 3 6 9 3 7 6 6 8 3 *